D1236996

VIBRATION ANALYSIS

Charles M. Haberman

Professor
Department of Mechanical Engineering
California State College at Los Angeles

Charles E. Merrill Publishing Company

Columbus, Ohio

A Bell & Howell Company

To B.V.M.

Consulting Editor

Richard J. Grosh
Purdue University

Library of Congress Catalog Card Number: 68-10502

PRINTED IN THE UNITED STATES OF AMERICA

Preface

The purpose of this text is to present the fundamentals of vibration analysis. Basic concepts are emphasized to provide the reader with a sufficient background for later study of more advanced topics in this area. Because of the extensiveness of vibration analysis, and because of the increased use of mathematics and computer applications in engineering education, the choice of topics to include for a first course is a compromise; and some of the included topics can only be covered in an introductory manner. Because computers are extensively used in professional practice to solve today's vibration problems (especially the large ones), and because today's engineering students are learning to use computers, the author has incorporated analog and digital computer solution methods into the last two chapters of this text. The instructor may choose to omit these topics if his students have not previously had courses in computer applications, or he may choose to integrate these topics throughout the course instead of covering them at the end.

The reader is assumed to have completed courses in dynamics, strength of materials, differential equations, and digital computer programming with elementary numerical analysis. Usually, the dynamics course will contain an introduction to vibration analysis; and the applied differential equations course will cover Laplace transforms, the solution of partial differential equations by separation of variables, and, possibly, electromechanical analogies. Such topics as Laplace transforms, matrix arithmetic, numerical-solution techniques, and Fortran programming are briefly reviewed in the text, and these review sections are denoted by an asterisk in the Table of Contents. It was decided to integrate the review material in the body of

iii

the text because, from his teaching experience, the author has found that this is much more effective than either brief appendix summaries or the use of other references, which often have a different nomenclature and a different emphasis or application. To increase the amount of material covered, it is assumed that the reader will cover the review topics on his own without the added benefit of a lecture.

The author wishes to express his appreciation to Professors Balise and Bhaumik for their helpful comments and suggestions. The author hopes that the reader will find the text useful, and that it achieves the stated objectives; he would appreciate comments, suggestions, and notifications of errors.

C. M. HABERMAN
California State College at Los Angeles

Table of Contents

TABLE OF CONTENTS

1 Introductory Concepts and Single-Degree-of-Freedom Free Vibrations

1.1   Introduction

If the motion of a system repeats itself, or if the system moves in a rapid back-and-forth oscillatory manner, we call that type of motion a *vibration*. All bodies or systems that possess both mass and elasticity are capable of undergoing vibration. A vibration may be undesirable (e.g., cause poor operation of machinery, generate disturbing noise, affect safety, cause excess wear and machine-part breakdown, etc.), may be of no concern, or may be needed in order to perform a specific task (e.g., vibration testers, shakers, vibration-measuring instruments, etc.).

Thus, an engineer may be called upon to minimize the adverse effects of undesirable vibrations, or, at other times, to enchance desirable vibrations. If the vibration repeats itself in equal intervals of time, it is a *periodic vibration*. The minimum time in which this periodic motion repeats itself is called the *period*, and the motion during this period is called a *cycle*. The *frequency*, which equals the reciprocal of the period, is the number of cycles per unit time, and the *radian or angular frequency* is measured in radians per unit time.

There are two general types of vibration: free and forced. In the case of a *free vibration*, there are no externally applied forces *during* vibration, but an external force may have caused an initial displacement or velocity in the system. If the external force is then removed, the body continues to vibrate, not because of an external force, but because of the action of elastic and other forces within the system itself. A *natural frequency*, which is a property of an elastic system, is a frequency of the system when it undergoes a free vibration without friction. A *forced vibration* is the vibration that takes place when one or more forces externally act on the system. This type of vibration is discussed in Chap. 2 and later chapters. The *degrees of freedom* for a system is the minimum number of independent coordinates needed to mathematically define a system. Figures 1–1a, b and 1–2a, b in Sec. 1.2 denote single-degree-of-freedom systems, since they are all constrained to move in only one manner (i.e., only one independent coordinate is needed to specify the geometric location of the mass or disk). If each mass in a system is constrained to move in only one direction (e.g., vertical or horizontal), then the degrees-of-freedom for that system equals the number of masses in that system. Figure 1–3 in Sec. 1.2 illustrates a two-degree-of-freedom system because either mass, m_b or m, can be allowed to move while the other mass is held motionless. Except for Sec. 1.2, the first two chapters of this text will be concerned with only single-degree-of-freedom systems. This chapter discusses simplified representations of more complex systems, undamped and damped free vibrations, types of damping, and an introduction to nonlinear vibrations.

1.2   *Some Design and Mathematical Model Considerations*

When the reader reads this text, he will note that most of the systems analyzed will consist of point masses connected by springs and dampers or of point masses connected by elastic beam segments. He will probably wonder how he can utilize the results or the analysis techniques for the control of vibrations in actual physical systems.

We shall postpone friction and other damping effects until later in this chapter. It should be noted that the analysis of a specific, complicated physical system requires a thorough knowledge of the details of this system in order to know what simplifying assumptions can be made. Engineers usually learn how to design by doing actual designing, instead of by doing problems or by studying textbooks. The design of many of today's machines involves quite a wide range of knowledge and a variety of different disciplines. This section will deal with a few considerations of how complex structures may be simplified to another system that lends to easier analysis than the actual one.

The spring-mass system in Fig. 1-1a has only one degree-of-freedom if the mass of the spring is neglected (or lumped with mass m), and if mass m is constrained to move in the vertical direction. This is because only one coordinate, x, is needed to define (that is, to specify the position of) this system. Though a coil spring possesses both mass and elasticity, in this text we shall either neglect the spring-mass or we shall lump its mass with another mass. In like manner, the disk-shaft system in Fig. 1-1b has only one degree-of-freedom, where θ is the only coordinate needed to define the system, if the mass of the shaft is neglected. Thus, in Fig. 1-1a, b, we neglect the mass of the spring and the shaft, but consider their properties of elasticity. The elasticity of the mass and the disk are assumed to be negligible, and hence are ignored. The system in Fig. 1-2a, consisting of an elastic cantilever beam segment, a spring, and a mass, has only one degree-of-freedom if the mass of the beam segment is neglected and if mass m is constrained to move vertically. Since they are both massless elastic bodies, the massless beam segment can be replaced by a spring, of spring constant k_b, as shown in Fig. 1-2b, so that we now have two springs in series. If we wanted to consider the mass of the cantilever beam segment, the two-degree-of-freedom system in Fig. 1-3 (consisting of two springs and two masses) is one of many ways in which it can be approximated.

For more general and practical applications, the mass and elasticity of a physical structure are usually distributed continuously, and possibly not uniformly, throughout this structure. If these properties of mass and elasticity were not separated, the calculations would probably involve a set of partial differential equations, as will be shown later in Chap. 5. In order to approximate such a system by a set of ordinary differential equations, a common practice is to divide such a structure into a number of discrete point masses connected by massless elastic elements (i.e., springs or beam segments). The original system is called a *distributed system* because the mass and elasticity

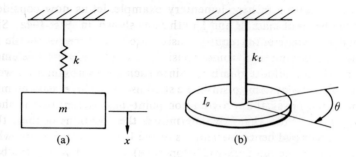

Fig. 1-1

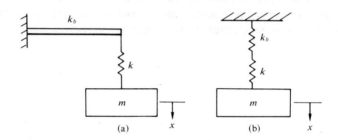

Fig. 1-2

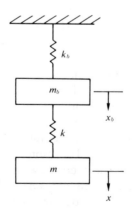

Fig. 1-3

properties are not separated, while the approximation system is called a *lumped-parameter system* because the two physical properties are lumped into separate elements.

To start with a rather elementary example, let us now consider a uniform beam of mass M and length L, as shown in Fig. 1–4a. Since the mass is distributed continuously throughout the beam, the best physical approximation would consist of very many mass elements (i.e., nearly an infinite number). Since such an approximation would be an unwieldly system to solve, we shall use only three mass elements to obtain the physical-approximation point-mass system that is shown in Fig. 1–4b. Note that we determined the locations of these three mass elements and hence the lengths of the four beam segments (which all have the same unit elastic properties) by first dividing the beam into three equal portions and then locating the mass elements at the centers of these three equal portions. The reader should be able to see easily that the approximation for the non-uniform beam in Fig. 1–4c would require a series of unequal masses connected by beam segments of differing elastic properties if the masses are equally spaced.

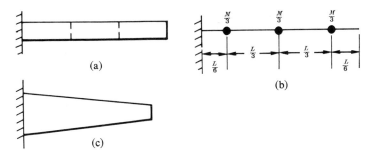

Fig. 1-4

An automobile is a complex system that has many degrees-of-freedom. Let us devise an approximation system for vehicle suspension studies. The three principal motions of the system are the vertical up-and-down motion of the automobile body, the angular pitching motion of the body about its center of gravity, and the vertical up-and-down motion of the wheels. Because of such factors as road roughness, etc., the vertical wheel motion has a much higher oscillation frequency (i.e., moves up and down with much greater rapidity) than the other two motions. Since the vertical wheel motions are too rapid to have a significant effect on the body motion, and since the vertical and pitching body motion is too slow to affect the vertical wheel motion, we can separate these almost independent motions in an approximation analysis. We can thus represent the automotive system of Fig. 1–5 by the simplified system of Fig. 1–6, where W is the automobile body weight, I_g is the body moment of inertia about the center of gravity, and k_1 and k_2 are the stiffnesses of the front and rear springs. Since the simplified system can move up and down vertically and can also rotate about the center of gravity, it is a two-degree-of-freedom system. The position of this system may be specified by two coordinates, such as: θ, the angle of rotation, and x, the vertical deflection of the center of gravity.

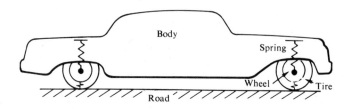

Fig. 1-5

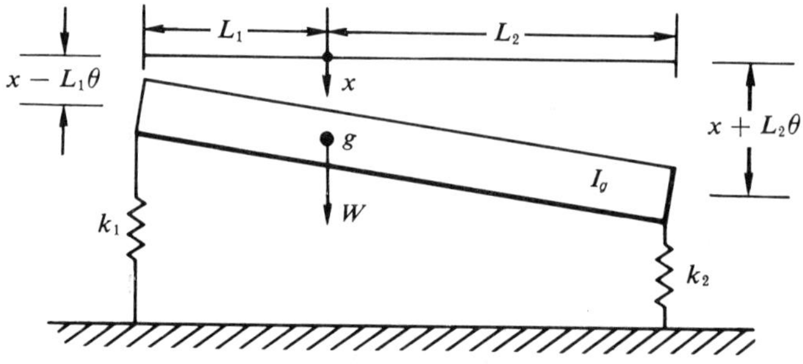

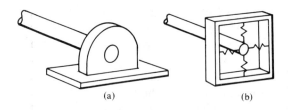

Fig. 1-6

As another example, the elasticity of a bearing support is often not the same in the lateral and vertical directions. Thus, the bearing support for the shaft in Fig. 1-7a may be represented by the system of Fig. 1-7b, where the four springs can have different spring constants.

(a)   (b)

Fig. 1-7

As for a three-dimensional example, it should not be difficult for the reader to see how a square table may be represented by the system of Fig. 1-8. Note that each table leg is represented by three masses and two beam segments, and that the horizontal masses and beam segments represent the table top and horizontal support members. The four corner horizontal masses must represent portions of the affected table leg, table top, and support member masses.

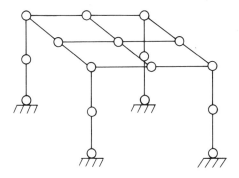

Fig. 1-8

Let us next take a structural engineering example. If a building is to be designed and located in a geographic region in which earthquakes are common, there should be an analysis of the building structure to make sure that it can withstand the forces caused by the horizontal ground motion in an earthquake excitation. Because of their usual massiveness, we shall represent the floors by rigid bodies of given mass. The floors of multi-story buildings are usually supported by steel or reinforced concrete columns. We may represent these columns by massless springs, since the mass of these columns is usually small when compared to a floor mass. When calculating a column spring constant, we may consider the column to be rigidly mounted to the upper or lower floors or to be connected by spring hinges at its ends. Since walls often do not add much to the structural strength of modern buildings, we may add a portion of the wall mass to the mass of the floor at which this wall portion is attached. Thus, the two-story building in Fig. 1-9 may be represented by the vertical spring-mass system of Fig. 1-10, where the ground motion is considered to be vertical and where the vertical motions of the two masses in Fig. 1-10 represent the horizontal motions of the two floors in Fig. 1-9. Note that each spring represents all of the columns between two floors, and that each mass represents the mass of a floor plus the equipment supported by this floor plus a portion of the mass of the walls.

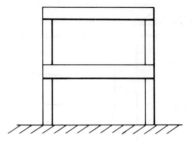

Fig. 1-9

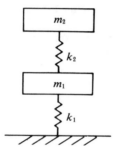

Fig. 1-10

1.3 Single-Degree-of-Freedom Free Vibrations without Damping

In a free vibration without damping, the *amplitude* (i.e., the maximum value of the displacement of the oscillating body from its equilibrium or neutral position) is the same for each repeated motion. In other words, the amplitude is constant. We shall also find that, for a free undamped single-degree-of-freedom vibration, the magnitude of the acceleration is proportional (but opposite in direction) to the magnitude of the displacement, and that the proportionality constant equals the square of the natural frequency expressed in angular velocity dimensions (i.e., radians per unit time). Again, a free vibration takes place when an elastic system that has no external forces or moments is disturbed from its equilibrium position, and this vibration is sustained by gravity forces and internal elastic forces and moments.

A classic example of a free vibration is that of a weight that is

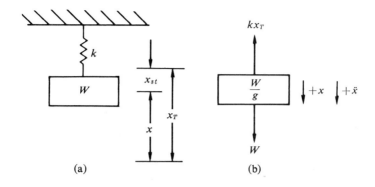

Fig. 1-11

suspended on a vertical linear spring. For a *linear spring*, the magnitude of the force F_s in the spring is proportional to the displacement x of the lower spring end from its equilibrium (or unstretched) position. More correctly, this spring force is proportional to the relative displacement (elongation or compression) between the two ends of the spring. The proportionality constant is called the *spring constant k*. Thus, F_s equals kx in magnitude. Consider Fig. 1–11 which shows a weight W suspended vertically from a spring whose spring constant is k. The *static deflection* x_{st} is the distance that the spring is pulled downward due to the weight. Since x_{st} is the relative displacement (stretch) between the spring ends

$$kx_{st} = W$$

Now, let the weight be displaced downward a distance x from the new equilibrium position. The total amount x_T that the spring is stretched is $(x + x_{st})$. Since the force in a deformed spring is directed to the equilibrium position of the spring, the spring force kx_T is directed upward (as shown in the free-body diagram in Fig. 1–11) when the lower end of the spring is stretched a distance x_T downward. The only other force is the weight W which acts in a downward direction. Using the free-body diagram and Newton's second law equation (i. e., $\Sigma F_x = m\ddot{x}$), we obtain

$$W - kx_T = \frac{W}{g}\left(\frac{d^2 x_T}{dt^2}\right)$$

where downward forces were considered to be positive and upward forces negative. Since x_T equals $(x + x_{st})$, and since x_{st} is equal to the constant W/k,

$$x_T = x + x_{st} = x + \frac{W}{k}$$

$$\frac{d^2 x_T}{dt^2} = \frac{d^2 x}{dt^2}$$

Upon substitution, we have

$$W - kx_T = W - kx - W = \frac{W}{g}\left(\frac{d^2 x}{dt^2}\right)$$

Denoting d^2x/dt^2 by $\ddot{x}$, we obtain the following differential equation with constant coefficients

$$\frac{W}{g}\ddot{x} + kx = 0 \tag{1-1}$$

which can be rewritten as

$$\ddot{x} + \omega_n^2 x = 0 \tag{1-2}$$

where $\omega_n^2 = kg/W$. From the reader's course in differential equations, we know that the solution of this homogeneous differential equation is

$$x = A \cos \omega_n t + B \sin \omega_n t \tag{1-3}$$

The solution procedure is shown in Ill. Ex. 1.1. We shall evaluate the two arbitrary constants A and B in terms of the initial displacement x_o and the initial velocity v_o. Since $x = x_o$ when time $t = 0$, Eq. (1-3) gives

$$x_o = A \cos 0 + B \sin 0$$

so that $A = x_o$. Differentiation of Eq. (1-3) gives

$$\dot{x} = -A\omega_n \sin \omega_n t + B\omega_n \cos \omega_n t$$

Since $\dot{x} = v_o$ when time $t = 0$, the previous equation gives

$$v_o = -A\omega_n \sin 0 + B\omega_n \cos 0$$

Thus, $B = v_o/\omega_n$. Substitution of these values of A and B into Eq. (1-3) gives

$$x = x_o \cos \omega_n t + \frac{v_o}{\omega_n} \sin \omega_n t \tag{1-4}$$

where $\omega_n = \sqrt{kg/W}$. Illustrative Example 1.1 shows that Eq. (1-3) may be rewritten as

$$x = M \sin (\omega_n t + \alpha) \tag{1-5}$$

where $M = \sqrt{A^2 + B^2}$ and $\alpha = \tan^{-1}(A/B)$. Thus, Eq. (1-5) shows that the vibration is a sinusoidal one where the amplitude is M and the phase angle is α. The oscillation frequency f_n, which is also the natural frequency of the system in cycles per unit time, is given by

$$f_n = \frac{\omega_n}{2\pi} = \frac{1}{2\pi}\sqrt{\frac{kg}{W}} \tag{1-6}$$

Note that ω_n is the angular natural frequency and is measured in radians per unit time. Since period T equals $1/f_n$, $T = 2\pi\sqrt{W/kg}$. The amplitude and phase angle values are given by

$$M = \sqrt{A^2 + B^2} = \sqrt{x_o^2 + \left(\frac{v_o}{\omega_n}\right)^2} \tag{1-7}$$

$$\alpha = \tan^{-1}\left(\frac{A}{B}\right) = \tan^{-1}\left(\frac{\omega_n x_o}{v_o}\right) \tag{1-8}$$

Illustrative Examples 1.2 to 1.6 furnish other examples of free, undamped vibrations having one degree-of-freedom. Since their equations of motion all have the form of Eq. (1–2), their solutions have the form of Eqs. (1–3) and (1–5). Their natural frequencies can also be easily evaluated from the equation of motion.

Illustrative Example 1.1. Show that Eqs. (1-3) and (1-5) are solutions of Eq. (1-2).

Solution. Since Eq. (1-2) is a homogeneous differential equation, we let $x = e^{nt}$. Thus, $\ddot{x} = n^2 e^{nt}$. Substitution of these values for x and $\ddot{x}$ into Eq. (1-2) and factoring the result gives

$$e^{nt}(n^2 + \omega_n^2) = 0$$

Thus, $n = \pm i\omega_n$ and the solution is

$$x = C_1 e^{i\omega_n t} + C_2 e^{-i\omega_n t}$$

Using Euler's formula, $e^{i\theta} = \cos\theta + i\sin\theta$, we have upon substitution

$$x = C_1(\cos\omega_n t + i\sin\omega_n t) + C_2(\cos\omega_n t - i\sin\omega_n t)$$

Thus, Eq. (1-3) is a soultion where A equals $(C_1 + C_2)$ and B equals $i(C_1 - C_2)$. Now let us rewrite Eq. (1-3) in the form

$$x = M\left(\frac{A}{M}\cos\omega_n t + \frac{B}{M}\sin\omega_n t\right)$$

Defining angle α such that $\sin\alpha = A/M$ and $\cos\alpha = B/M$, we have upon substitution

$$x = M(\sin\alpha\cos\omega_n t + \cos\alpha\sin\omega_n t)$$

From trigonometry we see that the previous equation is equivalent to Eq. (1-5). It should be noted that the definitions of $\sin\alpha$ and $\cos\alpha$ are in agreement with the definitions of M and $\tan\alpha$ (e.g., $\sin^2\alpha + \cos^2\alpha = (A^2 + B^2)/M^2 = 1$). The equivalence of Eqs. (1-3) and (1-5) exemplifies that the sum of two harmonic (i.e., sine or cosine) motions at the same frequency (even at different phase angles) is harmonic at that same frequency.

Illustrative Example 1.2. Consider the U-tube manometer, illustrated in Fig. 1-12, whose internal diameter is D. Let the specific weight of the fluid be γ and let the length of the fluid in the tube be L. If the lower fluid level is displaced downward a distance x from the equilibrium position, determine the equation of motion for the oscillation of the fluid in the U-tube.

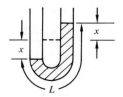

Fig. 1-12

Solution. Since the volume of the fluid is $\pi D^2 L/4$, the fluid weight W equals $\pi D^2 L\gamma/4$. The restoring force during displacement x is $\gamma\,(\pi D^2/4)(2\,x)$, which is in the upward, or negative, direction for the lower level. Upon applying Newton's second law (i.e., $\sum F_x = m\ddot{x}$), we obtain

$$\left(\frac{\pi D^2 L}{4}\right)\left(\frac{\gamma}{g}\right)\frac{d^2 x}{dt^2} = -\left(\frac{\gamma\pi D^2}{4}\right)(2\,x)$$

which reduces to

$$L\ddot{x} + 2\,gx = 0 \qquad\qquad (1\text{-}9)$$

Illustrative Example 1.3. Let a body of constant cross-sectional area A and weight W float in a fluid of specific weight γ. During equilibrium, h units of the float height are immersed. Let the float be displaced downward a distance x from equilibrium (i. e., its normal floating position). Determine the equation of motion for the bobbing (i. e., oscillating) float.

Solution. Archimedes' principle states that the upward buoyant force of the fluid on the floating body equals the weight of the fluid displaced by the body. During equilibrium, the downward weight force W is balanced by the upward buoyant force which then equals γAh. Thus, we have

$$W = \gamma Ah$$

When the body is displaced downward a distance x, the upward buoyant force is $\gamma A(h + x)$, Now, by applying Newton's second law, we obtain

$$\sum F = W - \gamma A(h + x) = \frac{W}{g}\frac{d^2 x}{dt^2}$$

Since $W = \gamma Ah$, we have

$$\frac{W}{g}\ddot{x} + \gamma Ax = 0 \qquad\qquad (1\text{-}10)$$

Illustrative Example 1.4. Consider the simple pendulum of weight W illustrated in Fig. 1-13. A *simple pendulum* is a particle or concentrated weight that oscillates in a vertical arc and is supported by a weightless cord. The only forces acting are those of gravity and cord tension (i. e., frictional resistance is neglected). If the cord length is l, determine the equation of motion if the maximum oscillation angle θ is small.

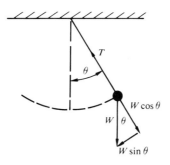

Fig. 1-13

Solution. The linear acceleration in the tangential direction is $l(d^2\theta/dt^2)$ or $l\ddot{\theta}$. If the weight and chord are displaced a positive angle θ from the vertical position, it can be seen, from Fig. 1-13, that there is a restoring force in the tangential direction of magnitude $W \sin \theta$ that is directed opposite to that of the direction of increasing θ. Thus, from Newton's second law, we have

$$\frac{W}{g} l\ddot{\theta} = -W \sin \theta$$

Division by W/g gives

$$l\ddot{\theta} + g \sin \theta = 0 \qquad (1\text{-}11)$$

Because of the term $\sin \theta$, the above equation is a nonlinear differential equation. If the maximum angle through which the pendulum oscillates is small, however, the approximation $\sin \theta = \theta$ can be used to get the following linear differential equation:

$$l\ddot{\theta} + g\theta = 0 \qquad (1\text{-}12)$$

Since $\ddot{\theta} + (g/l)\theta = 0$, ω_n, the natural frequency in rad/sec, equals $\sqrt{g/l}$. Using Eq. (1-4), the solution is

$$\theta = \theta_o \cos \sqrt{\frac{g}{l}} t + \dot{\theta}_o \sqrt{\frac{l}{g}} \sin \sqrt{\frac{g}{l}} t \qquad (1\text{-}13)$$

where θ_o and $\dot{\theta}_o$ are the angular displacement and angular velocity values at time zero.

Illustrative Example 1.5. A *compound pendulum* is a pendulum whose weight cannot be considered to be concentrated at one point. Consider the compound pendulum, illustrated in Fig. 1-14, whose center of gravity CG is at a distance $\bar{r}$ from the point of rotation. Write the equation of motion for the compound pendulum and show how it can be made equivalent to a simple pendulum.

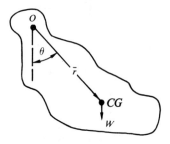

Fig. 1-14

Solution. Applying the rotation equation, $\sum M_o = I_o \ddot{\theta}$, about the point of rotation, we have

$$I_o \ddot{\theta} = -(W \sin \theta)\bar{r}$$

From physics, we know that $I_0 = I_{CG} + (W/g)\bar{r}^2$. If the maximum value of θ is small, we can use the approximation $\sin \theta = \theta$ to get the following linear differential equation:

$$I_o \ddot{\theta} + W\bar{r}\theta = 0$$

The radius of gyration k_o about the point of rotation is defined such that I_o equals $(W/g)k_o^2$. Upon substitution in the previous equation, and multiplying by $g/\bar{r}$, we obtain

$$\frac{k_0^2}{\bar{r}} \ddot{\theta} + g\theta = 0 \tag{1-14}$$

Thus, it is seen, by comparing Eqs. (1-12) and (1-14), that a compound pendulum can be treated as a simple pendulum supported by a cord of length $k_o^2/\bar{r}$.

Illustrative Example 1.6. Consider a rigid body that is suspended at its center of gravity by a light, thin elastic shaft of length L, which is in a vertical position. The other end of this circular shaft is rigidly attached, as shown in Fig. 1-1b in Sec. 1.2. In our strength of materials course, it is given that if such a shaft is twisted an angle θ, there will be a resisting torque exerted along the length of the shaft of magnitude $JG\theta/x$. J is the polar moment of inertia of the shaft cross-section, G is the shear modulus of the shaft, and x is the distance from the rigidly mounted end of the shaft. Write the equation of motion for the oscillation that will result.

Solution. Since the axis of rotation goes through the center of gravity of the body, we shall use I_{CG}, the moment of inertia about the center of gravity. The magnitude of the moment acting on the rigid body is $JG\theta/L$, and this moment is a resisting moment. Applying the dynamics equation $I_{CG}\ddot\theta = \sum M_{CG}$, we obtain the linear differential equation:

$$I_{CG}\ddot\theta + \left(\frac{JG}{L}\right)\theta = 0 \tag{1-15}$$

Denoting the term JG/L as the *shaft stiffness constant* k_t, where $J = (\pi/32)D^4$ for a circular shaft, we have

$$I_{CG}\ddot\theta + k_t\theta = 0 \tag{1-16}$$

1.4   The Energy Method

For bodies that vibrate without a loss of mechanical energy (i. e., have no friction or damping forces that dissipate mechanical energy), it is sometimes convenient to use energy techniques. Since there is no energy loss, and since the only energies involved in a mechanical system are kinetic energy KE and potential energy PE, the law of conservation of energy gives us the result that (PE + KE) is constant, so that

$$\frac{d}{dt}(\text{KE} + \text{PE}) = 0 \tag{1-17}$$

Applying this to the spring-mass system in Fig. 1-11 in Sec. 1.3, the kinetic energy of the mass is given by

$$\text{KE} = \frac{1}{2}\frac{W}{g}\dot x^2$$

Letting the equilibrium position be the zero reference for the potential energy, the potential energy of position PE_p equals $-Wx$ for a downward displacement x of the mass. The elastic potential energy of the spring PE_s for the same position is as follows, since the total spring force equals $k(x + x_{st})$ or $(kx + W)$,

$$\text{PE}_s = \int_0^x (W + kx)dx = Wx + \frac{1}{2}kx^2$$

The total potential energy $(\text{PE}_p + \text{PE}_s)$ equals $kx^2/2$, and the total energy is

$$\text{KE} + \text{PE} = \frac{1}{2}\frac{W}{g}\dot x^2 + \frac{1}{2}kx^2$$

The same result will be obtained when the mass is displaced upward a distance x. Application of Eq. (1-17) gives

$$\frac{d}{dt}(\text{KE}+\text{PE}) = \left(\frac{W}{g}\ddot{x} + kx\right)\dot{x} = 0$$

and division of the previous result by $\dot{x}$ gives us Eq. (1-1) for the equation of motion for this system.

We can also utilize another energy method, called *Rayleigh's method*, to calculate natural frequencies. Though this method is exemplified in Ill. Ex. 1.11, it will be much more fully discussed and illustrated in Sec. 4.2. In a vibrating system, the minimum values of the kinetic energy and the potential energy are zero. Since (KE + PE) is constant, a zero value of PE corresponds to a maximum value of KE, and a zero value of KE corresponds to a maximum value of PE. Thus, the maximum potential energy value equals the maximum kinetic energy value, and the equation that is the basis for Rayleigh's method is given by

$$(\text{KE})_{\max} = (\text{PE})_{\max} \tag{1-18}$$

Illustrative Example 1.7. What is the natural frequency and period of the system given in Fig. 1-11 if the mass has a weight of 20 lb and the spring has a spring constant of 13 lb/in.?

Solution. In vibration problems, displacements are usually given in inches. That is, the in. -lb_f-sec system is used. Thus, we shall use a gravitational acceleration value of $32.2 \times 12 = 386$ in./sec^2 in our calculations. Use of Eq. (1-6) gives

$$f_n = \frac{1}{2\pi}\sqrt{\frac{kg}{W}} = \frac{1}{2\pi}\sqrt{\frac{(13)(386)}{20}} = 15.8 \text{ cycles/sec}$$

Period T equals $1/f_n$, or 0.0633 sec.

Illustrative Example 1.8. Consider a cantilever beam of length L supporting a load of weight W at its end. In our strength of materials course it is given that the magnitude of the force F_y at the end of a cantilever beam that is required to deflect the free end a distance y is equal to $3EIy/L^3$. Assuming that the deflection shape of the oscillating beam is the same as that under static loading, and neglecting the mass of the beam, use the energy method to write the equation of motion for this oscillating elastic beam.

Solution. Since the potential energy PE is equal to $\int_0^y F_y dy$, we obtain

$$\text{PE} = \int_0^y \left(\frac{3EI}{L^3}\right) y\, dy = \frac{3}{2}\frac{EIy^2}{L^3}$$

The kinetic energy KE of the weight is

$$\text{KE} = \frac{1}{2}\left(\frac{W}{g}\right)(\dot{y})^2$$

Since the sum of the potential and kinetic energies of a conservative system (i.e., no friction or damping) is constant, we have

$$\frac{3}{2}\left(\frac{EIy^2}{L^3}\right) + \frac{1}{2}\left(\frac{W}{g}\dot{y}^2\right) = \text{constant}$$

Upon differentiating the above equation with respect to time to apply Eq. (1-17), we obtain

$$\frac{W}{g}\ddot{y} + \frac{3EI}{L^3}y = 0 \qquad (1\text{-}19)$$

Illustrative Example 1.9.　　Repeat Ill. Ex. 1.4 using the energy method.

Solution.　　The moment of inertia I_o of the mass about the point of rotation is equal to $(W/g)l^2$. The kinetic energy is given by

$$\text{KE} = \frac{1}{2}I_0\dot{\theta}^2 = \frac{1}{2}\frac{W}{g}l^2\dot{\theta}^2$$

When the pendulum is raised an angle θ, the mass is raised a distance of $l(1 - \cos\theta)$ in the vertical direction, and the potential energy is given by

$$\text{PE} = Wl(1 - \cos\theta)$$

Substitution in Eq. (1-17) gives

$$\frac{d}{dt}\left[\frac{1}{2}\frac{W}{g}l^2\dot{\theta}^2 + Wl(1 - \cos\theta)\right] = \left(\frac{W}{g}l^2\ddot{\theta} + Wl\sin\theta\right)\dot{\theta} = 0$$

Division by $Wl\dot{\theta}$ gives Eq. (1-11) and also Eq. (1-12) when the maximum value of θ is small.

Illustrative Example 1.10.　　Figure 1-15 shows a cylinder of mass m and radius r that rolls without slipping on a cylindrical surface of radius R. Determine the equation of motion and the natural frequency for small oscillations.

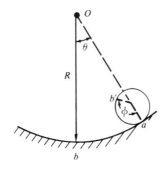

Fig. 1-15

Solution.   The translational kinetic energy of the center of the cylinder is equal to $(m/2)[(R - r)\dot\theta]^2$ since the translational velocity equals $(R - r)\dot\theta$ at that point. Because the cylinder is rolling, we must also consider the rotational kinetic energy, which equals $(I_g/2)(\dot\phi - \dot\theta)$ since the absolute angular velocity of the cylinder equals $(\dot\phi - \dot\theta)$. The centroidal moment of inertia for a cylinder is equal to $mr^2/2$. Since point b' upon rolling will touch point b, arc length ab equals both $R\theta$ and $r\phi$. Thus, we can use the equations $\phi = R\theta/r$ and $\dot\phi = R\dot\theta/r$ to obtain the kinetic energy equation:

$$\text{KE} = \frac{m}{2}\left[(R - r)\dot\theta\right]^2 + \frac{1}{2}\left(\frac{1}{2}mr^2\right)\left(\frac{R}{r} - 1\right)^2\dot\theta^2$$

It may be noted that the angular velocity equals $(\dot\phi - \dot\theta)$, and not $\dot\phi$, because point b is vertically below the center of the cylinder when the cylinder touches that point. The vertical through the center of the cylinder, when at the position shown in Fig. 1-15, is at an angle of $(\phi - \theta)$ from point b'. Thus, a vertical line, and any other line on the rigid cylinder, rotates an angle $(\phi - \theta)$ in rolling from any tangent point a to tangent point b, where point b is the zero angle reference point for both θ and ϕ. When the mass attains an angle θ, its center of gravity is vertically raised a distance of $(R - r)(1 - \cos\theta)$, so that the potential energy is given by the following equation:

$$\text{PE} = mg(R - r)(1 - \cos\theta)$$

Substitution of the KE and PE results into Eq. (1-17) gives

$$[\tfrac{3}{2}m(R - r)^2\ddot\theta + mg(R - r)\sin\theta]\dot\theta = 0$$

Use of the approximation $\sin\theta = \theta$ and division by $m\dot\theta$ gives us the following equation of motion:

$$\ddot\theta + \frac{2g}{3(R-r)}\theta = 0 \qquad\qquad (1\text{-}20)$$

From the previous equation, we see that the natural frequency in rad/sec is given by

$$\omega_n = \sqrt{\frac{2g}{3(R - r)}} \qquad\qquad (1\text{-}21)$$

Illustrative Example 1.11.   Suppose that the mass of the spring in Fig. 1-11 in Sec. 1.3 were not negligible. Use Rayleigh's method to determine the natural frequency of the system.

Solution.   Let the spring be of length L and let it have a unit mass of ρ lb per inch of length. Consider an infinitesimal length de of the spring that is located e inches from its top end. If weight W is moved a distance x, the bottom of the spring moves a distance x; the top of the spring does not move; and this infinitesimal element, whose mass is $\rho\,de$, moves a distance $(e/L)x$. Thus, the maximum velocity of this element is $(e/L)\dot x_{\max}$, where $\dot x_{\max}$ is the maximum velocity of weight W. Since the total kinetic energy of the system equals the kinetic energy of the mass plus that of the spring, the maximum kinetic energy is given by

$$(KE)_{max} = \frac{1}{2} \frac{W}{g} \dot{x}^2_{max} + \int_0^L \left(\frac{1}{2} \rho \, de \right) \left[\left(\frac{e}{L} \right) \dot{x}_{max} \right]^2$$

$$= \frac{1}{2} \frac{W}{g} \dot{x}^2_{max} + \left(\frac{\rho}{2} \right) \left(\frac{\dot{x}_{max}}{L} \right)^2 \int_0^L e^2 \, de$$

$$= \frac{1}{2} \left(\frac{W}{g} + \frac{1}{3} \rho L \right) \dot{x}^2_{max}$$

The maximum potential energy is given by $(PE)_{max} = \frac{1}{2} k x^2_{max}$. The application of Rayleigh's method requires an estimate of the deflection equation. From Sec. 1.3, we know that $x = A \sin \omega_n t$ is a valid assumption. Thus, $x_{max} = A$ and $\dot{x}_{max} = A\omega_n$, and the equation $(KE)_{max} = (PE)_{max}$ gives

$$\frac{1}{2} \left(\frac{W}{g} + \frac{1}{3} \rho L \right) (A\omega_n)^2 = \frac{1}{2} k A^2$$

from which we obtain

$$f_n = \frac{\omega_n}{2\pi} = \frac{1}{2\pi} \sqrt{\frac{k}{W/g + \rho L/3}} \tag{1-22}$$

Comparison of Eqs. (1-6) and (1-22) shows that we can account for the effect of the mass of the spring in Fig. 1-11 when we calculate the natural frequency of this spring-mass system, by adding one-third of the spring-mass (which is ρL) to that of the given mass and then considering the spring to be massless.

1.5 Equivalent Spring Constants

Often, combinations of springs or other types of elastic elements can be replaced by one elastic spring. Consider the two-spring, one-mass systems in Figs. 1-16a, b, where the two springs in Fig. 1-16a are close together. When the mass in either system is pulled down a distance x, the total force F_T equals $k_1 x + k_2 x$. Thus, the *equivalent spring constant*, which is defined to equal the total force divided by the total displacement, for either system is given by

$$k_{eq} = \frac{F_T}{x} = \frac{k_1 x + k_2 x}{x} = k_1 + k_2 \tag{1-23}$$

Thus, we could replace the two systems in Figs. 1-16a, b by the system in Fig. 1-11 in Sec. 1.3, if we let k equal $(k_1 + k_2)$. From Eq. (1-6), we see that the natural frequency of either system is equal to $(1/2\pi)\sqrt{(k_1 + k_2)g/W}$. In a similar manner, the equivalent spring constant for the disk-shaft system in Fig. 1-16c is equal to $k_{t1} + k_{t2}$, where k_t is defined in Ill. Ex. 1.6 in Sec. 1.3. Thus, we could replace this two-shaft system by the one-shaft system in Fig. 1-1b in Sec. 1.2 if we let $k_t = k_{t1} + k_{t2}$.

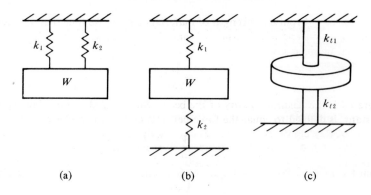

(a)   (b)   (c)

Fig. 1-16

Illustrative Example 1.12.   Determine the equivalent spring constant of the system in Fig. 1-17 which has two springs in series.

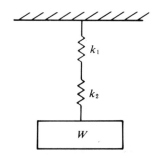

Fig. 1-17

Solution.   If we apply a force F_T to pull the mass down a distance x, then $x = x_1 + x_2$, where x_1 is the elongation between the two ends of the first spring and x_2 is the elongation between the two ends of the second spring. Since the equivalent spring constant for the system is defined to equal F_T/x and since $x_1 = F_T/k_1$ and $x_2 = F_T/k_2$, we have

$$k_{eq} = \frac{F_T}{x} = \frac{F_T}{x_1 + x_2} = \frac{F_T}{(F_T/k_1) + (F_T/k_2)}$$

from which we obtain

$$k_{eq} = \frac{1}{(1/k_1) + (1/k_2)} = \frac{k_1 k_2}{k_1 + k_2} \tag{1-24}$$

Illustrative Example 1.13.   Suppose that a weight W was attached to the free end of a cantilever beam of negligible mass. Determine the equivalent spring constant and the natural frequency of vibration.

Solution. From our strength of materials course, we find that the static deflection at the end of a cantilever beam of length L, when there is a weight W at that point, is given by

$$x_{st} = \frac{WL^3}{3EI}$$

where EI is the flexural rigidity of the beam. Since k_{eq}, the equivalent spring constant, is defined to equal the force per unit deflection, we have

$$k_{eq} = \frac{W}{x_{st}} = \frac{3EI}{L^3} \qquad (1\text{-}25)$$

From Eq. (1-6) in Sec. 1.3, the natural frequency of this beam system is given by

$$f_n = \frac{1}{2\pi} \sqrt{\frac{k_{eq}g}{W}} = \frac{1}{2\pi} \sqrt{\frac{3EIg}{WL^3}} \qquad (1\text{-}26)$$

We can use the same procedure to obtain the equivalent spring constants for beam systems that have a weight W at the center of the beam. If the beam is simply supported at both ends, the equivalent spring constant is given by

$$k_{eq} = \frac{48EI}{L^3} \qquad (1\text{-}27)$$

If the beam is fixed at both ends, the equivalent spring constant is given by

$$k_{eq} = \frac{192EI}{L^3} \qquad (1\text{-}28)$$

Illustrative Example 1.14. Determine the equivalent spring constant for the system shown in Fig. 1-18.

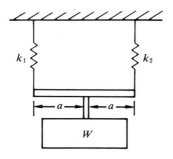

Fig. 1-18

Solution. Since the weight is attached to the midpoint of the bar, it moves a distance $(x_1 + x_2)/2$, where x_1 is the displacement of the bottom of the first spring and x_2 is the displacement of the bottom of the second spring when a vertical force F_T is applied to weight W. Since the force in each spring is $F_T/2$, $x_1 = (F_T/2)/k_1$ and $x_2 = (F_T/2)/k_2$. Thus, the equivalent spring constant is given by

$$k_{eq} = \frac{F_T}{x} = \frac{F_T}{(x_1 + x_2)/2} = \frac{F_T}{(F_T/4k_1) + (F_T/4k_2)}$$

from which we obtain

$$k_{eq} = \frac{1}{(1/4k_1) + (1/4k_2)} = \frac{4k_1 k_2}{k_1 + k_2}$$

1.6   Types of Damping and Equivalent Viscous Damping

Most vibrating systems encounter friction or some other type of resistance force that is called *damping*. Damping occurs whenever there is energy dissipation in the system. This energy dissipation, which causes the vibration amplitude to decrease with time, and which can cause the system to come to rest, is generally related to the relative motion between components in a system. The energy input to a damping device is converted to heat; hence, a damper is a non-conservative device.

There are several types of damping, and it should be noted that most are nonlinear and that all types cause a force that opposes the velocity or direction of motion. From our dynamics course, we are familiar with *coulomb* or *dry damping*. This type of damping results when there is a sliding motion between two touching surfaces, and the magnitude of the friction or damping force, which is nearly constant, is given by

$$F_d = \mu N \tag{1-29}$$

where μ is the coefficient of friction and N is the normal force. Another type is *viscous damping* where the magnitude of the damping force F_d is directly proportional to the relative velocity and is given by

$$F_d = c\dot{x} \tag{1-30}$$

where c is called the *damping coefficient* and $\dot{x}$ is the relative velocity between the two ends of the damper. Eq. (1–30) may be used as an approximation for the type of damping that is encountered by the sliding motion between lubricated surfaces and for certain components, such as a dashpot, that are immersed in oil. A viscous damper will be represented schematically by a dashpot, as shown in Fig. 1–19 in Sec. 1.7. For *velocity-squared damping*, the magnitude of the damping force F_d is proportional to the square of the relative velocity and is given by

$$F_d = a|\dot{x}|\dot{x} \tag{1-31}$$

The frictional damping force that results when a body moves through a fluid is nearly proportional to the square of the velocity, but the exact velocity exponent (which may be closer to 1.8) depends on many factors. Thus, we may prefer to use the following more general equation [which includes both Eqs. (1-30) and (1-31) if exponent b equals zero or unity]:

$$F_d = a|\dot{x}|^b \dot{x} \qquad (1\text{-}32)$$

where $0 \leq b \leq 1$ and where $|\dot{x}|^b$ was used so that force F_d can oppose velocity $\dot{x}$ (i.e., oppose the direction of motion). Other types of damping are *magnetic damping* and *structural damping*, both of which are rather complex. For the latter type, the friction is internal within the material and varies with the maximum stress of the vibration cycle. Often several types of damping exist within a system. Of those types which have been briefly described, only viscous damping is linear. That is, only viscous damping will result in a differential equation of motion that is both linear and adaptable to analytical mathematical solution techniques without involvement of complexities. If the nonlinear damping in a system is small compared to other forces, it may be approximated by linear damping in order to ease the solution. The numerical solutions of the nonlinear differential equations that result from nonlinear damping are covered in Secs. 2.14 and 2.15 of this text.

Since viscous damping is an idealized condition, let us consider how an *equivalent viscous damping coefficient, c_{eq},* may be calculated. From our dynamics course, we know that the energy ΔE that is dissipated due to damping during one cycle (i.e., during period T) is given by

$$\Delta E = \int F_d \, dx = \int_0^T F_d \, \dot{x} \, dt \qquad (1\text{-}33)$$

where F_d is the damping force and where the relation $dx = (dx/dt) \, dt$ was used. Thus, we must know both the equation for F_d and the variation of velocity $\dot{x}$ with time in order to calculate ΔE. Let us assume the sinusoidal motion $x = A \sin \omega t$ which can occur in a free vibration and in a vibration that has a sinusoidal external force. For viscous damping, $F_d = c\dot{x}$, and substitution in Eq. (1-33) gives

$$\Delta E = \int_0^T (c\dot{x})\dot{x} \, dt = \int_0^T c\omega^2 A^2 \cos^2 \omega t \, dt = c\omega\pi A^2$$

Thus, for sinusoidal motion, the equivalent viscous damping coefficient, c_{eq}, for non-viscous damping may be computed from the following equation :

$$c_{eq} = \frac{\Delta E}{\pi \omega A^2} \tag{1-34}$$

where A and ω are the amplitude and radian frequency of the sinusoidal motion and ΔE is the actual energy loss due to the non-viscous damping.

Illustrative Example 1.15. Determine the equivalent viscous damping coefficient for coulomb damping during a sinusoidal oscillation.

Solution. Since $F_d = \mu N$, and since F_d is constant, Eq. (1-33) gives $\Delta E = \mu N \int dx$ where $\int dx$ is the total displacement during one vibration cycle. Since the total displacement during one period of vibration equals $4A$, where A is the vibration amplitude, $\Delta E = 4A\mu N$. Thus, use of Eq. (1-34) gives

$$c_{eq} = \frac{4A\mu N}{\pi \omega A^2} = \frac{4\mu N}{\pi \omega A} \tag{1-35}$$

Illustrative Example 1.16. Determine the equivalent viscous damping coefficient for velocity-squared damping during a sinusoidal oscillation.

Solution. Since $F_d = a|\dot{x}|\dot{x} = \pm a\dot{x}^2$, $\omega = 2\pi/T$, $x = A \sin \omega t$, and $\dot{x} = \omega A \cos \omega t$, substitution in Eq. (1-33) gives

$$\Delta E = 2 \int_{-\pi/2}^{\pi/2} F_d \dot{x} \; d(\omega t)/\omega = 2 \int_{-\pi/2}^{\pi/2} (a\dot{x}^2)(\dot{x}/\omega) \; d(\omega t)$$

$$= 2a\omega^2 A^3 \int_{-\pi/2}^{\pi/2} \cos^3 \omega t \; d(\omega t) = 8a\omega^2 A^3/3$$

where the integration was performed over the half-cycle where velocity $\dot{x}$ was positive. Substitution of this ΔE result in Eq. (1-34) gives

$$c_{eq} = \frac{8a\omega A}{3\pi} \tag{1-36}$$

1.7   Single-Degree-of-Freedom Free Vibrations with Viscous Damping

Consider the system in Fig. 1-19 that consists of a spring, mass, and viscous damper. When the mass is displaced downward a distance x from the equilibrium position, the total stretch of the spring is $x + x_{st}$. Since both the spring and the viscous damper oppose the downward motion of the system, application of Newton's second law (i.e., $m\ddot{x} = \sum F_x$) gives

$$\frac{W}{g}\frac{d^2}{dt^2}(x + x_{st}) = -c\frac{d}{dt}(x + x_{st}) - k(x + x_{st}) + W$$

Since $kx_{st} = W$, and since static displacement x_{st} is a constant (so that its derivatives equal zero), the previous equation of motion simplifies to

$$\frac{W}{g}\ddot{x} + c\dot{x} + kx = 0 \tag{1-37}$$

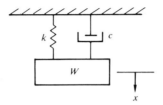

Fig. 1-19

Since Eq. (1-37) is a homogeneous differential equation, we can obtain a solution by letting $x = e^{nt}$. Thus, $\dot{x} = ne^{nt}$ and $\ddot{x} = n^2 e^{nt}$. Substitution of these values for x, $\dot{x}$, and $\ddot{x}$ into Eq. (1-37) and factoring the result gives

$$\left(\frac{W}{g}n^2 + cn + k\right)e^{nt} = 0$$

Thus, $(W/g)n^2 + cn + k = 0$, and use of the quadratic formula gives

$$n_{1,2} = \frac{g}{2W}(-c \pm \sqrt{c^2 - 4kW/g}) \tag{1-38}$$

The solution of Eq. (1-37) is given by

$$x = C_1 e^{n_1 t} + C_2 e^{n_2 t} \tag{1-39}$$

where C_1 and C_2 are arbitrary constants that are to be determined from the initial conditions and where the exponent coefficients n_1 and n_2 are given by Eq. (1-38).

Let the term c_c, which we shall call the *critical damping coefficient*, be the value of c that will cause the square root in Eq. (1-38) to equal zero. Thus, we have

$$c_c = 2\sqrt{\frac{kW}{g}} = \frac{2k}{\omega_n} \tag{1-40}$$

Let ζ, which is called the *damping ratio*, be defined to equal c/c_c. Since Eq. (1-38) can be rewritten as

$$n_{1,2} = -\frac{cg}{2W} \pm \sqrt{\left(\frac{cg}{2W}\right)^2 - \frac{kg}{W}}$$

and since $\omega_n^2 = kg/W$ (see Sec. 1.3) and $cg/W = 2\zeta\omega_n$, substitution of the last two equations in the previous $n_{1,2}$-equation gives us the following form of Eq. (1–38)

$$n_{1,2} = (-\zeta \pm \sqrt{\zeta^2 - 1})\omega_n \qquad (1\text{–}41)$$

There are three physical cases to consider: $\zeta > 1$, $\zeta < 1$, and $\zeta = 1$. If $\zeta > 1$, the system is said to be *overdamped* and Eq. (1–41) shows that exponents n_1 and n_2 are both real and that n_1 and n_2 are also both negative since $\sqrt{\zeta^2 - 1} < \zeta$. Thus, Eq. (1–39) shows that the motion is the sum of two exponential terms, each of which decreases with time since their exponents are negative. The initial displacement equals $C_1 + C_2$ for this case, which has very large damping and for which the mass creeps back to the equilibrium position. Since no vibration takes place (i.e., the motion is not periodic), this type of motion, which is illustrated in Fig. 1–21, is called *aperiodic motion*.

If $\zeta < 1$, the system is said to be *underdamped* and since the square root in Eq. (1–41) is imaginary, exponents n_1 and n_2 are complex conjugates and are given by

$$n_{1,2} = -\zeta\omega_n \pm i\omega_n\sqrt{1 - \zeta^2}$$

where $i = \sqrt{-1}$. Let us define ω_d, which will later be seen to be the *damped natural frequency* in rad/sec, by

$$\omega_d = \omega_n\sqrt{1 - \zeta^2} \qquad (1\text{–}42)$$

so that $n_{1,2} = -\zeta\omega_n \pm i\omega_d$. Substitution in Eq. (1–39) gives

$$x = e^{-\zeta\omega_n t}(C_1 e^{i\omega_d t} + C_2 e^{-i\omega_d t})$$

From Ill. Ex. 1.1 in Sec. 1.3, we see that the previous equation can be rewritten in the following two more useful forms:

$$x = e^{-\zeta\omega_n t}(A \cos \omega_d t + B \sin \omega_d t) \qquad (1\text{–}43)$$
$$x = M e^{-\zeta\omega_n t} \sin (\omega_d t + \alpha) \qquad (1\text{–}44)$$

where $M = \sqrt{A^2 + B^2}$ and $\alpha = \tan^{-1}(A/B)$. Thus, Eq. (1–44) shows that the motion is oscillatory and sinusoidal where the amplitude, which is $M e^{-\zeta\omega_n t}$, decreases exponentially with time. This motion is illustrated in Fig. 1–20, and Eq. (1–42) shows that its frequency ω_d is less than ω_n, the undamped angular natural frequency which was discussed in Sec. 1.3, by a factor of $\sqrt{1 - \zeta^2}$.

If $\zeta = 1$, the system is said to be *critically damped*. Since Eq. (1–41) shows that root $n_1 = $ root $n_2 = -\zeta\omega_n = -\omega_n$ when $\zeta = 1$, the solution of Eq. (1–37) for this equal-root case is given by

$$x = (C_1 + C_2 t)e^{-\omega_n t} \qquad (1\text{–}45)$$

Equation (1–45) shows that the motion for this case is aperiodic. This

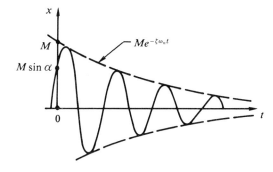

Fig. 1-20

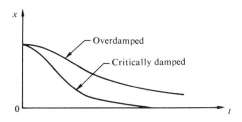

Fig. 1-21

case, which is the transition case between vibratory and non-vibratory motion, is the one in which the mass comes to rest at the equilibrium position in the shortest amount of time, as shown in Fig. 1–21.

Illustrative Example 1.17. Write the equation of motion and determine the damped vibration frequency for the system shown in Fig. 1-22. Consider the bar, which is of length L, to be massless.

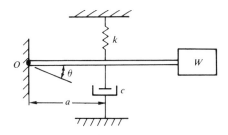

Fig. 1-22

Solution.   Let us displace the bar a small angle θ from the equilibrium position. The moment of weight W is balanced by the moment of the static spring force. The added spring force, when the bar is displaced an angle θ, equals $ka\theta$ and its moment about point O is $(ka\theta)a$. The damping force in $ca\dot{\theta}$ and its moment about point O is $(ca\dot{\theta})a$. Application of the dynamics equation $I_o\ddot{\theta} = \sum M_o$ gives

$$\frac{W}{g} L^2\ddot{\theta} = -ka^2\theta - ca^2\dot{\theta}$$

where $I_o = WL^2/g$. The previous equation of motion can be rewritten as

$$\ddot{\theta} + \frac{cga^2}{WL^2}\dot{\theta} + \frac{kga^2}{WL^2}\theta = 0$$

Thus, the undamped angular natural frequency ω_n for this system equals $\sqrt{ka^2g/WL^2}$. Use of steps similar to those used for the spring-mass-damper example in this section will show that the damped vibration frequency in rad/sec is given by

$$\omega_d = \sqrt{\frac{kga^2}{WL^2} - \left(\frac{cga^2}{2WL^2}\right)^2}$$

For comparison purposes, the damped vibration frequency for the system in Fig. 1-19 may be calculated from Eq. (1-42) to be

$$\omega_d = \sqrt{\frac{kg}{W} - \left(\frac{cg}{2W}\right)^2}$$

Thus, the two frequencies are equal when $a = L$.

1.8   The Logarithmic Decrement

The *logarithmic decrement* δ is defined to be the natural logarithm of the ratio of any two successive amplitudes in a damped vibration. It is shown in Ill. Ex. 1.18 in this section that this term can be used to calculate the amount of damping in a vibrating system. For a vibration of the spring-mass-damper system in Fig. 1–19, the vibrational motion is given by Eq. (1–44) and is illustrated in Fig. 1–20. In this figure, we see that the amplitudes occur approximately at the points where $\sin(\omega_d t + \alpha) = 1$, and that these points are spaced at time intervals of T, where T is the period of vibration. Using these results along with Eq. (1–44), the ratio of two successive amplitudes is given by

$$\frac{x_1}{x_2} = \frac{Me^{-\zeta\omega_n t_1}}{Me^{-\zeta\omega_n(t_1+T)}} = e^{\zeta\omega_n T}$$

where amplitude x_1 occurs at time t_1, where amplitude x_2 occurs at time $t_2 = t_1 + T$, and where $\sin(\omega_d t + \alpha) \approx 1$ at both amplitudes. Thus, the logarithmic decrement δ is given by

$$\delta = \ln\left(\frac{x_1}{x_2}\right) = \zeta\omega_n T = \frac{2\pi\zeta}{\sqrt{1-\zeta^2}} \tag{1-46}$$

where the equations $T = 2\pi/\omega_d$ and $\omega_d = \omega_n\sqrt{1 - \zeta^2}$ where used to obtain the last term. If damping coefficient $c \ll c_c$, $\zeta \ll 1$, and we have the approximation

$$\delta \approx 2\pi\zeta \qquad (1\text{-}47)$$

Since the ratio of any two successive amplitudes is given by

$$\frac{x_1}{x_2} = \frac{x_2}{x_3} = \frac{x_3}{x_4} = \cdots = \frac{x_{n-1}}{x_n} = e^\delta$$

where δ is given by Eq. (1-46), the ratio x_1/x_n is given by

$$\frac{x_1}{x_n} = \left(\frac{x_1}{x_2}\right)\left(\frac{x_2}{x_3}\right)\left(\frac{x_3}{x_4}\right)\cdots\left(\frac{x_{n-1}}{x_n}\right) = (e^\delta)^{n-1}$$

Thus, since $x_1/x_n = e^{(n-1)\delta}$, we can calculate the logarithmic decrement from the equation

$$\delta = \frac{1}{n-1}\ln\frac{x_1}{x_n} \qquad (1\text{-}48)$$

Illustrative Example 1.18. In a damped vibration of the system in Fig. 1-19, the ratio of the first amplitude to the fourth amplitude is 2.00. Determine the viscous damping coefficient c if $k = 20$ lb/in. and $W = 38.6$ lb.

Solution. Since $\zeta = c/c_c$, Eq. (1-40) gives $c = 2\zeta\sqrt{kW/g}$. From Eq. (1-48), we have

$$\delta = \frac{1}{3}\ln 2.00 = 0.231$$

From Eq. (1-47), we have the approximation

$$\zeta = \frac{\delta}{2\pi} = \frac{0.231}{6.28} = 0.0369$$

Note that ζ is small enough to make this approximation valid. Thus, we have

$$c = 2\zeta\sqrt{\frac{kW}{g}} = 0.0738\sqrt{\frac{(20)(38.6)}{386}} = 0.104 \text{ lb-sec/inch}$$

1.9   Nonlinear Free Vibrations

Many engineering vibration problems are nonlinear. For example, we saw in Sec. 1.6 that most damping is nonlinear. Suppose for the spring-mass-damper system in Fig. 1-19 of Sec. 1.7 that we had velocity-squared damping. Then from Eq. (1-31) the equation of motion is

$$\frac{W}{g}\ddot{x} + a|\dot{x}|\dot{x} + kx = 0 \qquad (1\text{-}49)$$

which is a nonlinear differential equation because of the $a|\dot{x}|\dot{x}$ (or $\pm a\dot{x}^2$)

term. If we used the more general nonlinear damping force given by Eq. (1–32), the equation of motion would be

$$\frac{W}{g}\ddot{x} + a|\dot{x}|^b\dot{x} + kx = 0 \qquad (1\text{–}50)$$

which is also a nonlinear differential equation when $b \neq 0$. We saw in Ill. Ex. 1.4 of Sec. 1.3 that a pendulum-oscillation problem has a nonlinear differential equation of motion when the angular amplitude is large. Nonlinear differential equations of the form of Eq. (1–11) can be solved analytically by integration using elliptic integrals, as shown in Ill. Ex. 1.23, and as discussed later in this section.

The spring force is usually linear and its magnitude equals kx within its proper elastic limits. The spring may, however, vary nonlinearly with x, or may have a different value for spring constant k outside these limits. Hysteresis effects can also be important. Many other system components are linear only in certain ranges. The nonlinear spring problem and the pendulum-oscillation problem are two examples of nonlinear restoring forces. A *bilinear spring* is a spring that has two different spring constants. That is, $k = k_1$ if $x \leq x_1$ and $k = k_2$ if $x \geq x_1$, and the graph of the spring force versus x curve consists of two connected straight lines. The stress-strain curve of many elastic materials have similar bilinear relationships. If the spring-mass system in Fig. 1–11 of Sec. 1.3 had a bilinear spring, the equations of motion are linear in two regions, but the system itself must be considered to be nonlinear since it does not have the same linear relationship throughout its whole range of motion. Two other systems that are linear in two regions are the spring-mass example in Ill. Ex. 1.19 and the coulomb-damping example in Ill. Ex. 1.20.

A nonlinear free vibration of a single-degree-of-freedom system can be expressed, using Newton's second law, as $m\ddot{x} = F(\dot{x}, x)$ which can be rewritten as the following functional equation-of-motion form where $f(\dot{x}, x)$ denotes a nonlinear relationship with velocity $\dot{x}$ or displacement x (or with both terms) and where $f(\dot{x}, x) = -F(\dot{x}, x)/m$.

$$\ddot{x} + f(\dot{x}, x) = 0 \qquad (1\text{–}51)$$

Using the relationship $dt = dx/\dot{x}$ so that $\ddot{x} \equiv d\dot{x}/dt = \dot{x}\, d\dot{x}/dx$, we can modify the previous equation to the following simpler forms:

$$\dot{x}\frac{d\dot{x}}{dx} + f(\dot{x}, x) = 0$$

$$\frac{d\dot{x}}{dx} = g(\dot{x}, x) \qquad (1\text{–}52)$$

where $g(\dot{x}, x) = -f(\dot{x}, x)/\dot{x}$.

Solution of the nonlinear differential equations that represent a nonlinear vibration by analytical means are usually difficult, and usually require advanced mathematics. That is, there is no general analytical method for the exact solution of nonlinear differential equations. Each case must be treated separately, since solution methods vary with equation type, and the exact solutions are known for relatively few case types. The motion given by Eq. (1–51) and simplified to Eq. (1–52) may be graphically studied by the *phase-plane method*, which can be laborious and which is very briefly described in the next paragraph. It gives a visual picture of the vibration behavior. Perturbation and other special analytical methods are usually applied only when the magnitude of the nonlinear term is relatively small, and most of these analytical methods give only approximate solutions. Though the perturbation method is exemplified in Ill. Ex. 1.24, the reader is referred to the R. Bellman reference or the third W. T. Thomson reference in Appendix A-3 for a much more thorough discussion of this method. It should be noted that superposition cannot be used in the solution of nonlinear differential equations, as it can with linear differential equations. Fortunately, nonlinear differential equations, and, hence, nonlinear vibration problems, can be solved by analog computers and by digital computers using numerical methods. The use of numerical methods and analog and digital computers is further discussed in Secs. 2.13 and 2.14 and in Chaps. 6 and 7. Analog and digital computers are available at both college and in industry.

The phase-plane method for solving Eq. (1–52) is an approximation method that results in a plot (i.e., graph) of velocity $\dot{x}$ versus displacement x. Starting with the initial velocity $\dot{x}(0)$ and the initial displacement $x(0)$ to give the first point of the approximation curve, a short line segment is drawn to represent the slope $d\dot{x}/dx$ for each $(\dot{x}, x)$ point that is chosen. Thus, the resultant curve consists of a series of connected straight-line segments that approximate the variation of velocity $\dot{x}$ with displacement x for the given set of initial conditions, which is the starting point for the curve. Since this method is very laborious, its usage is generally confined to stability analyses and other specific system-behavior studies, rather than being used as a general method to obtain the complete solution to a problem for a given set of initial conditions. Phase-plane methods are discussed much more thoroughly in the N. Minorsky reference in Appendix A-3.

A nonlinear free vibration whose equation of motion has the functional form $m\ddot{x} = f(x)$, so that the acceleration varies in a nonlinear manner only with displacement x, can be solved by *two integrations*. Rewriting the equation of motion as

$$m\ddot{x} = m\left(\frac{d\dot{x}}{dt}\right) = m\dot{x}\left(\frac{d\dot{x}}{dx}\right) = f(x)$$

we can integrate the equation $m\dot{x}\,d\dot{x} = f(x)\,dx$ to obtain

$$\dot{x}^2 \equiv \left(\frac{dx}{dt}\right)^2 = \left(\frac{2}{m}\right)\int_X^x f(x)\,dx$$

where X is the displacement when velocity $\dot{x}$ equals zero (which occurs at the amplitude points of a vibration). Thus, the initial velocity $\dot{x}_o$ (if $x = 0$ when $t = 0$) can be computed from

$$\dot{x}_o^2 = \left(\frac{2}{m}\right)\int_X^0 f(x)\,dx \tag{1-53}$$

We can also integrate the equation $m\dot{x}\,d\dot{x} = f(x)\,dx$ with different limits to obtain

$$\dot{x}^2 \equiv \left(\frac{dx}{dt}\right)^2 = \dot{x}_o^2 + \left(\frac{2}{m}\right)\int_{x_o}^x f(x)\,dx$$

where $\dot{x}_o$ and x_o are the values of the velocity and displacement at $t = 0$. Solving the previous equation for dt, and integrating the result, gives

$$t = \int_{x_o}^x \frac{dx}{\sqrt{\dot{x}_o^2 + (2/m)[F(x) - F(x_o)]}} \tag{1-54}$$

where $[F(x) - F(x_o)]$ represents the value of the integral $\int_{x_o}^x f(x)\,dx$ in the last equation for $\dot{x}^2$. If we let the limits for Eq. (1–54) be $x_0 = 0$ and $x = X$, where X is the amplitude of the nonlinear vibration, then the calculated value of t will equal $T/4$, where T is the period of the vibration, because a vibration goes from $x = 0$ to $x = X$ in a quarter of a cycle.

Many nonlinear and linear systems can result in unstable oscillations, but some nonlinear systems are unstable at small magnitudes but stabilize when the motion magnitudes become larger. A vibration whose amplitude increases with time is unstable and it can cause damage to the system. The vibration shown in Fig. 1–20 of Sec. 1.7 is stable since its amplitude, which equals $Me^{-\zeta\omega_n t}$, decreases with time. Note that this oscillation would be unstable if the amplitude envelope were equal to $Me^{\zeta\omega_n t}$ since the amplitude would now increase exponentially with time. Thus, if the real part of either root n_1 or n_2, as given by Eq. (1–41), is positive, an unstable oscillation results. Courses in servo-mechanisms and control systems include a discussion of Nyquist, Bode, root-locus, and other methods that determine the stable and unstable oscillation ranges for given linear mechanical and electrical systems. More will be said about the solution of nonlinear vibrations in Secs. 2.13 to 2.15 and in Sec. 3.11.

Illustrative Example 1.19. Write the equation of motion for the spring-mass system in Fig. 1-23 assuming that the friction is negligible, and show that this system can be represented by the spring-mass system in Fig. 1-11 in Sec. 1.3 with a bilinear spring.

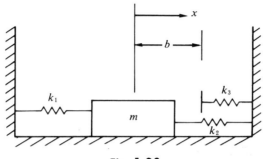

Fig. 1-23

Solution. This system is one that is linear in two regions because its equations of motion are

$$m\ddot{x} + (k_1 + k_2)x = 0 \qquad \text{if } x \leq b$$
$$m\ddot{x} + (k_1 + k_2 + k_3)x = 0 \qquad \text{if } x \geq b$$

Thus, this system can be represented by the system in Fig. 1-11 with a bilinear spring that has the properties

$$k = k_1 + k_2 \qquad \text{if } x \leq b$$
$$k = k_1 + k_2 + k_3 \qquad \text{if } x \geq b$$

Illustrative Example 1.20. Write the equation of motion for the spring-mass system in Fig. 1-24. Include the effect of coulomb damping.

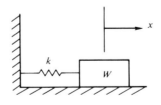

Fig. 1-24

Solution. From Eq. (1-29) in Sec. 1.6, we see that the magnitude of the coulomb damping force is μW, where μ is the coefficient of friction between the mass and the horizontal surface. This force acts to the left when the mass moves to the right (i.e., when velocity $\dot{x} > 0$). The coulomb damping force acts to the right when the mass moves to the left (i.e., when velocity $\dot{x} < 0$). Thus, this system is one that is linear in two regions because its equations of motion are

$$\frac{W}{g}\ddot{x} + kx = -\mu W \qquad \text{if } \dot{x} > 0$$

$$\frac{W}{g}\ddot{x} + kx = \mu W \qquad \text{if } \dot{x} < 0$$

The homogeneous part of the solution for either differential equation is given by Eq. (1-3) or (1-5) in Sec. 1.3, since the homogeneous equation is $(W/g)\ddot{x} + kx = 0$. The particular solution can be obtained by the method of undetermined coefficients. Substitution of $x = A$, where A is a constant, in both equations of motion yields $A = -\mu W$ and $A = \mu W$. Thus, the solutions are

$$x = M \sin(\omega_n t + \alpha) - \mu W \qquad \text{if } \dot{x} > 0$$
$$x = M \sin(\omega_n t + \alpha) + \mu W \qquad \text{if } \dot{x} < 0$$

where $\omega_n = \sqrt{kg/W}$ and constants M and α can be computed from the initial conditions. If the mass is given an initial displacement without an initial velocity, a plot of displacement x versus time t will show a damped oscillation where the amplitude decreases an amount of $4\mu W/k$ per cycle.

Illustrative Example 1.21. Write the equation of motion for the spring-mass system shown in Fig. 1-25, when the lower end of the spring is attached to a fixed pin joint. Assume the sliding surface to be sufficiently lubricated so that coulomb damping may be ignored.

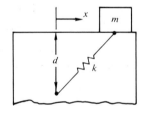

Fig. 1-25

Solution. When $x = 0$, the spring length is d. Since the spring length equals $\sqrt{x^2 + d^2}$ at position x, the magnitude of the spring force F_s is

$$F_s = k(\sqrt{x^2 + d^2} - d)$$

and the component of the spring force in the x-direction is

$$F_x = -k(\sqrt{x^2 + d^2} - d)\left(\frac{x}{\sqrt{x^2 + d^2}}\right) = -k\left(1 - \frac{d}{\sqrt{x^2 + d^2}}\right)x$$

Application of Newton's second law (i.e., $m\ddot{x} = F_x$) gives the following non-linear differential equation of motion:

$$m\ddot{x} + k\left(1 - \frac{d}{\sqrt{x^2 + d^2}}\right)x = 0$$

Illustrative Example 1.22. Sometimes a nonlinear differential equation can be transformed into a linear differential equation, which can be solved analytically. Briefly describe how this is done in orbital mechanics.

Solution. The motion of a space vehicle about the earth (which is, in essence, a periodic motion, as is planetary motion about the sun) is governed by the following nonlinear differential equation:

$$m(\ddot{r} - r\dot{\theta}^2) = -\frac{GMm}{r^2}$$

where m is the mass of the space vehicle, M is the mass of the earth, r is the distance of the space vehicle from the center of the earth, and G is the universal gravitational constant whose value is 3.44×10^{-8} ft^4/lb-sec^4. This equation is based on the fact that after rocket burnout, space vehicles are subjected only to the force of the gravitational pull of the earth. That is, $(\ddot{r} - r\dot{\theta}^2)$ is the acceleration component that is directed from the center of the earth (i.e., perpendicular to the curvilinear flight path), and GMm/r^2 is the Newton force of attraction between the earth and the space vehicle. In addition, $h = r^2\dot{\theta}$ is constant, where h is the angular momentum per unit mass. Substitution of the following equations

$$\dot{\theta} = \frac{h}{r^2}$$

$$\ddot{r} = -\frac{h^2}{r^2}\frac{d^2}{d\theta^2}\left(\frac{1}{r}\right)$$

transforms the nonlinear differential equation into the following linear one, after some mathematical manipulation is performed:

$$\frac{d^2u}{d\theta^2} + u = \frac{GM}{h^2}$$

The solution is

$$u = \frac{1}{r} = A\cos(\theta - \theta_0) + \frac{GM}{h^2}$$

After evaluation of the arbitrary constants A and θ_0, the solution can be expressed in the following compact form:

$$r = \frac{h^2/GM}{1 + \epsilon\cos\theta}$$

which, from analytic geometry, is the equation for a conic section, where, r is the distance from the focus and ϵ is the eccentricity. That is, the orbit path is elliptic if $\epsilon < 1$, parabolic if $\epsilon = 1$, and hyperbolic if $\epsilon > 1$. The center of the earth is the focus of the conic orbit path. The constant h can be evaluated from the value of $r^2\dot{\theta}$ at time zero (i.e., at rocket burnout). It can be shown that the orbit eccentricity ϵ can be determined from the conditions at rocket burnout as follows:

$$\epsilon = \sqrt{\left(\frac{r_0 v_0^2}{GM} - 1\right)^2 \cos^2 \beta_0 + \sin^2 \beta_0}$$

where $v_0, \beta_0,$ and r_0 are the velocity, flight path angle (measured from the perpendicular to r), and distance from the center of the earth at burnout. The value of angle θ at burnout can be determined from the following equation:

$$\tan \theta_0 = \frac{\sin \beta_0 \cos \beta_0}{\cos^2 \beta_0 - GM/r_0 v_0^2}$$

Illustrative Example 1.23. For the simple pendulum in Fig. 1-13, suppose that the maximum oscillation angle θ is 90°. Determine the period of this nonlinear vibration by use of integration methods.

Solution. Since Ill. Ex. 1.4 shows that the equation of motion is

$$l\ddot{\theta} = -g \sin \theta$$

we can use the equation $\ddot{\theta} = \dot{\theta}(d\dot{\theta}/d\theta)$ to obtain the following equation after a multiplication by $d\theta$

$$(l\dot{\theta})d\dot{\theta} = -(g \sin \theta) \, d\theta$$

Integration of the previous equation gives

$$\dot{\theta}^2 \equiv \left(\frac{d\theta}{dt}\right)^2 = \dot{\theta}_0^2 - \frac{2}{l} \int_{\theta_0}^{\theta} (g \sin \theta) \, d\theta$$

$$= \dot{\theta}_0^2 + \frac{2g}{l} (\cos \theta - \cos \theta_0)$$

where $\dot{\theta}_0$ and θ_0 are the angular velocity and angular displacement when $t = 0$. Solving the previous equation for dt gives

$$dt = \frac{d\theta}{\sqrt{\dot{\theta}_0^2 + (2g/l)(\cos \theta - \cos \theta_0)}}$$

If $\theta_0 = 0$ when $t = 0$, $\cos \theta - \cos \theta_0 = \cos \theta - 1 = -2(\sin^2 \theta/2)$. Substituting the preceding result in the preceding equation and integrating gives

$$t = \frac{1}{\dot{\theta}_0} \int_0^{\theta} \frac{d\theta}{\sqrt{1 - (4g/l\dot{\theta}_0^2) \sin^2 (\theta/2)}}$$

If $\theta_{\max}$ is the maximum oscillation angle and if $4g/l\dot{\theta}_0^2 > 1$, it can be shown that the preceding equation can be rewritten in the following form when the upper limit of the integral is $\theta_{\max}$

$$t \equiv \frac{T}{4} = \sqrt{\frac{l}{g}} \int_0^{\theta_{\max}} \frac{d\theta}{\sqrt{1 - \sin^2 (\theta_{\max}/2) \sin^2 \theta}}$$

The integral on the right side of the previous equation is an elliptic integral of the first kind whose values are tabulated in mathematics tables. For $\theta_{\max} = \pi/2$, the period T equals $7.42 \sqrt{l/g}$.

Illustrative Example 1.24. For the spring-mass system in Fig. 1-1a, suppose that the spring force is nonlinear and equals $kx + bx^2$, where b is a very small term and x is the displacement of the mass from the equilibrium position. Use the perturbation method to approximate the variation of displacement x with time when the initial displacement and velocity of the mass are X and zero, respectively.

Solution. The equation of motion for this system is

$$m\ddot{x} = -kx - bx^2$$

which can be rewritten as

$$\ddot{x} + \omega_n^2 x + ax^2 = 0$$

where $\omega_n^2 = k/m$ and $a = b/m$. Note that a is a very small term (unless m is very small) and that the ax^2-term is always positive. The perturbation method is useful for solving nonlinear vibration problems whose mathematical form, when there is no applied force on the system, is as follows:

$$\ddot{x} + \omega_n^2 x + a f(x, \dot{x}) = 0$$

where a must have a very small value and where the function $f(x, \dot{x})$ includes all of the nonlinear terms. Note that the solution is known when $a = 0$. The perturbation method seeks a series solution whose coefficients are a^i as follows, where the x_i-terms are functions of time t.

$$x = x_0 + ax_1 + a^2 x_2 + a^3 x_3 + \cdots = \sum_{i=0}^{n} a^i x_i$$

We shall use only the first two terms as follows for our approximate solution for the given problem

$$x = x_0 + ax_1$$

We would also guess that the frequency ω of this nonlinear vibration should not equal the natural frequency ω_n, so we shall also use a two-term approximation for the square of frequency ω as follows:

$$\omega^2 = \omega_n^2 + a\omega_1^2$$

Substitution of these two equations for x and ω^2 into the equation of motion for the nonlinear spring-mass system gives

$$\ddot{x}_0 + a\ddot{x}_1 + (\omega^2 - a\omega_1^2)(x_0 + ax_1) + a(x_0^2 + 2ax_0 x_1 + a^2 x_1^2)$$
$$= (\ddot{x}_0 + \omega^2 x_0) + a(\ddot{x}_1 + \omega^2 x_1 - \omega_1^2 x_0 + x_0^2)$$
$$+ a^2(2x_0 x_1 - \omega_1^2 x_1) + a^3(x_1^2) = 0$$

Since a^2 and a^3 are very small numbers, we shall utilize only the first two terms of the preceding equation. If we make both the first term and the second term (whose coefficient is a) equal to zero, we obtain the two following equations:

$$\ddot{x}_0 + \omega^2 x_0 = 0$$
$$\ddot{x}_1 + \omega^2 x_1 = \omega_1^2 x_0 - x_0^2$$

Let us use the following initial conditions for these two differential equations: $x_0(0) = X$ and $\dot{x}_0(0) = \dot{x}_1(0) = 0$. The solution for x_0 in the first differential equation is $X \cos \omega t$ and substitution in the second equation gives

$$\ddot{x}_1 + \omega^2 x_1 = \omega_1^2 X \cos \omega t - X^2 \cos^2 \omega t$$

$$= \omega_1^2 X \cos \omega t - \frac{X^2}{2}(\cos 2\omega t + 1)$$

whose solution is as follows:

$$x_1 = A \cos \omega t + B \sin \omega t + \frac{\omega_1^2 X}{2\omega} t \sin \omega t + \frac{X^2}{6\omega^2} \cos 2\omega t - \frac{X^2}{2\omega^2}$$

The initial condition $x_1(0) = 0$ gives $A = X^2/3\omega^2$ and the initial condition $\dot{x}_1(0) = 0$ gives $B = X^2/2\omega^2$. The third term, $(\omega_1^2 X/2\omega)t \sin \omega t$, is a large term (because it has the factor t) which we shall set equal to zero because the term ax_1 must be a small correction to be added to x_0 for the perturbation method to be properly applied. Thus, $\omega_1^2 \approx 0$ and $\omega \approx \omega_n$, and this perturbation method approximation for displacement x is

$$x = x_0 + ax_1 = X \cos \omega_n t + \frac{X^2}{3\omega_n^2} \cos \omega_n t + \frac{X^2}{2\omega_n^2} \sin \omega_n t$$

$$+ \frac{X^2}{6\omega_n^2} \cos 2\omega_n t - \frac{X^2}{2\omega_n^2}$$

PROBLEMS

1.1. Represent an airplane wing, of decreasing width to the tip, by a system composed of beam segments and four concentrated masses m_1, m_2, m_3, and m_4 that are equally spaced.

1.2. Represent a propeller by a system composed of beam segments, a central mass whose value is m (the mass of the central hub), and two other masses of value M (the mass of a propeller half).

1.3. In Fig. 1–1b, suppose that the mass of the shaft is not negligible. Divide this shaft into three segments, represent this disk-shaft system by an inertia disk at the center of each segment, and connect massless elastic shafts to the four disks, in order to obtain an equivalent system composed of inertia disks and massless elastic shafts.

1.4. Repeat Prob. 1.3, except connect the four disks by massless springs, to represent the elasticity of the shaft segments.

1.5. A mass M is located at the center of a horizontal elastic string that is firmly tied at both ends. Attached to the bottom of mass M is a spring-mass system as shown in Fig. 1–1b. Draw an equivalent system composed of only massless springs and concentrated masses.

1.6. A massless horizontal lever of length L is pivoted at its center, and is rotated by a torque $T \cos \omega t$. If each end of the lever has a concentrated mass m and is restrained by a vertical spring of spring constant k that is connected to the lever end and to an upper ceiling, draw an equivalent spring-mass system composed of only concentrated masses and massless springs.

1.7. Represent a complete airplane by a system composed of beam segments and four concentrated masses, where m_1 is the mass of the fuselage, m_2 is the mass of the tail section, and m_3 is the mass of each wing.

1.8. If the spring in the spring-mass system in Fig. 1–1a is stretched a distance x_{st} due to the weight of mass m, determine the natural frequency of the system as a function of this stretched distance.

1.9. Find the frequency and period of a harmonic motion whose equation is $x = 3 \sin (11\pi t + \pi/3)$, where x is in inches, when time t is in seconds and when the angle is in radians. Also find the maximum displacement, velocity, and acceleration, and the values of the displacement, velocity, and acceleration at $t = 0$ and at $t = 0.4$ sec.

1.10. For a harmonic motion of the form of Eq. (1–5), where $\omega_n = 50$ rad/sec, find the values of M and α if the initial conditions are $x(0) = 0.200$ in. and $\dot{x}(0) = 30$ in./sec. Also determine the constants A and B for this motion to also be represented by Eq. (1–3).

1.11. Should a U-tube manometer that has a column of mercury nine inches long be connected to the discharge side, for measuring the pressure of a one-cylinder blower that makes 100 strokes per minute?

1.12. The weight of a floating hydrometer for measuring the specific gravity of a liquid is 0.0671 lb. If the diameter of the upper section of the hydrometer, which protrudes above the surface, is 0.350 in., find the frequency of vibration when this hydrometer bobs up and down in a fluid whose specific gravity is 1.73.

1.13. Suppose that we had a compound pendulum composed of a semi-circular disk that is pivoted at the center of its straight boundary. Write the equation of motion and determine the natural frequency of this system and the string length for an equivalent simple pendulum.

1.14. When an 8–lb connecting rod is supported at the wrist pin end and given an initial displacement so that it oscillates, the measured frequency is 54 cycles/min. If the center of gravity is 13 in. from the pivot point, what is the moment of inertia of this connecting rod about its center of gravity?

1.15. If a rotating machine (whose moment of inertia about the axis of rotation is I_o) is supported atop four vertical springs of equal spring constant k, where each spring is mounted between the machine and a foundation, and is at a distance b from the axis of rotation, write the equation of motion and determine the natural frequency of this system.

1.16. A three-ton elevator is moving downward at a constant velocity of 4 ft/sec. An accident occurs when the cable length is 70 ft, which stops the drum rotation. The cross-sectional area of the metal-wire cable is 1.5 in². and its modulus of elasticity is 8×10^6. Neglecting the rope weight, determine the amplitude and frequency of the ensuing vibration and the maximum stress in the cable due to this accident.

1.17. Calculate the natural frequencies of the two systems in Fig. 1–26 when the bar is considered to be rigid and weightless.

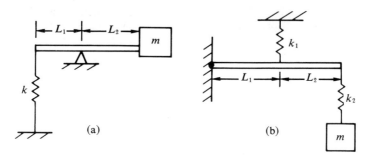

Fig. 1-26

1.18. Calculate the natural frequencies of the three systems in Fig. 1–27.

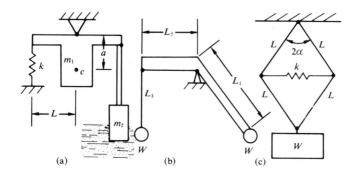

Fig. 1-27

1.19. Use the energy method to obtain the equation of motion and the natural frequencies of the four systems in Fig. 1–28. Verify your results using Newton's second law.

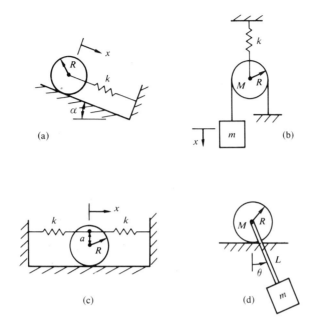

Fig. 1-28

1.20. Use the energy method to obtain the natural frequencies of the three systems in Fig. 1–29.

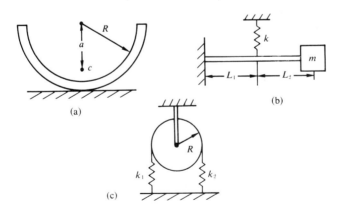

Fig. 1-29

1.21. Use the energy method to determine the frequency of roll for a boat in still water. Let θ denote the angle of rotation about the center of gravity from the vertical position and let d represent the distance of the metacenter above the center of gravity. The *meta-center* is defined to be the point at which the buoyant force intersects a line through the center of gravity which is vertical when the boat is not rolling.

1.22. Use the energy method to determine the natural frequency of a beam that is simply supported at both ends, has a concentrated mass m at its midspan, and weighs w lb per unit length. For vibration, use the same deflection equation as that for the static deflection of a simply-supported beam that has a concentrated mass at its midspan.

1.23. Verify Eqs. (1–27) and (1–28).

1.24. Determine the equivalent spring constant for an elastic massless beam that is simply supported at both ends and which supports a concentrated mass m that is located at a distance a from the left end. Also determine the equivalent spring constant for a massless beam that is rigidly supported at one end and simply supported at the other end and which supports a concentrated mass m at its midspan.

1.25. Suppose that the upper shaft in Fig. 1–16c is 1.5 in. in diameter and 9 in. long and that the lower shaft, which is of the same material, is 3 in. in diameter and 13 in. long. Specify the two lengths required if we wish to replace these two shafts with two 2.5–in. diameter shafts of the same material and having the same values for k_{t1} and k_{t2}, as before. The equation for k_t is given in Ill. Ex. 1.6.

1.26. Calculate the equivalent spring constant and the natural frequency of the system in Fig. 1–18 when a vertical spring, k_3 replaces the rigid support that connects weight W to the rigid horizontal bar of length 2a.

1.27. Calculate the equivalent spring constant for the system in Fig. 1–30.

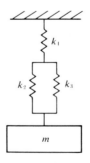

Fig. 1-30

1.28. Calculate the equivalent viscous damping coefficient during a sinusoidal oscillation of 0.100–in. amplitude and 15–rad/sec frequency for coulomb damping when the normal force is 38 lb and the coefficient of friction is 0.18.

1.29. Repeat Prob. 1.28 for a damping force that equals $0.123 \, |\dot{x}|\dot{x}$ lb for $\dot{x}$ in in./sec.

1.30. For the spring-mass-damper system in Fig. 1–19, suppose that $W = 60$ lb, $k = 45$ lb/in., and $c = 0.45$ lb–sec/in. Calculate the displacement and velocity of the mass as a function of time if the mass is given an initial displacement of 2 in. and released without imparting an initial velocity. Also calculate the acceleration, velocity, and displacement of the mass at 2 sec.

1.31. Repeat Prob. 1.30 for a zero initial displacement and an initial velocity of 3 in./sec.

1.32. Repeat Prob. 1.30, for an initial velocity of 2 in./sec.

1.33. Compute the critical damping coefficient, damping ratio, logarithmic decrement, and the ratio of any two consecutive amplitudes during a free vibration of the spring-mass-damper system in Prob. 1.30. Also compute the ratio of the damped natural frequency to the undamped natural frequency, and the amplitude of the vibration after one cycle.

1.34. A 30–lb machine is suspended from a spring, whose spring constant is 15 lb/in., and a linear damper, whose damping coefficient is 0.04 lb–sec/in., is attached to this weight and to the factory floor. Compute the undamped and damped natural frequencies, the critical damping coefficient, the logarithmic decrement, and the vibration amplitude five cycles after a 2–in. initial displacement.

1.35. Write the equation of motion for small oscillations, and compute the critical damping coefficient and the damped natural frequency for the system in Fig. 1–31, where the weightless bar is pivoted at its left end.

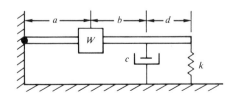

Fig. 1-31

1.36. Write the equation of motion for the spring-mass system in Fig. 1–32 and state the spring requirements for representing this system by the system in Fig. 1–11.

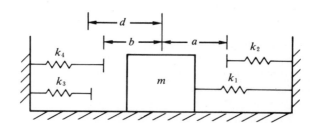

Fig. 1-32

1.37. In Fig. 1–24, suppose that $W = 12$ lb, $k = 20$ lb/in., and $\mu = 0.18$. Compute the variation of the velocity and displacement with time when there is an initial displacement of 6 in. without an initial velocity.

1.38. For Prob. 1.37, compute the maximum velocity, the amplitude decrease per cycle, and the displacement where the mass stops moving (i.e., comes to rest).

1.39. In Fig. 1–24, suppose that $W = 8$ lb, $k = 18$ lb/in., and $\mu = 0.14$. If the mass is pulled a distance of 9 in., compute the vibration frequency, the amplitude at the start of the fourth cycle, and the displacement and number of cycles when the mass comes to rest.

1.40. Verify the fourth equation in Ill. Ex. 1.22.

1.41. Verify Eq. (1–54).

1.42. Use Eq. (1–53) to show that the maximum veloicty equals $X\sqrt{k/m}$, and use Eq. (1–54) to compute the period of vibration for the spring-mass system in Fig. 1–1a.

1.43. For the system in Fig. 1–15, suppose that angle θ has a maximum value of 90° for the oscillation of the rolling cylinder. Determine the period of this nonlinear oscillation.

1.44. Verify the solution for x_1 in Ill. Ex. 1.24.

1.45. For the spring-mass system in Fig. 1–1a, suppose that the spring force is nonlinear and equals $kx + bx^3$, where b is a very small term. Use the perturbation method to approximate the variation of displacement x with time t when the initial conditions are $x(0) = X$ and $\dot{x}(0) = 0$.

2 Single-Degree-of-Freedom Forced Vibrations

2.1 Introduction

This chapter deals with the analysis of forced vibrations of single-degree-of-freedom systems. That is, the previous chapter analyzed the vibrations of one-degree-of-freedom systems that do not have any applied forces, while this chapter will study the effect on such systems when external forces are applied. Several different types of external forces (harmonic, self-excited, general machinery, shock, random, etc.) are considered (some in a very introductory manner). Because of the differing nature of these forces, some require special solution techniques. Various applications, such as vibration-measuring instruments and reciprocating and rotating unbalance, are included. The method of undetermined coefficients, impedance method, Fourier series, Laplace transform, and numerical solution techniques are presented; and it will be seen that each solution technique has its own particular advantages and limitations.

2.2   *Forced Harmonic Vibrations without Damping*

Consider a weight W suspended vertically from a spring, as shown in Fig. 2–1. The damping is negligible, and the weight is subjected to a periodic external disturbing force, $F \sin \omega t$. Application of Newton's second law gives

$$\frac{W}{g}\ddot{x} = F \sin \omega t - kx$$

which can be rewritten as

$$\ddot{x} + \frac{kg}{W}x = \frac{Fg}{W}\sin \omega t \qquad (2\text{–}1)$$

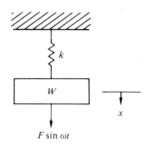

$F \sin \omega t$

Fig. 2-1

Since the homogeneous equation is $\ddot{x} + (kg/W)x = 0$, the homogeneous solution x_h is given by Eq. (1–3) and Eq. (1–5), as follows:

$$x_h = A \cos \omega_n t + B \sin \omega_n t = M \sin (\omega_n t + \alpha)$$

where $\omega_n = \sqrt{kg/W}$. Using the method of undetermined coefficients, which was presented in the reader's differential equations course and which is easier to apply than the more general variation-of-parameters method, the particular solution is of the form

$$x_p = C \sin \omega t + D \cos \omega t$$

Substitution of x_p in Eq. (2–1) gives

$$\ddot{x}_p + \frac{kg}{W}x_p = (-C\omega^2 \sin \omega t - D\omega^2 \cos \omega t) + \frac{kg}{W}(C \sin \omega t + D \cos \omega t)$$

$$= \frac{Fg}{W}\sin \omega t$$

Collecting the coefficients of $\sin \omega t$ and $\cos \omega t$, we have

$$\left[-\frac{1}{g}(C\omega^2 W) + Ck\right]\sin \omega t + \left[Dk - \frac{1}{g}(D\omega^2 W)\right]\cos \omega t = F \sin \omega t$$

Since the coefficients of sin ωt give

$$-\frac{1}{g}(C\omega^2 W) + Ck = C\left[k - \frac{\omega^2 W}{g}\right] = F$$

the solution for C is

$$C = \frac{F}{k - (W/g)\omega^2} = \frac{F/k}{1 - (\omega/\omega_n)^2} = \frac{X_0}{1 - r^2}$$

where X_0 is the *static deflection* due to a steady force of magnitude F and equals F/k, and where r is the *frequency ratio* (i.e., the applied frequency ω divided by the natural frequency ω_n). Since the coefficients of cos ωt give

$$Dk - \frac{1}{g}(D\omega^2 W) = D\left[k - \frac{\omega^2 W}{g}\right] = 0$$

then $D = 0$, and the complete solution for x is

$$x = x_h + x_p = A \cos \omega_n t + B \sin \omega_n t + \frac{F/k}{1 - r^2} \sin \omega t \qquad (2\text{-}2)$$

where arbitrary constants A and B are evaluated from the initial conditions. Section 2.5 presents an easier method to apply, than that of undetermined coefficients, to obtain the steady-state solution when the applied force is harmonic (e.g, $F \sin \omega t$ or $F \cos \omega t$). Since the homogeneous solution may also be expressed as $M \sin (\omega_n t + \alpha)$, the solution is the sum of two sinusoids that have different frequencies (i.e., ω_n and ω). The last term of Eq. (2–2) shows that displacement x approaches infinity when the applied frequency ω equals the natural frequency ω_n, which is a condition of *resonance*. This is why the determination of the natural frequency of a system is important to designers, who are usually more interested in maximum displacements than in the timewise variation of the displacement. It will be seen in Sec. 2.3 that damping causes finite displacements at resonance, though these displacements are still large. If $x = x_0$ and $\dot{x} = v_0$ when time $t = 0$, calculation shows that $A = x_0$ and $B = (v_0/\omega_n) - Fr/k(1 - r^2)$, so that Eq. (2–2) becomes

$$x = x_0 \cos \omega_n t + \left(\frac{v_0}{\omega_n} - \frac{Fr/k}{1 - r^2}\right) \sin \omega_n t + \frac{F/k}{1 - r^2} \sin \omega t \qquad (2\text{-}3)$$

Illustrative Example 2.1.   Discuss what happens when the system in Fig. 2-1 has zero initial displacement and velocity and when impressed frequency ω almost equals the natural frequency ω_n.

Solution.   Substitution of $x_0 = 0$, $v_0 = 0$, and $\omega/\omega_n = r \approx 1$ into Eq. (2-3) gives the following solution:

$$x = \frac{F/k}{1 - r^2}(\sin \omega t - \sin \omega_n t)$$

Letting $\omega_n = \omega + \epsilon$, where ϵ has a very small value, the previous equation may be rewritten as

$$x = 2\left(\frac{F/k}{1 - r^2}\right)\cos\frac{\epsilon}{2}t \sin\left(\omega + \frac{\epsilon}{2}\right)t$$

At times the two sinusoids add to each other, and at other times they cancel, resulting in a *beating phenomenon*, as shown in Fig. 2-2. The preceding equation shows that the beating phenomenon can be considered to be a sine wave of frequency $\omega + \epsilon/2$ and varying amplitude $[2F/k(1 - r^2)] \cos (\epsilon t/2)$. This example also shows that the sum of two harmonic (i.e., sine or cosine) motions at different frequencies is not harmonic.

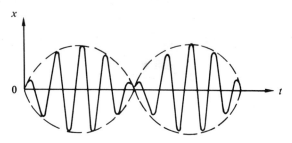

Fig. 2-2

Illustrative Example 2.2.　At resonance, where $\omega = \omega_n$, Eq. (2-2) is not a proper solution of Eq. (2-1) since the assumed solution for x_p (i.e., $C \sin \omega_n t + D \cos \omega_n t$) is now contained in the homogeneous solution x_h, which is still given by Eq. (1-3). Obtain the proper particular solution x_p.

Solution.　When $\omega = \omega_n$, the particular solution must be of the form

$$x_p = Ct \sin \omega_n t + Dt \cos \omega_n t$$

Substituting Eq. (2-1) and solving for C and D, as before, gives $C = 0$ and $D = -F/2\sqrt{kW/g}$, so that

$$x_p = -\frac{F/2}{\sqrt{kW/g}}t \cos \omega_n t$$

Thus, since the steady-state amplitude varies directly with time, it takes a while for this amplitude to build up to a large unsafe value while at resonance. It is safe, then, for a machine or system to quickly pass through the resonance condition in order to attain some other frequency.

Illustrative Example 2.3.　Suppose, for the spring-mass system in Fig. 2-1, that $W = 38.6$ lb, $k = 40$ lb/in, $F = 60$ lb, and $\omega = 12$ rad/sec. Find the displacement and velocity variation with time if the initial displacement is 3 in. and the initial velocity is 5 in./sec.

Solution.   The angular natural frequency is

$$\omega_n = \sqrt{\frac{kg}{W}} = \sqrt{\frac{(40)(386)}{38.6}} = 20 \text{ rad/sec}$$

Thus, $r = \omega/\omega_n = 12/20 = 0.600$, and

$$\frac{F/k}{1 - r^2} = \frac{60/40}{1.00 - 0.36} = 2.35$$

Substitution in Eq. (2-3) gives

$$x = 3.00 \cos 20t + \left[\frac{5.00}{20.0} - 2.35(0.600)\right] \sin 20t + 2.35 \sin 12t$$

$$= 3.00 \cos 20t - 1.16 \sin 20t + 2.35 \sin 12t$$

Differentiation of the previous displacement equation gives the following equation for velocity:

$$\dot{x} = -60.0 \sin 20t - 23.2 \cos 20t + 28.2 \cos 12t$$

2.3   Forced Harmonic and Self-Excited Vibrations with Viscous Damping

Consider the spring-mass-viscous-damper system in Fig. 2–3, in which the mass is subjected to a periodic, external disturbing force $F \sin \omega t$. Using Newton's second law, its equation of motion is

$$\frac{W}{g} \ddot{x} + c\dot{x} + kx = F \sin \omega t \tag{2-4}$$

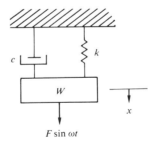

Fig. 2-3

where $c\dot{x}$ is the viscous damping force. The homogeneous solution x_h is given by Eq. (1–45), Eq. (1–43), or Eq. (1–39) in Sec. 1.7 depending upon whether damping ratio ζ equals, is less than, or exceeds unity. An oscillation results only if $\zeta < 1$. The homogeneous solution x_h is called the *transient solution* in vibration analysis. It should be noted that the transient solution dies out with time since it approaches zero (i.e., the amplitude exponentially decreases with increasing time and soon disappears) for all three ζ-cases, as shown by Eqs. (1–39), (1–43), and (1–45). For this reason, engineers are often only interested in the particular or *steady-state solution* since this part of the solution remains with time. The particular solution can be obtained by using the method of undetermined coefficients as we did in Sec. 2.2. Thus, the particular solution is of the form

$$x_p = C \sin \omega t + D \cos \omega t \tag{2-5}$$

Substituting in Eq. (2–4) and calculating coefficients C and D, as we did in Sec. 2.2, we obtain

$$C = \frac{F[k - (W/g)\omega^2]}{[k - (W/g)\omega^2]^2 + (c\omega)^2} \tag{2-6}$$

$$D = \frac{-Fc\omega}{[k - (W/g)\omega^2]^2 + (c\omega)^2} \tag{2-7}$$

Equation (2–5) for the particular solution x_p can be rewritten in the following more useful form:

$$x_p = X \sin (\omega t - \phi) \tag{2-8}$$

where the *steady-state amplitude* X is given by

$$X = \sqrt{C^2 + D^2} = \frac{F}{\sqrt{[k - (W/g)\omega^2]^2 + (c\omega)^2}} \tag{2-9}$$

Thus, the particular solution is

$$x_p = \frac{F}{\sqrt{[k - (W/g)\omega^2]^2 + (c\omega)^2}} \sin (\omega t - \phi) \tag{2-10}$$

where phase angle ϕ is given by

$$\phi = \tan^{-1}\left[\frac{c\omega}{k - (W/g)\omega^2}\right] \tag{2-11}$$

After substitution of the dimensionless terms $X_0 = F/k$, $r = \omega/\omega_n = \omega/\sqrt{kg/W}$, and $\zeta = c\omega_n/2k$, into Eqs. (2–10) and (2–11) and after doing some algebraic manipulation, we obtain the following equations:

$$\text{M.F.} = \frac{X}{X_0} = \frac{1}{\sqrt{(1 - r^2)^2 + (2\zeta r)^2}} \tag{2-12}$$

$$\phi = \tan^{-1}\left(\frac{2\zeta r}{1 - r^2}\right) \tag{2-13}$$

where amplitude ratio X/X_0 (or M.F.) is called the *magnification factor* (which is a dimensionless way of expressing the steady-state amplitude), r is the frequency ratio, and ζ is the damping ratio. Equations (2–12) and (2–13), which are plotted in Figs. 2–4 and 2–5, show that both the magnification factor and the phase angle can be expressed only as functions of the two dimensionless terms r and ζ.

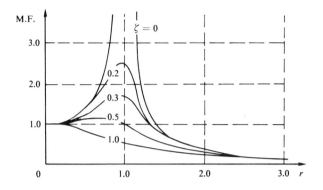

Fig. 2-4

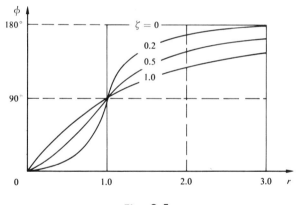

Fig. 2-5

Note from Eq. (2–8) that the steady-state motion of the mass (i.e., displacement response x_p) is harmonic (i.e., sinusoidal), has the same frequency as the applied force, and lags the applied force by an angle ϕ. Also note that the steady-state solution is not a function of the initial conditions. That is, only the transient part of the solution, which dies out with time, varies with the initial displacement and initial velocity. Note from Eq. (2–12) that M.F. $= 1/(1 - r^2)$ when ζ, and hence, c, equals zero (i.e., when there is no damping). From Eqs. (2–9) and (2–11), we see that amplitude X equals $F/c\omega$ and that phase angle ϕ equals $90°$ at resonance, where $\omega = \omega_n = \sqrt{kg/W}$. Also note from this result and in Fig. 2–4 that at resonance when $r = 1$ (i.e., $\omega = \omega_n$), the magnification factor approaches infinity when $c = \zeta = 0$. Note further how increased values of damping (i.e., c or ζ) reduces M.F. and hence the steady-state amplitude at resonance where $r = 1$. In Fig. 2–5, where phase angle ϕ is the angle between the force and the motion, note the effect of damping and note that with no damping (i.e., $\zeta = 0$), the force and motion are in phase for $\omega < \omega_n$ but are 180° out of phase for $\omega > \omega_n$. That is, when ω is larger than natural frequency ω_n, the weight W moves in the opposite direction to the motion of the applied force $F \sin \omega t$. Also note in Fig. 2–5 that for all values of ζ, phase angle ϕ approaches zero as r approaches zero and that $\phi = 90°$ at resonance where $r = 1$.

It should be noted that the equations obtained so far [i.e., Eqs. (2–10) to (2–13)] apply only for *harmonic* (i.e., sine and cosine) external forces, where it can be shown that an external force of $F \cos \omega t$ gives similar results. In some of the applications given later in this chapter, it will be seen that the harmonic external force can be applied to the mass (as shown in Fig. 2–3), can be applied to the support of the system, or can occur as the result of an unbalanced machine component. Self-excited vibrations will be discussed in the next paragraph. Other types of external forces (general, shock, random, etc.) and their effects on vibratory motion will be discussed later in this chapter.

In most forced vibration applications, the external forces will vary only with time t. For a *self-excited vibration*, the external or exciting force is a linear or nonlinear function of the displacement, velocity, or acceleration of a system mass, two examples of which are aircraft wing flutter and automobile wheel shimmy. The motion of the system will cause the energy of the system, and hence the amplitude, to increase unless this energy is balanced by energy dissipation due to damping (e.g., heat caused by friction). Thus, for stable systems, the total or net energy decreases with time. For unstable systems, there is a continuous increase in vibration amplitude due to a continuous net increase of energy to the system. In other words, the amplitude will increase with time if the system motion causes the net energy to increase.

Illustrative Example 2.4. Suppose that the system in Fig. 2–3 is excited by a force that is proportional to the velocity of the mass, rather than by $F \sin \omega t$. Discuss the stability of this system.

Solution. Use of Newton's second law gives the following linear differential equation of motion:

$$\frac{W}{g}\ddot{x} + c\dot{x} + kx = F_0\dot{x}$$

which can be rewritten as

$$\frac{W}{g}\ddot{x} + (c - F_0)\dot{x} + kx = 0$$

By analogy to Eqs. (1-37) and (1-38) in Sec. 1.7, the solution is given by Eq. (1-39) where

$$n_{1,2} = \frac{g}{2W}(F_0 - c \pm \sqrt{(c - F_0)^2 - 4kW/g})$$

Thus, the solution may be expressed as

$$x = e^{(F_0 - c)gt/2W}(C_1 e^{bt} + C_2 e^{-bt})$$

where $b = (g/2W)\sqrt{(c - F_0)^2 - 4kW/g}$. Thus, from the previous equation, it is seen that the amplitude of the motion exponentially increases with time when $F_0 > c$, so that the motion is unstable, and that the motion is stable when $F_0 < c$ since the amplitude exponentially decreases with time. Note that if $F_0 > c$, we essentially have a negative damping force $(c - F_0)\dot{x}$ which will cause both the net system energy and the vibration amplitude to increase with time. Also, by analogy to Sec. 1.7, the motion is oscillatory when $4kW/g > (c - F_0)^2$; otherwise, the motion is aperiodic.

2.4   Comparison of Rectilinear and Rotational Systems

Many different types of physical systems are mathematically analogous. This is true of rectilinear and rotational mechanical systems. That is, from our dynamics course, we know that their differential equations of motion have the same mathematical form. For example, the equation of motion for the rotational system shown in Fig. 2-6, using the dynamics equation $I\ddot{\theta} = \sum M$, is

$$I\ddot{\theta} = T \sin \omega t - c_t \dot{\theta} - k_t \theta$$

which can be rewritten as

$$I\ddot{\theta} + c_t \dot{\theta} + k_t \theta = T \sin \omega t \qquad \qquad (2\text{-}14)$$

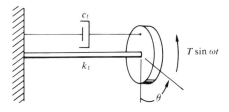

Fig. 2-6

where $c_t \dot{\theta}$ is the damping moment and c_t is the *rotational damping coefficient* in in.-lb-sec/rad. Note the mathematical analogy between

Eqs. (2–4) and (2–14). The individually analogous terms are given in Table 2–1, where the units are enclosed in parentheses beside the symbol for the rectilinear or rotational physical term. Though this text will concentrate mostly on rectilinear motion examples, the use of this analogy table should allow the reader to easily extrapolate the results to equivalent rotational systems.

Physical Term	Rectilinear System	Rotational System
Time	t (sec)	t (sec)
Displacement	x (in.)	θ (rad)
Inertia	m (lb-sec^2/in.)	I (in.-lb-sec^2)
Force or Torque	F (lb)	T (in.-lb)
Spring Constant	k (lb/in.)	k_t (in.-lb/rad)
Damping Coefficient	c (lb-sec/in.)	c_t (in.-lb-sec/rad)

Table 2-1

2.5   The Mechanical Impedance Method

Rather than use such classical solution techniques as the method of undetermined coefficients or variation of parameters, it is easier to use a complex algebra form of the mechanical impedance method (which shall be called the *mechanical impedance method* in this text), for obtaining the steady-state solutions of harmonic forced vibrations. The three limitations of this complex algebra method are that it only gives the steady-state response (i.e., only the particular solution and not the transient solution, that the differential equation of motion must be linear, and that the externally applied force must be harmonic (e.g., $F \sin \omega t$ or $F \cos \omega t$). This method is based on the fact, shown in Sec. 2.3, that if the differential equation of motion is linear, and if the applied force is harmonic, the steady-state motion of the mass is harmonic at the same frequency with a lagging phase angle. That is, the steady-state motion of a single-degree-of-freedom system has the form $X \sin (\omega t - \phi)$ when the applied force is $F \sin \omega t$, and has the form $X \cos (\omega t - \phi)$ when the applied force is $F \cos \omega t$. The mechanical impedance method evaluates amplitude X and phase angle ϕ for the steady-state motion. Since *Euler's formula* is

$$e^{i\theta} = \cos \theta + i \sin \theta$$

where $i = \sqrt{-1}$, then an external force $Fe^{i\omega t}$ may be expressed as

$$Fe^{i\omega t} = F \cos \omega t + iF \sin \omega t$$

Thus, $F \cos \omega t = Re[Fe^{i\omega t}]$, where the symbol Re denotes the real part of $Fe^{i\omega t}$, and $F \sin \omega t = Im[Fe^{i\omega t}]$, where the symbol Im denotes the imaginary part of $Fe^{i\omega t}$. If the applied force has the complex exponential form $Fe^{i\omega t}$, then the displacement x has the form

$$x = Xe^{i(\omega t - \phi)} = \bar{X}e^{i\omega t} \qquad (2\text{--}15)$$

where X is the amplitude of the motion, ϕ is the lagging phase angle, and $\bar{X} = Xe^{-i\phi}$. Differentiation gives the following equations for velocity $\dot{x}$ and acceleration $\ddot{x}$:

$$\dot{x} = i\omega Xe^{i(\omega t - \phi)} = i\omega x = i\omega \bar{X}e^{i\omega t} \qquad (2\text{--}16)$$
$$\ddot{x} = i^2\omega^2 Xe^{i(\omega t - \phi)} = -\omega^2 x = -\omega^2 \bar{X}e^{i\omega t} \qquad (2\text{--}17)$$

Now let us consider the spring-mass-damper system in Fig. 2–3 in Sec. 2.3 with an external force of $Fe^{i\omega t}$. The equation of motion is

$$\frac{W}{g}\ddot{x} + c\dot{x} + kx = Fe^{i\omega t}$$

Substitution of Eqs. (2–15), (2–16), and (2–17) and cancellation of $e^{i\omega t}$ gives

$$\left(-\frac{W}{g}\omega^2 + ic\omega + k\right)\bar{X} = F$$

where $\bar{X}$ is the complex amplitude $Xe^{-i\phi}$. Thus,

$$\bar{X} = Xe^{-i\phi} = \frac{F}{[k - (W/g)\omega^2] + ic\omega}$$

and amplitude X is given by

$$X = \left|\frac{Fe^{i\phi}}{[k - (W/g)\omega^2] + ic\omega}\right| = \frac{F}{\sqrt{[k - (W/g)\omega^2]^2 + (c\omega)^2}} \qquad (2\text{--}18)$$

since $|e^{i\phi}| = |\sin \phi - i \cos \phi| = 1$. The phase angle ϕ is given by

$$\phi = \tan^{-1}\left[\frac{c\omega}{k - (W/g)\omega^2}\right] \qquad (2\text{--}19)$$

The solutions for amplitude X and phase angle ϕ also hold when the external force is $F \sin \omega t$ or $F \cos \omega t$ because

$$F \sin \omega t = Im[Fe^{i\omega t}]$$
$$X \sin(\omega t - \phi) = Im[Xe^{i(\omega t - \phi)}]$$
$$F \cos \omega t = Re[Fe^{i\omega t}]$$
$$X \cos(\omega t - \phi) = Re[Xe^{i(\omega t - \phi)}]$$

Note that these results for $\sin \omega t$ [i.e., Eqs. (2–18) and (2–19) substituted in $x_p = X \sin(\omega t - \phi)$] agree with Eqs. (2–10) and (2–11) in Sec. 2.3.

The name "mechanical impedance" for this method comes from the fact that this method can be started by substitution of $-(W/g)\omega^2 x$ for each inertia force, $ic\omega x$ for each damping force, and kx for each spring force. As a comparison to electrical impedance, which is the voltage drop across an electrical element per unit current, *mechanical impedance* is defined to be force per unit displacement. Thus, the mechanical impedance for a mass equals $-(W/g)\omega^2 x/x$ or $-(W/g)\omega^2$, the mechanical impednace for a damper is $ic\omega x/x$ or $ic\omega$, and the mechanical impedance for a spring is k. The equation of motion can be written using these impedance values when the mechanical impedance method is applied.

Illustrative Example 2.5. Suppose that the system in Fig. 2-3 of Sec. 2.3 is excited by a force of $F_1 \sin \omega_1 t + F_2 \cos \omega_2 t$. Find the steady-state response.

Solution. Since the applied force consists of two harmonics, the impedance method tells us that the steady-state solution is

$$x_p = X_1 \sin(\omega_1 t - \phi_1) + X_2 \cos(\omega_2 t - \phi_2)$$

where the superposition property for a linear differential equation was applied. Use of Eqs. (2-18) and (2-19) give the following values for the amplitudes and phase angles:

$$X_1 = \frac{F_1}{\sqrt{[k - (W/g)\omega_1^2]^2 + (c\omega_1)^2}}$$

$$\phi_1 = \tan^{-1}\left[\frac{c\omega_1}{k - (W/g)\omega_1^2}\right]$$

$$X_2 = \frac{F_2}{\sqrt{[k - (W/g)\omega_2^2]^2 + (c\omega_2)^2}}$$

$$\phi_2 = \tan^{-1}\left[\frac{c\omega_2}{k - (W/g)\omega_2^2}\right]$$

2.6 Solution of Vibrations with General Forces by Fourier Series

Let us now be more general and consider periodic external forces that are not harmonic, such as those that often occur in machine oscillations. From the reader's course in differential equations, we know that any periodic function can be expressed by an infinite series of sine and cosine terms. Thus, a periodic external force $F(t)$ of period T can be represented by the following Fourier series:

$$F(t) = \frac{a_0}{2} + \sum_{n=1}^{\infty} (a_n \cos n\omega t + b_n \sin n\omega t) \qquad (2\text{-}20)$$

where

$$a_0 = \frac{2}{T}\int_0^T F(t)\, dt \qquad\qquad (2\text{--}20a)$$

$$a_n = \frac{2}{T}\int_0^T F(t)\cos n\omega t\, dt \qquad\qquad (2\text{--}20b)$$

$$b_n = \frac{2}{T}\int_0^T F(t)\sin n\omega t\, dt \qquad\qquad (2\text{--}20c)$$

Now let us suppose that such a periodic, non-harmonic external force is applied to the spring-mass-damper system in Fig. 2-3 in Sec. 2.3. The equation of motion for this system can be written as

$$\frac{W}{g}\ddot{x} + c\dot{x} + kx = F(t) = \frac{a_0}{2} + \sum_{n=1}^{\infty}(a_n\cos n\omega t + b_n\sin n\omega t)$$

where Fourier series theory was used to represent the non-harmonic force by an infinite number of hamonic (i.e., sine and cosine) forces. Since the differential equation of motion is linear, we can use super-position and the impedance method to obtain the steady-state response. Use of Eqs. (2–18) and (2–19) gives the following equation for the steady-state response:

$$x_p = \frac{a_0}{2k} + \sum_{n=1}^{\infty}\frac{a_n\cos(n\omega t - \phi_n) + b_n\sin(n\omega t - \phi_n)}{\sqrt{[k - (W/g)n^2\omega^2]^2 + (cn^2\omega^2)^2}} \qquad (2\text{--}21)$$

where phase angle ϕ_n is given by

$$\phi_n = \tan^{-1}\left[\frac{cn\omega}{k - (W/g)n^2\omega^2}\right] \qquad\qquad (2\text{--}22)$$

Illustrative Example 2.6. Suppose that the external force for the spring-mass-damper system in Fig. 2-3 is a rectangular wave, as shown in Fig. 2-7. Find the steady-state response.

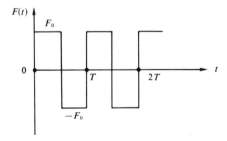

Fig. 2-7

Solution. We wish to determine the values of a_0, a_n, and b_n for Eq. (2–21). Since $F(-t) = -F(t)$, force $F(t)$ is an odd periodic function, and we know from Fourier series theory that $a_0 = a_n = 0$. Use of Eq. (2–20c) gives

$$b_n = \frac{2}{T}\left[\int_0^{T/2} F_0 \sin n\omega t\, dt + \int_{T/2}^T (-F_0) \sin n\omega t\, dt\right]$$

$$= \frac{2}{T}\left(\frac{TF_0}{2\pi n}\right)\left\{\left[-\cos\frac{2\pi n}{T}t\right]_0^{T/2} + \left[\cos\frac{2\pi n}{T}t\right]_{T/2}^T\right\}$$

where $\omega = 2\pi/T$ and $\int \sin\theta\, d\theta = -\cos\theta$. Thus, $b_n = 4F_0/n\pi$ if n is odd (i.e., 1, 3, 5, etc.), and $b_n = 0$ if n is even. Therefore, the steady-state response is

$$x_p = \sum_{n=1}^{\infty} \frac{b_n \sin(n\omega t - \phi_n)}{\sqrt{[k - (W/g)n^2\omega^2]^2 + (cn^2\omega^2)^2}}$$

where phase angle ϕ_n is given by Eq. (2-22), and where

$$b_n = \begin{cases} 4F_0/n\pi & \text{if } n \text{ is odd} \\ 0 & \text{if } n \text{ is even} \end{cases}$$

2.7   Vibration Isolation and Transmissibility

When a machine is attached to a supporting structure or foundation, it is often preferable to mount the machine on springs and dampers, as shown in Fig. 2–8, or on some other type of flexible support in order to minimize both the transmitted vibration and the forces transmitted to the support. Springs and dampers, besides being used to isolate vibrating machinery from their foundations, are also used to isolate delicate instruments from the motion of their surroundings. Since we see from Fig. 2–8 that the total force transmitted to the support is the sum of the spring and damping forces, the transmitted force F_T is as follows when the support is considered to be rigid enough so that its deflection may be neglected:

$$F_T = kx + c\dot{x} \qquad (2\text{-}23)$$

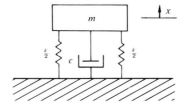

Fig. 2-8

If the machine is excited by a harmonic sinusoidal force, we know from Sec. 2.3 that its steady-state motion may be expressed as $X \sin(\omega t - \phi)$. Because the transient forces die out with time, we are usually interested only in the transmitted steady-state force which is

$$F_T = kX \sin(\omega t - \phi) + c\omega X \cos(\omega t - \phi)$$

Since $A \sin bt + B \cos bt = D \sin(bt - \gamma)$,

$$F_T = X\sqrt{k^2 + c^2\omega^2} \sin(\omega t - \phi + \gamma) \tag{2-24}$$

where $\gamma = \tan^{-1}(c\omega/k) = \tan^{-1}(2\zeta r)$, since $c\omega/k = 2\zeta r$. Thus, the amplitude of this transmitted force is

$$|F_T| = X\sqrt{k^2 + c^2\omega^2} = kX\sqrt{1 + \left(\frac{c\omega}{k}\right)^2} = kX\sqrt{1 + (2\zeta r)^2}$$

From Eq. (2–12) in Sec. 2.3, it is seen that amplitude X for the steady-state motion equals (M.F.)(F/k) during an $F \sin \omega t$ external force excitation. Thus,

$$|F_T| = F(\text{M.F.})\sqrt{1 + (2\zeta r)^2} \tag{2-25}$$

Note from Eq. (2–25) that if the support were rigid (i.e., had no springs or dampers), the amplitude of the transmitted force would equal the amplitude F of the impressed force. The ratio of the amplitude of the transmitted force to the amplitude of the external or impressed force is defined as the *transmissibility* TR. Thus, we have

$$TR = \frac{|F_T|}{F} = \text{M.F.}\sqrt{1 + (2\zeta r)^2} \tag{2-26}$$

where M.F. is given by Eq. (2–12) as a function of ζ and r. Equation (2–26) is plotted as TR versus r and ζ in Fig. 2–9. From this figure we see that $TR = 1.0$ for all values of damping ratio ζ when $\omega/\omega_n = \sqrt{2}$. Also note that added damping increases the transmitted force when $r < \sqrt{2}$ and decreases this force when $r > \sqrt{2}$. Since $TR > 1$ for $r < \sqrt{2}$ and $TR < 1$ for $r > \sqrt{2}$, a successful vibration isolation (i.e., reduction of the force transmitted to the support) is only possible when the ratio ω/ω_n exceeds $\sqrt{2}$. Since $r = \omega/\sqrt{kg/W}$, we can use Fig. 2–9 for design purposes to determine the values of the spring constant k for a given set of values for W, ω, and ζ that will give specified values of TR, the value of the transmitted force compared to the applied force. When we wish to consider the isolation of the motion of the surroundings from a delicate instrument and other types of motion isolation, rather than a force isolation as discussed previously, we can also use Fig. 2–9. The effectiveness of such a motion isolation is defined to be the ratio of the vibration amplitude of the body to that of its support. Since this amplitude ratio is also given by Eq. (2–26), Fig. 2–9 can be used for both force and motion isolation design purposes.

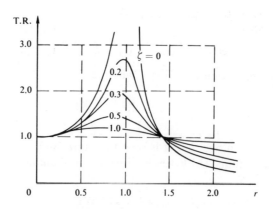

Fig. 2-9

2.8   Vibration-Measuring Instruments

A *seismic instrument* consists of a mass, spring, and supporting base where the base is attached to the body when a body motion (i.e., x_1 versus t) is to be measured. A seismic instrument is schematically shown in Fig. 2–10. Since the displacement motion of mass m is x_2, the pen records the relative displacement $(x_2 - x_1)$ on a rotating drum inside the instrument. It will be shown that this relative displacement between the mass and the base indicates the motion of the base and hence also the motion of the vibrating body. The damper also serves to cause the transient motion to die out with time.

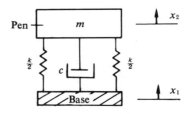

Fig. 2-10

Since the spring force is proportional to the relative displacement between its two ends, and since the damping force is proportional to the relative velocity, the spring and damping forces are $k(x_2 - x_1)$ and $c(\dot{x}_2 - \dot{x}_1)$, respectively. Thus, Newton's second law gives

$$m\ddot{x}_2 = -c(\dot{x}_2 - \dot{x}_1) - k(x_2 - x_1) \qquad \text{(2-27)}$$

which can be rewritten as follows, where $z = (x_2 - x_1)$ so that $\ddot{x}_2 = \ddot{z} + \ddot{x}_1$.

$$m(\ddot{z} + \ddot{x}_1) = -c\dot{z} - kz$$

If the motion x_1 of the base (or vibrating body) is $X_1 \sin \omega t$, then the previous equation becomes

$$m\ddot{z} + c\dot{z} + kz = -m\ddot{x}_1 = m\omega^2 X_1 \sin \omega t \qquad \text{(2-28)}$$

Since this equation has the same mathematical form as Eq. (2-4), we can use Eqs. (2-8) and (2-12), where $F = m\omega^2 X_1$, to get the steady-state solution

$$z_p = (x_2 - x_1)_p = \left(\frac{m\omega^2 X_1}{k}\right)(\text{M.F.}) \sin(\omega t - \phi)$$
$$= r^2(\text{M.F.})X_1 \sin(\omega t - \phi) = Z \sin(\omega t - \phi) \qquad \text{(2-29)}$$

where phase angle ϕ is given by Eq. (2-13) and is plotted in Fig. 2-5 and where magnification factor M.F. is given by Eq. (2-12). From Eq. (2-29), we see that the amplitude ratio Z/X_1 is

$$\frac{Z}{X_1} = r^2(\text{M.F.}) = \frac{r^2}{\sqrt{(1 - r^2)^2 + (2\zeta r)^2}} \qquad \text{(2-30)}$$

which is plotted in Fig. 2-11.

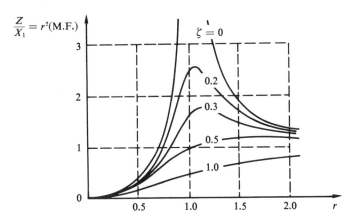

Fig. 2-11

A *vibrometer*, which has a relatively low natural frequency, and hence, has a high value of ω/ω_n or r, is used to measure the displacement x_1 of the vibrating body. The low natural frequency ω_n can be obtained by use of a weak spring (i.e., low spring constant k) or a large mass m. In Fig. 2–11, note that the amplitude ratio Z/X_1 approaches unity for large values of r, regardless of the value of the damping ratio ζ. Since Fig. 2–5 in Sec. 2.3 shows that phase angle ϕ approaches 180° for large values of r and for all values of ζ, the measured relative motion $z = (x_2 - x_1)$ and the body or base motion x_1 are 180° out of phase. Thus, $x_1 = -(x_2 - x_1)$ at steady-state, and we can easily determine the motion x_2 of the vibrating body from the measured motion $(x_2 - x_1)$.

The *seismograph* is used to measure earthquake vibrations, and the *torsiograph* is a special and elaborate instrument that is used to measure torsional vibrations. An *accelerometer* has a high natural frequency (obtainable by the use of hard or strong springs) and is used to measure the acceleration $\ddot{x}_1$ of a vibrating body. If the body or base motion is $X_1 \sin \omega t$, its acceleration is $-\omega^2 X_1 \sin \omega t$. Combining Eqs. (2–29) and (2–30), we have

$$(x_2 - x_1)_p = \frac{(\omega^2/\omega_n^2)X_1 \sin(\omega t - \phi)}{\sqrt{(1 - r^2)^2 + (2\zeta r)^2}} \tag{2–31}$$

Since the high natural frequency of an accelerometer means a low value for r, and since Fig. 2–5 shows that phase angle ϕ approaches zero for small values of r, we have the following approximation for Eq. (2–31) for small values of r:

$$(x_2 - x_1)_p \approx \frac{\omega^2}{\omega_n^2} X_1 \sin \omega t = -\frac{\ddot{x}_1}{\omega_n^2} \tag{2–32}$$

which shows that the measured steady-state motion $(x_2 - x_1)_p$ is nearly proportional to the body acceleration $\ddot{x}_1$ when r is small. By use of integrating circuits, the accelerometer output, which is usually converted to an electrical signal, can be integrated to obtain the velocity or displacement variation with time. Since most motions to be measured consist of several harmonics, phase and amplitude distortion of the recorded harmonics for an accelerometer can occur. For this reason, the damping ratio ζ should be between 0.65 and 0.70 to minimize these distortion problems.

Illustrative Example 2.7.   Discuss the acceleration measurement error when the body motion is given by $0.100 \sin 60\,t$ in., and where $\zeta = 0.700$ and $\omega_n = 120$ rad/sec for an instrument that is used as an accelerometer.

Solution.   The frequency ratio r is

$$r = \frac{\omega}{\omega_n} = 0.500$$

which shows that the natural frequency should be much higher if this instrument is to be used generally as an accelerometer. This could be done by using stronger springs. From Eqs. (2-12) and (2-13) (or from Figs. 2-4 and 2-5), M.F. $= 1.00$ and $\phi = 45°$ when $\zeta = 0.700$ and $r = 0.500$. Use of Eq. (2-29) shows that the measured relative displacement is

$$(x_2 - x_1)_p = r^2(\text{M.F.})X_1 \sin(\omega t - \phi)$$
$$= (0.250)(1.00)(0.100)\sin(\omega t - 45°)$$
$$= 0.0250 \sin(\omega t - 45°)$$

The body acceleration $\ddot{x}_1$ is

$$\ddot{x}_1 = -\omega^2 X_1 \sin \omega t = -\omega_n^2 r^2 X_1 \sin \omega t$$
$$= -\omega_n^2(0.0250)\sin \omega t$$

Thus, the accelerometer formula $\ddot{x}_1 = -\omega_n^2(x_2 - x_1)$ can be used to determine the amplitude variation of the body acceleration if the 45° phase lag is accounted for. Almost exact amplitude values will result for this particular case, because M.F. $= 1.00$ when $\zeta = 0.700$ and $r = 0.500$.

2.9   Other Applications

Illustrative Exs. 2.8 to 2.11 will consider the following four forced-vibration applications: rotating and reciprocating unbalance, critical speed of a rotating shaft, base excitation, and shaft support by flexible bearings. These examples will further show that a forced vibration can be caused not only by an external force applied to a mass, but also by a motion excitation applied to a system support, and by an unbalance in a machine component. The equation of motion for a single-degree-of-freedom forced-sinusoidal vibration with viscous damping is

$$m\ddot{x} + c\dot{x} + kx = F_{eq} \sin \omega t \qquad (2\text{-}33)$$

and from Eqs. (2-10) to (2-13) in Sec. 2.3, we see that its solution is

$$x = x_h + \frac{F_{eq}}{k}(\text{M.F.})\sin(\omega t - \phi) \qquad (2\text{-}34)$$

where M.F. is given by Eq. (2-12), and ϕ is given by Eq. (2-13), both as a function of r and ζ. The transient solution x_h, which dies out with time, is given by Eq. (1-39), (1-43), or (1-45) in Sec. 1.7, depending upon the value of ζ.

Illustrative Example 2.8.   Discuss the analysis of the rotating and reciprocating unbalance that may occur in machines with rotors (e.g., turbines and electric motors) and reciprocating engines.

Solution. If the center of gravity of a rotor does not coincide with its axis of rotation, there is an unbalance which may be represented by an equivalent point mass m that has an eccentricity e. Figure 2-12 is a schematic of a rotating machine of total mass M and moment unbalance me that is constrained to move in the vertical direction. If the eccentric mass m rotates at angular velocity ω, its vertical displacement is $(x + e \sin \omega t)$, where x is the displacement of mass $(M - m)$. Use of Newton's second law gives the following equation of motion:

$$(M - m)\frac{d^2 x}{dt^2} + m\frac{d^2}{dt^2}(x + e \sin \omega t) = -kx - c\frac{dx}{dt}$$

which may be rewritten as

$$M\ddot{x} + c\dot{x} + kx = (me\omega^2) \sin \omega t \qquad (2\text{-}35)$$

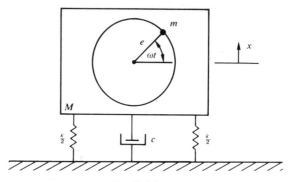

Fig. 2-12

If F_{eq} is set to equal $me\omega^2$, Eqs. (2-33) and (2-35) are identical, and use of Eq. (2-34) gives the following steady-state solution

$$x_p = \frac{me\omega^2}{k}\,(\text{M.F.}) \sin(\omega t - \phi) \qquad (2\text{-}36)$$

where M.F. and ϕ are given by Eqs. (2-12) and (2-13), and are plotted in Figs. 2-4 and 2-5. From Eq. (2-36), we see that the amplitude X is

$$X = me\omega^2(\text{M.F.})/k = mer^2(\text{M.F.})/M \qquad (2\text{-}37)$$

where $r = \omega/\omega_n$ and $\omega_n = \sqrt{k/M}$. Thus, ratio MX/me is

$$\frac{MX}{me} = r^2(\text{M.F.}) \qquad (2\text{-}38)$$

By comparing Eqs. (2-30) and (2-38), we see that Fig. 2-11 can be used to show the variation of both MX/me and Z/X_1 with r and ζ.

This analysis can also include the unbalance in a reciprocating engine. Let mass m represent the mass of the piston and a portion of the connecting rod. The inertia force of the reciprocating mass m approximately equals $(me\omega^2) \sin \omega t$ when the ratio e/L is small. The term e denotes the radius of the crank, L denotes the connecting rod length, and ω is the angular velocity of the crank. Thus, the motion x for a spring-damper-mounted engine is represented by Eq. (2-33), where $F_{eq} = me\omega^2$ as in a rotational unbalance since the inertia force is $(me\omega^2) \sin \omega t$.

Illustrative Example 2.9.   Discuss the analysis of a rotating shaft during whirl.

Solution.   A rotating shaft tends to vibrate transversely when the rotative speed equals the *critical* or *whirling speed*, which coincides with the natural frequency of lateral beam vibration of the shaft. This vibration results from an unbalance in the rotating system. Consider the rotating shaft of negligible mass in Fig. 2-13 that has an unbalanced disk of mass m that is mounted at its midspan. The center of gravity G, geometric center O, and center of rotation C are shown in Fig. 2-13b. A viscous damping force due to air friction, and proportional to the velocity at the disk center, is assumed to exist for this example.

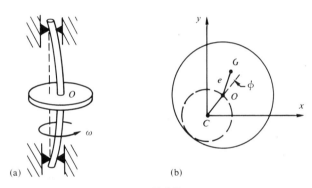

(a)   (b)

Fig. 2-13

Application of Newton's second law in the x-and y-directions, where k is the lateral stiffness of the shaft at the disk location, gives

$$m \frac{d^2}{dt^2} (x + e \cos \omega t) = -kx - c\dot{x}$$

$$m \frac{d^2}{dt^2} (y + e \sin \omega t) = -ky - c\dot{y}$$

which can be rewritten as

$$m\ddot{x} + c\dot{x} + kx = me\omega^2 \cos \omega t \qquad (2\text{-}39)$$

$$m\ddot{y} + c\dot{y} + ky = me\omega^2 \sin \omega t \qquad (2\text{-}40)$$

Since these two equations have the same form as Eq. (2-35), use of Eq. (2-36) shows the steady-state solutions to be

$$x_p = \frac{me\omega^2}{k} \text{(M.F.)} \cos (\omega t - \phi) \qquad (2\text{-}41)$$

$$y_p = \frac{me\omega^2}{k} \text{(M.F.)} \sin (\omega t - \phi) \qquad (2\text{-}42)$$

where ϕ is the angle between line OG and the extension of line CO past center O, as shown in Fig. 2-13b, and where M.F. and ϕ are given by Eqs. (2-12) and (2-13) and are plotted in Figs. 2-4 and 2-5. Since the lateral deflection r of the shaft at steady-state equals $\sqrt{x_p^2 + y_p^2}$ and since $m\omega^2/k = \omega^2/\omega_n^2 = r^2$, the amplitude ratio r/e is

$$\frac{r}{e} = \frac{\sqrt{x_p^2 + y_p^2}}{e} = m\omega^2(\text{M.F.})/k = r^2(\text{M.F.}) \tag{2-43}$$

Thus, comparison of Eqs. (2-30) and (2-43) shows that Fig. 2-11 can be used to show the variation of amplitude ratio r/e with r and ζ. Use of Eq. (2-13) or Fig. 2-5 shows that ϕ is very small when r is very small (which means physically that the disk rotates with center of gravity G outside the geometric center O), that $\phi = 90°$ at resonance where $r = 1$, and that ϕ approaches $180°$ for high values of r (and rotational speed ω). Figure 2-11 and Eq. (2-43) show that at very high speeds (i.e., high values of ω and r), ratio r/e approaches unity, which means physically that the center of gravity G tends to coincide with the center of rotation C (i.e., the system tends to rotate about its center of gravity).

Illustrative Example 2.10.   Discuss the analysis of flexibly-mounted systems where the excitation is applied to the support instead of to the mass.

Solution.   This has been partially discussed in Sec. 2.8 which deals with vibration-measuring instruments. The equation of motion for the system in Fig. 2-10 is given by Eq. (2-27) which can be rewritten as

$$m\ddot{x}_2 + c\dot{x}_2 + kx_2 = kx_1 + c\dot{x}_1 \tag{2-44}$$

If the motion x_1 of the support base is $X_1 \sin \omega t$, then the previous equation becomes

$$m\ddot{x}_2 + c\dot{x}_2 + kx_2 = kX_1 \sin \omega t + c\omega X_1 \cos \omega t$$

This equation could also be used to approximate the motion of a vehicle of mass m over a sinusoidally-shaped road of amplitude X_1. Since $A \sin bt + B \cos bt = D \sin (\omega t + \alpha)$, the previous equation becomes

$$m\dot{x}_2 + c\dot{x}_2 + kx_2 = X_1\sqrt{k^2 + c^2\omega^2} \,(\sin \omega t + \gamma) \tag{2-45}$$

where $\gamma = \tan^{-1}(c\omega/k) = \tan^{-1}(2\zeta r)$. Since Eqs. (2-33) and (2-45) have the same form if $F_{eq} = X_1\sqrt{k^2 + c^2\omega^2} = X_1 k\sqrt{1 + (2\zeta r)^2}$, use of Eq. (2-34) shows the steady-state solution to be

$$x_2 = X_1(\text{M.F.})\sqrt{1 + (2\zeta r)^2} \sin (\omega t - \phi + \gamma)$$

where M.F. and ϕ are given by Eqs. (2-12) and (2-13) as functions of r and ζ. Since amplitude X_2 equals $X_1(\text{M.F.})\sqrt{1 + (2\zeta r)^2}$, the maximum acceleration $\ddot{x}_{2_{\max}}$ is

$$\ddot{x}_{2_{\max}} = \omega^2 X_2 = \omega^2 X_1(\text{M.F.})\sqrt{1 + (2\zeta r)^2}$$

From Newton's second law, $F_{T_{\max}} = m\ddot{x}_{2_{\max}}$, and the maximum force transmitted to mass m is

$$F_{T_{\max}} = m\omega^2 X_1 (\text{M.F.})\sqrt{1 + (2\zeta r)^2} \tag{2-46}$$

Illustrative Example 2.11.   Discuss the analysis of the unbalanced diskshaft system in Ill. Ex. 2.9 when it is supported in flexible bearings, and where the flexibilities are different in the x-and y-directions. Assume here that damping is negligible.

Solution.   As discussed in Sec. 1.2, a flexible bearing can be represented by the four-spring system shown in Fig. 1-7b. Application of Newton's second law in the x-and y-directions, when not at a shaft critical speed, gives the following non-whirl equations of motion:

$$m \frac{d^2}{dt^2}(x + e \cos \omega t) + k_x x = 0$$

$$m \frac{d^2}{dt^2}(y + e \sin \omega t) + k_y y = 0$$

which can be rewritten as

$$m\ddot{x} + k_x x = me\omega^2 \cos \omega t \tag{2-47}$$
$$m\ddot{y} + k_y y = me\omega^2 \sin \omega t \tag{2-48}$$

where k_x and k_y are the bearing flexibilities in the x-and y-directions. Comparison of these two equations with Eq. (2-1) in Sec. 2.2 shows that the steady-state solutions may be obtained using Eq. (2-2), where $F_{eq} = me\omega^2$, to give

$$x_p = \frac{me\omega^2}{k_x - m\omega^2} \cos \omega t = X \cos \omega t \tag{2-49}$$

$$y_p = \frac{me\omega^2}{k_y - m\omega^2} \sin \omega t = Y \sin \omega t \tag{2-50}$$

Squaring Eqs. (2-49) and (2-50) and adding the result gives us the following equation of an ellipse:

$$\frac{x_p^2}{X^2} + \frac{y_p^2}{Y^2} = \cos^2 \omega t + \sin^2 \omega t = 1 \tag{2-51}$$

Thus, the center of the unbalanced disk, which is the point where the shaft passes through the disk, moves in an elliptical motion.

2.10   The Laplace Transformation

This method of solution, which should be a review for most readers, is applicable only to linear differential equations with constant or polynomial coefficients. It transforms differential equations (which may have integrals) into algebraic equations, and it does not separate the homogeneous and particular parts of the solution, as do classical

methods. A very advantageous feature of the Laplace transform solution technique is the ease with which it handles step functions, impulse
functions, periodic functions, and other special kinds of functions. In
Sec. 2.12, it will be seen that this method of solution is very useful for
solving vibration problems that have pulse and other types of discontinuous or short-duration excitations.

Let t be a real variable, and let the function $f(t)$ be defined for all
positive values of t. Then the function $F(s)$ defined by the equation

$$F(s) = \int_0^\infty e^{-st} f(t)\, dt \tag{2-52}$$

is called the *Laplace transform* of the function $f(t)$. This new function
$F(s)$ does not involve the variable t (since the integral has fixed limits),
but involves the parametric variable s, which is part of the exponent
of the exponential function e^{-st}.

The exponential function e^{-st} causes the integral to converge properly, provided that the function $f(t)$ does not increase too rapidly as
the variable t approaches infinity. Thus, some functions of t do not
have Laplace transforms. That is, the function $f(t)$ must possess certain
requirements if its Laplace transform is to exist. The function $f(t)$ can
have a finite number of discontinuities, but not an infinite number of
discontinuities between $t = 0$ and $t = \infty$. Another requirement is that
the limit of the function $f(t)/e^{st}$ must approach zero as the variable t
approaches infinity. These are not the only requirements, but the
requirements for $f(t)$ may by briefly summarized by saying that the
function $f(t)$ must be such that the integral given in Eq. (2–52) exists,
and that the requirements for the existence of an integral were given
in the reader's course in integral calculus.

It can easily be shown that the Laplace transform has the following
linearity property, where a_1 and a_2 are constants:

$$L\{a_1 f_1(t) + a_2 f_2(t)\} = a_1 L\{f_1(t)\} + a_2 L\{f_2(t)\} \tag{2-53}$$

which can be proved as follows:

$$L\{a_1 f_1(t) + a_2 f_2(t)\} = \int_0^\infty [a_1 f_1(t) + a_2 f_2(t)] e^{-st}\, dt$$
$$= a_1 \int_0^\infty f_1(t) e^{-st}\, dt + a_2 \int_0^\infty f_2(t) e^{-st}\, dt$$
$$= a_1 L\{f_1(t)\} + a_2 L\{f_2(t)\}$$

As seen from Eq. (2–52), the Laplace transform of a function can
be obtained by integration and the results of these integrations for
different functions can be tabulated in a table for future reference.
Table A–2 in the Appendix contains values for the Laplace transform

of some specific functions $f(t)$, which are designated in this table. The Laplace transforms of K, e^{at}, sin at, and cos at can be computed by integration as follows:

$$L\{K\} = \int_0^\infty Ke^{-st}\,dt = -\frac{Ke^{-st}}{s}\Big|_0^\infty = \frac{K}{s} \text{ if } s > 0$$

$$L\{e^{at}\} = \int_0^\infty e^{at}e^{-st}\,dt = \int_0^\infty e^{-(s-a)t}\,dt = -\frac{e^{-(s-a)t}}{s-a}\Big|_0^\infty = \frac{1}{s-a} \text{ if } s > a$$

$$L\{\sin at\} = \int_0^\infty e^{-st}\sin at\,dt = -\frac{e^{-st}}{s^2+a^2}(s\sin at + a\cos at)\Big|_0^\infty$$

$$= \frac{a}{s^2+a^2} \text{ if } s > 0$$

$$L\{\cos at\} = \int_0^\infty e^{-st}\cos at\,dt = \frac{-e^{-st}}{s^2+a^2}(a\sin at + s\cos at)\Big|_0^\infty$$

$$= \frac{s}{s^2+a^2} \text{ if } s > 0$$

In this section and the next, we shall see that the Laplace transform has certain properties that simplify the determination of the Laplace transform of some specific types of functions $f(t)$. We shall also denote the Laplace transform of the function $f(t)$ by $L\{f(t)\}$. That is, $F(s)$ and $L\{f(t)\}$ have the same meaning.

A most useful property of the Laplace transform is that the Laplace transform of a derivative of the function $f(t)$ of order n can be expressed as a function of the transform of $f(t)$ itself. This result makes the Laplace transform a very useful tool for solving differential equations. This usage is shown in Ill. Ex. 2.12 of this section. The Laplace transform of the first derivative of $f(t)$ is defined to be

$$L\left\{\frac{df(t)}{dt}\right\} = \int_0^\infty e^{-st}\frac{df(t)}{dt}\,dt$$

Since $d(uv) = (u\,dv + v\,du)$, integration gives

$$uv = \int u\,dv + \int v\,du$$

which can be rewritten as

$$\int u\,dv = uv - \int v\,du \tag{2-54}$$

We shall apply Eq. (2–54) to evaluate $L\{df(t)/dt\}$, where the method is called *integration by parts*. Letting $u = e^{-st}$ and $dv = df(t)$, then $v = f(t)$ and $du = -se^{-st}\,dt$. Substitution in Eq. (2–54) gives

$$\int_0^\infty e^{-st}\frac{df(t)}{dt}\,dt = e^{-st}f(t)\Big|_0^\infty + s\int_0^\infty e^{-st}f(t)\,dt$$

Because of the conditions required on the function $f(t)$, the term $e^{-st} f(t)$ approaches zero as t approaches infinity. Thus, we obtain

$$L\left\{\frac{df(t)}{dt}\right\} = sF(s) - f(0+) \tag{2-55}$$

Suppose that the function $f(t)$ is discontinuous at $t = 0$. The term $f(0+)$ tells us to use the value of $f(t)$ on the positive side of this discontinuity. If function $f(t)$ is continuous at $t = 0$, then we use the value of the function at $t = 0$ for $f(0+)$. Eq. (2-55) may be rewritten as

$$L\{f'(t)\} = sL\{f(t)\} - f(0+) \tag{2-56}$$

Using Eq. (2-52) the Laplace transform of the second derivative of the function $f(t)$ is defined to be

$$L\left\{\frac{d^2 f(t)}{dt^2}\right\} = \int_0^\infty e^{-st} \frac{d^2 f(t)}{dt^2} dt$$

Integration by parts, where $u = e^{-st}$ and $v = df(t)/dt$ in Eq. (2-54), gives

$$L\left\{\frac{d^2 f(t)}{dt^2}\right\} = e^{-st} \frac{df(t)}{dt}\bigg|_0^\infty + s \int_0^\infty e^{-st} \frac{df(t)}{dt} dt$$

$$L\left\{\frac{d^2 f(t)}{dt^2}\right\} = e^{-st} \frac{df(t)}{dt}\bigg|_0^\infty + sL\left\{\frac{df(t)}{dt}\right\}$$

For the Laplace transform to exist, the term $e^{-st} df(t)/dt$ must approach zero as t approaches infinity. Thus, we obtain

$$L\{f''(t)\} = sL\{f'(t)\} - f'(0+)$$

Upon substituting for $L\{f'(t)\}$, using Eq. (2-56), we obtain

$$L\{f''(t)\} = s^2 L\{f(t)\} - sf(0+) - f'(0+) \tag{2-57}$$

It can be shown, using mathematical induction, that the transform of the derivative of $f(t)$ of order n is given by the following general form:

$$L\{f^{(n)}(t)\} = s^n L\{f(t)\} - [s^{n-1} f(0+) + s^{n-2} f'(0+) \tag{2-58}$$
$$+ s^{n-3} f''(0+) + \ldots + f^{(n-1)}(0+)]$$

In Ill. Ex. 2.12, we shall find that we have to break the Laplace transform of the solution into partial fractions before we can solve the given differential equation. This can be done by the algebraic method that the reader has learned before, but simpler methods are desirable. Let us assume that $F(s)$, the Laplace transform of the solution of a specified differential equation, can be expressed as the quotient of two polynomials in s [i.e., $F(s) = N(s)/D(s)$] and that the denominator $D(s)$ is of a higher degree in s than the numerator $N(s)$.

Also assume that the denominator $D(s)$ may be factored into n roots $r_1, r_2, r_3, \ldots, r_n$, all of which are different. Then $F(s)$ is given by

$$F(s) = \frac{N(s)}{D(s)} = \frac{N(s)}{(s - r_1)(s - r_2) \ldots (s - r_n)}$$

which may be expanded by partial fractions as

$$F(s) = \frac{C_1}{s - r_1} + \frac{C_2}{s - r_2} + \ldots + \frac{C_n}{s - r_n} \qquad (2\text{--}59)$$

To determine the value of any constant C_i, multiply both sides of the above equation by $(s - r_i)$ and let s equal r_i. This can be expressed mathematically as

$$C_i = [(s - r_i)F(s)]_{s=r_i} \qquad (2\text{--}60)$$

The result is C_i, because all of the terms on the right-hand side of Eq. (2–59) equal zero, except the ith term, when these terms are multiplied by $(s - r_i)$ and s is set equal to r_i. The ith term equals C_i, since multiplication by $(s - r_i)$ cancels its denominator. If the denominator $D(s)$ has a root r_n of order k, then it will be of the form

$$D(s) = (s - r_1)(s - r_2) \ldots (s - r_n)^k \qquad (2\text{--}61)$$

and the partial fraction expansion of the Laplace transform $F(s)$ is given by

$$F(s) = \frac{C_1}{s - r_1} + \frac{C_2}{s - r_2} + \ldots + \frac{C_{n1}}{s - r_n} + \frac{C_{n2}}{(s - r_n)^2} + \ldots + \frac{C_{nk}}{(s - r_n)^k} \qquad (2\text{--}62)$$

The coefficients C_1 to C_{n-1} are obtained as before. The coefficient C_{nk} is obtained by multiplying $F(s)$ by $(s - r_n)^k$ and letting s equal r_n. That is,

$$C_{nk} = [(s - r_n)^k F(s)]_{s=r_n} \qquad (2\text{--}63)$$

The term $C_{n(k-1)}$ is obtained by multiplying $F(s)$ by $(s - r_n)^k$, differentiating the result with respect to s, and substituting s equals r_n. This can be expressed mathematically as follows:

$$C_{n(k-1)} = \left\{ \frac{d}{ds} [(s - r_n)^k F(s)] \right\}_{s=r_n} \qquad (2\text{--}64)$$

Continuing the process, the formula for the coefficient C_{n1} becomes

$$C_{n1} = \frac{1}{(k - 1)!} \left\{ \frac{d^{k-1}}{ds^{k-1}} [(s - r_n)^k F(s)] \right\}_{s=r_n} \qquad (2\text{--}65)$$

As a general result, for the term $C_{np}/(s - r_n)^p$, where $p \leq k$, the coefficient C_{np} can be obtained from the equation

$$C_{np} = \frac{1}{(k-p)!} \left\{ \frac{d^{k-p}}{ds^{k-p}} [(s - r_n)^k F(s)] \right\}_{s=r_n} \qquad (2\text{-}66)$$

where it is seen that the required differentiations can add a complication to the application of this method.

Now let us consider the case in which the denominator $D(s)$ has a factor of the form $(s^2 + a^2)$. This could be handled by factoring $(s^2 + a^2)$ into $(s + ia)(s - ia)$, where $i = \sqrt{-1}$, and applying Eq. (2-60), but this method can be somewhat awkward at times. Since

$$F(s) = \frac{N(s)}{D(s)} = \frac{N(s)}{(s - r_1)(s - r_2)\ldots(s^2 + a^2)}$$

we can partial fraction $F(s)$ as follows:

$$F(s) = \frac{C_1}{s - r_1} + \frac{C_2}{s - r_2} + \ldots + \frac{C_{n-1}s + C_n}{s^2 + a^2} \qquad (2\text{-}67)$$

If we multiply $F(s)$ by $(s^2 + a^2)$ and let $s = ia$, so that $(s^2 + a^2) = 0$, all of the terms on the right-hand side of Eq. (2-67), except the last term, will equal zero. Since its denominator is cancelled, the last term now equals $(C_{n-1}(ia) + C_n)$. Thus, we have

$$[(s^2 + a^2)F(s)]_{s=ia} = C_n + iaC_{n-1}$$

Let us suppose that the term $[(s^2 + a^2)F(s)]_{s=ia}$ equals $(M + iN)$ when it is evaluated. From the above equation it is seen that

$$C_{n-1} = \frac{N}{a} \qquad (2\text{-}68)$$

$$C_n = M \qquad (2\text{-}69)$$

where M is the real part and N is the imaginary part of $[(s^2 + a^2)F(s)]_{s=ia}$.

Illustrative Example 2.12.   Solve the following differential equation by the Laplace transform method, where $y(0) = 1$, $\dot{y}(0) = 0$, and $\ddot{y}(0) = 0$.

$$\frac{d^3y}{dt^3} - \frac{dy}{dt} = 1$$

Solution.   Let $Y = Y(s) = L\{y(t)\}$. Application of Eq. (2-58) gives

$$L\left\{\frac{d^3y}{dt^3}\right\} = s^3 Y - [s^2 y(0) + s\dot{y}(0) + \ddot{y}(0)] = s^3 Y - s^2$$

Application of Eq. (2-56) gives

$$L\left\{\frac{dy}{dt}\right\} = sY - y(0) = sY - 1$$

From Table A-2 in the Appendix, we see that $L\{1\} = 1/s$. Thus, the Laplace transform of the differential equation is

$$(s^3 Y - s^2) - (sY - 1) = \frac{1}{s}$$

which can be rewritten as

$$s^4 Y - s^2 Y - s^3 + s = 1$$

so that

$$Y(s^4 - s^2) = Y(s^2)(s + 1)(s - 1) = s^3 - s + 1$$

Thus, Y, the Laplace transform of $y(t)$, is the sum of the following partial fractions

$$Y = Y(s) = \frac{s^3 - s + 1}{s^2(s + 1)(s - 1)} = \frac{A}{s} + \frac{B}{s^2} + \frac{C}{s + 1} + \frac{D}{s - 1}$$

Application of Eq. (2-60) gives

$$C = [(s + 1)Y(s)]_{s=-1} = \left[\frac{s^3 - s + 1}{s^2(s - 1)}\right]_{s=-1} = -\frac{1}{2}$$

$$D = [(s - 1)Y(s)]_{s=1} = \left[\frac{s^3 - s + 1}{s^2(s + 1)}\right]_{s=1} = \frac{1}{2}$$

Application of Eqs. (2-63) and (2-64) gives

$$B = [s^2 Y(s)]_{s=0} = \left[\frac{s^3 - s + 1}{(s + 1)(s - 1)}\right]_{s=0} = -1$$

$$A = \left\{\frac{d}{ds}[s^2 Y(s)]\right\}_{s=0} = \left\{\frac{d}{ds}\left[\frac{s^3 - s + 1}{(s + 1)(s - 1)}\right]\right\}_{s=0} = 1$$

Thus, the Laplace transform of $y(t)$ is

$$Y(s) = L\{y(t)\} = \frac{1}{s} - \frac{1}{s^2} - \frac{1/2}{s + 1} + \frac{1/2}{s - 1}$$

Since we see from Table A-2 that $L\{1\} = 1/s$, then the *inverse transform* $L^{-1}\{1/s\} = 1$. From Table A-2 in the Appendix, we also see that $L^{-1}\{1/s^2\} = t$, $L^{-1}\{1/(s + 1)\} = e^{-t}$, and $L^{-1}\{1/(s - 1)\} = e^t$. Since $L^{-1}\{Y(s)\} = y(t)$, the solution of the differential equation is the inverse transform of the $Y(s)$ equation, and is given by

$$y(t) = 1 - t - \tfrac{1}{2} e^{-t} + \tfrac{1}{2} e^t$$

which can be checked by substitution in the differential equation and the three initial conditions.

Illustrative Example 2.13.   Suppose that $Y(s) = 3/[(s - 2)(s^2 + 4)]$. Find $y(t)$, the inverse transform of $Y(s)$.

Solution.   Using partial fractions, we have

$$Y(s) = \frac{3}{(s - 2)(s^2 + 4)} = \frac{A}{s - 2} + \frac{Bs + C}{s^2 + 4}$$

Use of Eq. (2-60) gives

$$A = [(s - 2)Y(s)]_{s=2} = \left[\frac{3}{s^2 + 4}\right]_{s=2} = \frac{3}{8}$$

Since

$$[(s^2 + 4)Y(s)]_{s=i2} = \left[\frac{3}{s-2}\right]_{s=i2} = \frac{3}{2i-2} = -\frac{3}{4} - i\frac{3}{4}$$

Equations (2-68) and (2-69) tell us that $B = (-3/4)/2 = -3/8$ and $C = -3/4$. Thus, we have

$$Y(s) = L\{y(t)\} = \frac{3/8}{s-2} - \frac{(3/8)s}{s^2+4} - \frac{3/4}{s^2+4}$$

Using Table A-2 in the Appendix, we find that the inverse transform $y(t) = L^{-1}\{Y(s)\}$ is

$$y(t) = \tfrac{3}{8} e^{2t} - \tfrac{3}{8} \cos 2t - \tfrac{3}{8} \sin 2t$$
$$= \tfrac{3}{8} (e^{2t} - \cos 2t - \sin 2t)$$

2.11 Other Properties of Laplace Transforms

Many of the properties of Laplace transforms are summarized in Table A-1 in the Appendix. Little will be said or proven about these properties, because it is assumed in this text that the reader has studied Laplace transforms in a previous course; therefore, only a brief review will be necessary. We see from Table A-1 that the Laplace transform of an integral is given by

$$L\left\{\int_a^t f(t)\,dt\right\} = \frac{1}{s}F(s) + \frac{1}{s}\int_a^0 f(t)\,dt \qquad (2\text{-}70)$$

and that

$$L\{tf(t)\} = -\frac{d}{ds}F(s) \qquad (2\text{-}71)$$

$$L\left\{\frac{f(t)}{t}\right\} = \int_s^\infty F(s)\,ds \qquad (2\text{-}72)$$

Thus, Eq. (2-71) shows that

$$L\{t \sin at\} = -\frac{d}{ds}\left(\frac{a}{s^2+a^2}\right) = \frac{2as}{(s^2+a^2)^2}$$

since $F(s) = L\{f(t)\} = L\{\sin at\} = a/(s^2+a^2)$. Table A-1 also contains the *first shifting theorem* which states that if $L\{f(t)\} = F(s)$, then

$$L\{e^{at} f(t)\} = F(s-a) \qquad (2\text{-}73)$$

The *second shifting theorem* states that if $a \geq 0$, then

$$L\{f(t-a)u(t-a)\} = e^{-as}L\{f(t)\} \qquad (2\text{-}74)$$

where the unit step function $u(t-a)$ is defined to equal unity if $t > a$ and zero if $t < a$, as shown in Fig. 2-14a. Thus, Eq. (2-73) shows that

$$L\{e^{at} \sin bt\} = \frac{b}{(s-a)^2 + b^2}$$

since $F(s) = L\{\sin bt\} = b/(s^2 + b^2)$. Equation (2–74) shows that

$$L\{\sin (t-a)u(t-a)\} = e^{-as}\left(\frac{1}{s^2+1}\right)$$

since $L\{f(t)\} = L\{\sin t\} = 1/(s^2 + 1^2)$. The second shifting theorem is very useful in the solution of vibration problems that have discontinuous or pulse force-or-motion excitations. The two shifting theorems are proven in Ill. Ex. 2.14.

A periodic function has been previously defined to be a function that repeats itself after a certain increment of time. That is, $f(t)$ is a periodic function if $f(t+T) = f(t)$ for all values of t. If T is the minimum value for which the previous relation holds, then T is the period of function $f(t)$. From Table A–1 in the Appendix, we see that the Laplace transform of a periodic function of period T is given by

$$L\{f(t)\} = \frac{1}{(1-e^{-Ts})}\int_0^T e^{-st} f(t)\, dt \tag{2–75}$$

The application of this very useful property, which requires integration only through one period of $f(t)$, is exemplified in Ill. Ex. 2.15. The *convolution theorem, initial-value theorem,* and the *final-value theorem* are given by Eqs. (2–76), (2–77), and (2–78), respectively, and in Table A–1. Many textbooks furnish examples for the application of these and other properties that are given in this section and in Table A–1.

$$F(s) = L\{g(t)\}L\{h(t)\} = L\left\{\int_0^t g(t-T)h(T)\, dT\right\} \tag{2–76}$$

$$f(0+) = \lim_{t\to 0+} f(t) = \lim_{s\to\infty} sL\{f(t)\} \tag{2–77}$$

$$\lim_{t\to\infty} f(t) = \lim_{s\to 0} sL\{f(t)\} \tag{2–78}$$

Let us now consider the Laplace transforms of some special functions that occur in engineering. The *unit step function,* which is plotted in Fig. 2–14a, is defined mathematically as follows:

$$u(t-a) = 0 \qquad \text{if } t < a$$
$$u(t-a) = 1 \qquad \text{if } t > a$$

From Table A–2, we see that its Laplace transform is given by

$$L\{u(t-a)\} = \frac{e^{-as}}{s} \tag{2–79}$$

which is proven in Ill. Ex. 2.16. The *unit gate function* $g(t)$, which is plotted in Fig. 2–14b, is defined mathematically as follows:

$$g(t) = 0 \qquad \text{if } t < a$$
$$g(t) = 1 \qquad \text{if } a < t < b$$
$$g(t) = 0 \qquad \text{if } t > b$$

Since it can be seen from Fig. 2–14a, b that $g(t)$ can be obtained by subtracting $u(t - b)$ from $u(t - a)$, then $g(t) = u(t - a) - u(t - b)$ and use of Eq. (2–79) gives

$$L\{g(t)\} = L\{u(t - a) - u(t - b)\} = \frac{e^{-as}}{s} - \frac{e^{-bs}}{s} = \frac{e^{-as} - e^{-bs}}{s} \qquad (2\text{–}80)$$

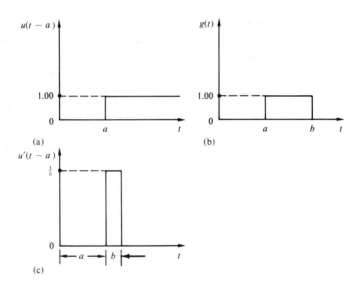

Fig. 2-14

The *unit impulse* or *delta function* $u'(t - a)$, which is located at $t = a$ and is plotted in Fig. 2–14c, is a gate function of width b and height $1/b$ (i.e., its area is unity), where width b approaches zero so that height $1/b$ approaches infinity. A unit impulse is denoted functionally by $u'(t - a)$ because it is often considered to be the first derivative of the step function $u(t - a)$, even though the derivative of $u(t - a)$ does not exist at $t = a$ because of the discontinuity at that point. Note, however, that the slope of step function $u(t - a)$ equals zero when $t \neq a$ (since this step function equals zero when $t < a$ and equals unity when $t > a$). From Table A–2, we see that the Laplace transform of a unit impulse is given by

$$L\{u'(t - a)\} = e^{-as} \qquad (2\text{–}81)$$

which is proven in Ill. Ex. 2.16.

As will be seen in Sec. 2.12, pulse functions occur in vibration problems as well as in electrical circuit problems, and such problems are very conveniently handled by use of Laplace transforms. A rectangular pulse, a triangular pulse, a sawtooth pulse, and a sinusoidal pulse are illustrated in Figs. 2-15a, b, c, and d, respectively. Their Laplace transforms are given in Table A-2 in the Appendix. The result for the rectangular pulse may be easily obtained from Eq. (2-80) since Fig. 2-15a shows that a rectangular pulse is a gate function of height M that starts at $t = 0$ and ends at $t = a$. Thus, the result is obtained by substitution of $a = 0$ and $b = a$ in Eq. (2-80), and by multiplication of the result by M.

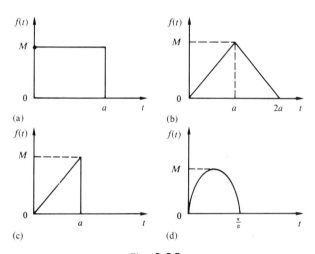

Fig. 2-15

Illustrative Example 2.14. Verify the two shifting property theorems (i.e., Eqs. (2-73) and (2-74)).

Solution. The proof of Eq. (2-73) is as follows:

$$L\{e^{at} f(t)\} = \int_0^\infty e^{at} f(t) e^{-st}\, dt = \int_0^\infty e^{-(s-a)t} f(t)\, dt = F(s - a)$$

Equation (2-74) can be proved as follows:

$$L\{f(t - a)u(t - a)\} = \int_0^\infty f(t - a)u(t - a)e^{-st}\, dt$$

Since $u(t - a) = 0$ for $t < a$ and $u(t - a) = 1$ for $t > a$, we have

$$L\{f(t - a)u(t - a)\} = \int_a^\infty f(t - a)e^{-st}\, dt$$

Let $t_1 = t - a$. Thus, $dt = dt_1$ and we obtain

$$\int_0^\infty f(t_1)e^{-s(t_1+a)}\, dt_1 = e^{-as} \int_0^\infty f(t_1)e^{-st_1}\, dt_1 = e^{-as} L\{f(t)\}$$

Illustrative Example 2.15.   Determine the Laplace transform of the periodic rectangular wave that is illustrated in Fig. 2-16.

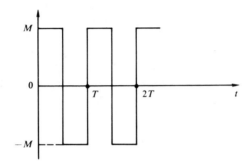

Fig. 2-16

Solution.   Since Fig. 2-16 shows that

$$f(t) = M \qquad \text{for } 0 \le t \le \frac{T}{2}$$

$$f(t) = -M \qquad \text{for } \frac{T}{2} \le t \le T$$

substitution in Eq. (2-75) gives

$$L\{f(t)\} = \frac{M}{1 - e^{-Ts}}\left[\int_0^{T/2} Me^{-st}\, dt + \int_{T/2}^{T} - Me^{-st}\, dt\right]$$

Evaluation of the integrals gives

$$L\{f(t)\} = \frac{M}{1 - e^{-Ts}}\left\{-\frac{1}{s}\,[e^{-st}]_0^{T/2} + \frac{1}{s}\,[e^{-st}]_{T/2}^{T}\right\}$$

$$= \left(\frac{M/s}{1 - e^{-Ts}}\right)(1 - 2e^{-sT/2} + e^{-sT})$$

Factoring the numerator and denominator of the above result gives

$$L\{f(t)\} = \frac{M}{s}\,\frac{(1 - e^{-sT/2})^2}{(1 - e^{-sT/2})(1 + e^{-sT/2})} = \frac{M(1 - e^{-sT/2})}{s(1 + e^{-sT/2})}$$

Multiplication of the numerator and denominator by $e^{Ts/4}$ gives

$$L\{f(t)\} = \frac{M(e^{Ts/4} - e^{-Ts/4})}{s(e^{Ts/4} + e^{-Ts/4})}$$

from which we obtain

$$L\{f(t)\} = \frac{M}{s}\tanh\frac{Ts}{4} \qquad\qquad (2\text{-}82)$$

Illustrative Example 2.16.   Verify Eqs. (2-79) and (2-81).

Solution.   Since $u(t - a) = 0$ if $t < a$ and $u(t - a) = 1$ if $t > a$, the proof of Eq. (2-79) is as follows:

$$L\{u(t - a)\} = \int_0^\infty e^{-st} u(t - a)\, dt$$

$$= \int_0^a e^{-st}(0)\, dt + \int_a^\infty e^{-st}(1)\, dt = \frac{e^{-as}}{s}$$

As for the proof of Eq. (2-81), note that the impulse function $u'(t - a)$ can be expressed mathematically as follows:

$$u'(t - a) = \lim_{b \to 0} \left[\frac{1}{b} u(t - a) - \frac{1}{b} u(t - a - b) \right]$$

Thus, its Laplace transform, using Eq. (2-79), is given by

$$L\{u'(t - a)\} = \lim_{b \to 0} \left[\frac{1}{b} \frac{e^{-as}}{s} - \frac{1}{b} \frac{e^{-(a+b)s}}{s} \right] = \lim_{b \to 0} \left[\frac{e^{-as}(1 - e^{-bs})}{bs} \right]$$

The bracketed term is an indeterminant quantity (i.e., $0/0$). Differentiating separately the numerator and denominator of the term within the brackets with respect to b and applying *L'Hospital's rule*, we obtain

$$L\{u'(t - a)\} = \lim_{b \to 0} \left[\frac{e^{-as}(se^{-bs})}{s} \right] = e^{-as}$$

2.12   Shock, Impact, and Random Vibrations

Up to this point, this chapter has been concentrating mainly on vibrating systems where the excitation is either harmonic or else can be resolved by Fourier series as an infinite sum of harmonic components of differing amplitudes and frequencies. This section will consider, in a very introductory and incomplete manner, some more involved types of force or motion excitations. Some systems are subjected to abrupt excitations. Physical examples are a punch press, the striking of a hammer, automotive travel on a rough road, the shooting of a gun, the gust force on an airplane, and the dropping of a package. Such abrupt excitations can cause strong transient vibrations, which can be important to analyze, if the system is elastic. *Shock* generally denotes a rapidly applied force, displacement, or velocity excitation whose time duration is quite short and which has some large first-derivative values. *Impact*, for which no general theory has been developed, implies the collision of two elastic or inelastic bodies (e.g., a bullet striking its target). To simplify the problem, the only suddenly applied excitations that we shall consider are those of simple mathematical form (e.g., the four types of pulses in Fig. 2–15). For a pulse excitation, the steady-state response is zero and we are only interested in the transient motion, whose amplitude variation can be rapid. The analysis of pulse and other discontinuous excitation problems cannot be easily done by classical methods; consequently, Laplace transform methods are usually employed for such problems. Table A–2 in the

Appendix furnishes the Laplace transforms for a step function, an impulse function, and the four pulses shown in Fig. 2–15. Three examples that illustrate the solution of vibration problems by Laplace transforms are given in Ill. Exs. 2.17, 2.18, and 2.19.

Some systems, such as jet engines, have a non-periodic excitation that is random in its variation with time. Since the resulting vibrational motion will also be random, this type of vibration must be analyzed by methods that are quite different from those discussed in this text so far. Since a random vibration implies that the instantaneous value of both the excitation and the resultant motion is not predictable with respect to time, statistical methods are used. Thus, a random vibration is a continuous oscillation whose instantaneous amplitude can be predicted only by probability methods, and it is often desirable to calculate such statistical properties as the mean, mean square value, probability distribution, variance, standard deviation, and root mean square for such random functions as the excitation and amplitude variations. Suppose that we had a single-degree-of-freedom system that is excited by a random external force $F(t)$ or by randomly applied motion that can be represented by a force $F(t)$. The transient response motion will die with time if damping is present, while the steady-state motion $x_p(t)$ will be random. The *mean-square value* $\bar{F}^2$ of the random excitation force $F(t)$, where τ is a multiple of period T, is defined to be as follows:

$$\bar{F}^2 = \frac{1}{\tau} \int_0^\tau [F(t)]^2 \, dt \qquad (2\text{--}83)$$

For a random vibration, engineers are often mainly interested in the average energy that can be computed from the mean-square value for displacement $x(t)$, which allows us to ignore phase angle ϕ. The *mean-square value* $\bar{x}^2$ for displacement $x(t)$ is defined to be

$$\bar{x}^2 = \frac{1}{\tau} \int_0^\tau [x(t)]^2 \, dt \qquad (2\text{--}84)$$

where τ is a multiple of period T, or more properly as

$$\bar{x}^2 = \lim_{\tau \to \infty} \frac{1}{\tau} \int_0^\tau [x(t)]^2 \, dt \qquad (2\text{--}85)$$

since the averaging time τ should be large when a mean-square value is determined for a random vibration.

If a displacement-versus-time solution is obtained for a given random vibration problem, the probability that displacement x will lie between x_0 and $x_0 + \Delta x$ can be determined graphically by drawing a horizontal line at x_0 and a horizontal line at $(x_0 + \Delta x)$ on a plot of x

versus time t. This probability will equal the sum of the measured time intervals for which $x_0 < x < (x_0 + \Delta x)$ divided by the total time considered. This probability value can be computed more easily by statistical means. That is, the probability that displacement x will attain a certain value can be determined from the mean value and the standard deviation, where the standard deviation is a measure of the width of the distribution of the x-values about the mean value. For damage considerations, peak amplitude values, and their probabilities, can be important. Because a random vibration can be considered to be composed of a continuous spectrum of frequencies whose amplitudes vary in a random manner, spectral density techniques are useful in the analysis of random vibrations. Since the amplitude of a single frequency component (when the frequency spectrum is divided into components or frequency intervals) varies randomly, the mean-square value is used. The *spectral density* of a function plotted against frequency ω is the ratio of the mean-square value of that function in a frequency interval $\Delta \omega$ divided by $\Delta \omega$, the length of the frequency interval. Thus, the area under a portion of this plot (i.e., curve) equals the mean-square-value of the function between the two indicated frequencies. In Sec. 2.6, we saw that by using Fourier series any periodic force or motion can be represented as a superposition of harmonic functions (i.e., sines and cosines) of different amplitudes, phase angles, and frequencies. It can be shown that the mean-square value for such a periodic function, which is composed of components at different frequencies, equals the sum of the mean-square values for each harmonic component. Random excitation and response functions can usually be considered to have a large number of frequency components. The frequency content of a recorded random motion can be determined from methods based on Fourier series if the motion is periodic, or from methods that utilize the Fourier integral when the motion is non-periodic. Because of its complicated nature, the requirement of advanced solution techniques, and the space that it would otherwise take, the subject of random vibrations is very sketchily covered in this text. More complete references, such as the two S. H. Crandall references and the third W. T. Thomson reference in Appendix A–3, should be consulted.

Illustrative Example 2.17.   Suppose for the spring-mass-damper system in Fig. 2-3 in Sec. 2.3 that the external excitation force is $F(t)$, the initial displacement is x_0, and the initial velocity is v_0. Find the Laplace transform of displacement $x(t)$.

Solution.   Let us denote the Laplace transform of $x(t)$ by $X(s)$ or X, and the Laplace transform of force $F(t)$ by $F(s)$. The equation of motion for the spring-mass-damper system is

$$\frac{W}{g}\ddot{x} + c\dot{x} + kx = F(t)$$

Using Eqs. (2-56) and (2-57), the Laplace transform of this equation of motion, where $m = W/g$, is

$$m(s^2 X - sx_0 - v_0) + c(sX - x_0) + kx = F(s)$$

which can be rewritten as

$$X(ms^2 + cs + k) - (ms + c)x_0 - mv_0 = F(s)$$

Thus, $X(s)$, the Laplace transform of $x(t)$, is

$$X(s) = \frac{[F(s) + mv_0 + (ms + c)x_0]}{ms^2 + cs + k} \qquad (2\text{-}86)$$

Illustrative Example 2.18.   Suppose that in Ill. Ex. 2.17 the damping is negligible, both the initial displacement and initial velocity equal zero, and the external force is a rectangular pulse of magnitude F_0 and duration a.

Solution.   Since $c = x_0 = v_0 = 0$, Eq. (2-86) becomes

$$X(s) = L\{x(t)\} = \frac{F(s)}{ms^2 + k} = \frac{F(s)/m}{s^2 + k/m}$$

where $m = W/g$. From Table A-2 in the Appendix, we see that $F(s)$, which is the Laplace transform of the rectangular pulse force, equals $F_0(1 - e^{-as})/s$. Substituting this, along with $\omega_n = \sqrt{k/m}$, gives

$$X(s) = \frac{F_0/m}{s(s^2 + \omega_n^2)} - \frac{F_0/m}{s(s^2 + \omega_n^2)} e^{-as}$$

Use of partial fractions gives

$$\frac{F_0/m}{s(s^2 + \omega_n^2)} = \frac{A}{s} + \frac{Bs + C}{s^2 + \omega_n^2}$$

Application of Eq. (2-60) gives

$$A = \left(\frac{F_0/m}{s^2 + \omega_n^2}\right)_{s=0} = \frac{F_0}{m\omega_n^2}$$

since

$$\left(\frac{F_0/m}{s}\right)_{s=i\omega_n} = \frac{F_0}{i\omega_n m} = -i\frac{F_0}{\omega_n m}$$

Application of Eqs. (2-68) and (2-69) show that $B = (-F_0/\omega_n m)/\omega_n = -F_0/m\omega_n^2$ and $C = 0$. Thus, we have

$$X(s) = L\{x(t)\} = \frac{F_0}{k}\left(\frac{1}{s} - \frac{s}{s^2 + \omega_n^2}\right) - \frac{F_0}{k} e^{-as}\left(\frac{1}{s} - \frac{s}{s^2 + \omega_n^2}\right)$$

where $m\omega_n^2 = m(k/m) = k$. Table A-2 shows that the inverse transforms of $1/s$ and $s/(s^2 + \omega_n^2)$ are 1 and $\cos \omega_n t$, respectively. For the second term of $X(s)$, we must apply Eq. (2-74), the second shifting theorem, in order to obtain its inverse transform. Usually we expect to apply this property when we have pulse excitations. The solution for $x(t)$ is the inverse transform of $X(s)$, which is as follows where $\omega_n = \sqrt{kg/W}$:

$$x(t) = \frac{F_0}{k}(1 - \cos \omega_n t) - \frac{F_0}{k}[1 - \cos \omega_n(t - a)]u(t - a)$$

Note the discontinuity in the $x(t)$-response which is caused by the discontinuity in the applied force at $t = a$. That is,

$$x(t) = \frac{F_0}{k}(1 - \cos \omega_n t) \qquad \text{if } t < a$$

$$x(t) = \frac{F_0}{k}[-\cos \omega_n t + \cos \omega_n(t-a)] \qquad \text{if } t > a$$

Illustrative Example 2.19.   Consider the vibration-measuring instrument illustrated in Fig. 2-10 of Sec. 2.8, and assume that damping may be neglected. If the support base of the instrument is suddenly displaced and maintained at distance h, determine the maximum acceleration and force to which the spring-mounted instrument is subjected, and the displacement variation of mass m with time.

Solution.   Application of Newton's second law gives

$$m\ddot{x}_2 = -k(x_2 - x_1)$$

where $x_1 = hu(t)$ because there is a step-function displacement excitation applied to the base of the instrument. Thus, the equation of motion can be written as

$$m\ddot{x}_2 + kx_2 = khu(t)$$

The values of $x_2(0)$ and $\dot{x}_2(0)$ are both zero, since mass m has no displacement or velocity at the time that the instrument base is initially displaced. Using Eqs. (2-57) and (2-79), the Laplace transform of the equation of motion, where $X = X(s) = L\{x_2(t)\}$, is as follows:

$$ms^2 X + kX = \frac{kh}{s}$$

From the previous equation, we see that

$$X(s) = L\{x_2(t)\} = \frac{kh}{s(ms^2 + k)} = \frac{hk/m}{s(s^2 + k/m)}$$

$$= \frac{h\omega_n^2}{s(s^2 + \omega_n^2)} = \frac{A}{s} + \frac{Bs + C}{s^2 + \omega_n^2}$$

Application of Eq. (2-60) gives

$$A = \left(\frac{h\omega_n^2}{s^2 + \omega_n^2}\right)_{s=0} = h$$

since

$$\left(\frac{h\omega_n^2}{s}\right)_{s=i\omega_n} = \frac{h\omega_n}{i} = -ih\omega_n$$

Application of Eqs. (2-68) and (2-69) show that $B = -h\omega_n/\omega_n = -h$ and $C = 0$. Thus, we have

$$X(s) = L\{x_2(t)\} = \frac{h}{s} - \frac{hs}{s^2 + \omega_n^2}$$

Using Table A-2 in the Appendix, we see that the inverse transform of $X(s)$ is

$$x_2 = h - h \cos \omega_n t = h\left(1 - \cos \sqrt{\frac{k}{m}} t\right)$$

which is the displacement variation of mass m with time. Let $z = (x_2 - x_1)$, so that for $t > 0$ we have

$$z(t) = -h \cos \sqrt{\frac{k}{m}} t = -h \cos \omega_n t$$

since $x_1 = h$ for $t > 0$. Because $\ddot{z} = -\omega_n^2 z$, the maximum acceleration subjected to the instrument is

$$\ddot{z}_{max} = (-\omega_n^2)(-h) = \omega_n^2 h = \frac{kh}{m}$$

and the maximum force is

$$F_{max} = m\ddot{z}_{max} = m\frac{kh}{m} = kh$$

where it may be noted that the maximum force can also be calculated by

$$F_{max} = k(x_2 - x_1)_{max} = kz_{max} = kh$$

Illustrative Example 2.20. For the damped spring-mass system in Fig. 2-3, suppose that $W = 38.6$ lb, $k = 6.4$ lb/in., and the damping ratio is 0.20. Determine the mean-square steady-state displacement of the mass when the applied force $F(t)$ is $100(\cos 4t + \cos 12t)$ lb, when t is in seconds. Also obtain the equation for this steady-state displacement.

Solution. Using superposition and Eq. (2-8) of Sec. 2.3, the form of the equation for the steady-state displacement x_p of the mass is

$$x_p = X_1 \cos(\omega_1 t - \phi_1) + X_2 \cos(\omega_2 t - \phi_2)$$

where $\omega_1 = 4$ rad/sec and $\omega_2 = 12$ rad/sec. From Eqs. (2-10) and (2-12), $X_{1_0} = X_{2_0} = F/k = 100/6.4 = 15.6$ in. Equations (2-12) and (2-13) give

$$X_i = \frac{15.6}{\sqrt{(1 - r_i^2)^2 + (2\zeta r_i)^2}}$$

$$\phi_i = \tan^{-1}\left(\frac{2\zeta r_i}{1 - r_i^2}\right)$$

where $\zeta = 0.20$ and $r_i = \omega_i/\omega_n$. Since $\omega_n = \sqrt{kg/W}$, the natural frequency is $\sqrt{(6.4)(386)/(38.6)}$ or 8.0 rad/sec. Thus, $r_1 = 4/8 = 0.50$ and $r_2 = 12/8 = 1.50$. Substitution of these values for r_1 and r_2 in the previous two equations gives $X_1 = 20.1$ in., $X_2 = 11.2$ in., $\phi_1 = \pi/12.0$ rad, and $\phi_2 = -\pi/7.05$ rad. Thus, the steady-state displacement x_p is

$$x_p = 20.1 \cos\left(4t - \frac{\pi}{12}\right) + 11.2 \cos\left(12t + \frac{\pi}{7.05}\right)$$

Rewriting Eq. (2-84) in a summation form, for n applied forces, gives

$$\bar{x}^2 = \frac{1}{n} \sum_{i=1}^{n} X_i^2$$

Substitution of the computed values X_1 and X_2 in the previous equation gives the following result for the mean-square displacement $\bar{x}^2$:

$$\bar{x}^2 = \frac{1}{2}(X_1^2 + X_2^2) = 262 \text{ in.}^2$$

2.13   Solution of Nonlinear and Other Vibration Problems by Numerical Methods

To illustrate how nonlinear and other types of vibration problems may be solved by numerical methods, let us consider the spring-mass-damper system in Fig. 2–3 of Sec. 2.3, which has an applied external force $F \sin \omega t$. To be more general, let the damping force be nonlinear, and be given by Eq. (1–32). Using Newton's second law, the equation of motion is

$$\frac{W}{g}\ddot{x} = F \sin \omega t - a|\dot{x}|^b \dot{x} - kx$$

which is a nonlinear differential equation (if $b \neq 0$) that can be rewritten as

$$\frac{W}{g}\ddot{x} + a|\dot{x}|^b \dot{x} + kx = F \sin \omega t \qquad (2\text{--}87)$$

We shall solve this differential equation by numerical means. As mentioned previously, such a nonlinear differential equation cannot be solved by Laplace transform or classical analytical methods. This section shall present two elementary numerical-solution methods, where one method is not very accurate. More accurate numerical methods for solving ordinary differential equations will be discussed in Sec. 2.15.

We shall first discuss the Euler method. As a background for the Euler method, consider the following equation from calculus:

$$f(x + \Delta x) - f(x) = \int_x^{x+\Delta x} f'(\xi)\, d\xi \qquad (2\text{--}88)$$

If the slope $f'(\xi)$ is constant in the interval from x to $(x + \Delta x)$, this slope equals $f'(x)$, and this equation gives the following formula when the integral is evaluated:

$$f(x + \Delta x) = f(x) + (\Delta x)f'(x) \qquad (2\text{--}89)$$

which is the equation for the *Euler method* and which is also the sum of the first two terms of a Taylor series. This equation can be rewritten in subscript form as follows:

$$f_{x+\Delta x} = f_x + (\Delta x)f'_x \qquad (2\text{-}90)$$

Essentially, this method says that if we know the values of $f(x)$ and $f'(x)$, which are at location x, we can *approximate* $f(x + \Delta x)$ by use of Eq. (2–89). The accuracy of this method, which *assumes* that $f'(x)$ is constant in each Δx-interval, improves as Δx is made smaller. This method is graphically illustrated in Fig. 2–17, where the sum of the unshaded areas under the curve is the error that results from using this method.

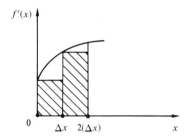

Fig. 2-17

The trapezoidal method is the other numerical method that we shall discuss in this section. Consider the following equation:

$$f(x + \Delta x) - f(x) = \int_{x}^{x+\Delta x} f'(\xi)\, d\xi \qquad (2\text{-}91)$$

If we assume that the average value of slope $f'(\xi)$ in the interval from x to $x + \Delta x$ is equal to $[f'(x) + f'(x + \Delta x)]/2$ and substitute this result in Eq. (2–91), we obtain the following approximation after the integral is evaluated:

$$f(x + \Delta x) = f(x) + \left(\frac{\Delta x}{2}\right)[f'(x) + f'(x + \Delta x)] \qquad (2\text{-}92)$$

The *trapezoidal method* is given by Eq. (2–92), which can be rewritten in subscript form as follows:

$$f_{x+\Delta x} = f_x + \left(\frac{\Delta x}{2}\right)(f'_x + f'_{x+\Delta x}) \qquad (2\text{-}93)$$

The trapezoidal method is graphically illustrated in Fig. 2–18, and comparison of the size of the unshaded area under this curve with that

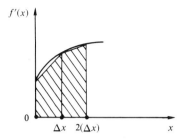

Fig. 2-18

of Fig. 2–17 clearly shows the superiority of the trapezoidal method over the Euler method. The trapezoidal method cannot be directly applied in the solution of first-order differential equations of the general type

$$f'(x) \equiv \frac{df}{dx} = F(x, f)$$

This is because we need the value of $f(x + \Delta x)$ in order to calculate $f'(x + \Delta x)$. The trapezoidal method, however, is very useful in the solution of higher-order differential equations. For a third-order differential equation written in the following general form:

$$f'''(x) \equiv \frac{d^3 f}{dx^3} = F(x, f, f', f'')$$

we can numerically obtain $f''(x + \Delta x)$ using the Euler method, and we can numerically obtain both $f'(x + \Delta x)$ and $f(x + \Delta x)$ by using the trapezoidal method [i.e., by use of Eq. (2–92)].

It is easiest to illustrate how numerical methods are used to solve an ordinary differential equation by first using a rather simple example. For this reason, Ill. Ex. 2.20 shows how to numerically solve the second-order differential equation $\ddot{y} + 4y = 0$ by use of the Euler and trapezoidal methods. After looking at this example, let us next make a comparison of analytical and numerical techniques.

Numerical-solution techniques consist of repetitive, step-by-step procedures that are tedious and laborious to apply, but are easy to set up for solution by a digital computer which can quickly do this labor. The numerical method advantages over classical and Laplace transform techniques are that they can be used to solve nonlinear differential equations and that they can handle almost any type of forcing or excitation function. Classical and Laplace transform techniques give a solution in the form of an equation, and values at different points may be obtained by substitution of numerical values for the independent variable. Thus, we can construct a table of values for the solution for various values of the independent variable. Numerical methods, on the other hand, furnish the solution in the form of a table of values, instead of by an equation. In other words, it is a step-by-step method by which the solution is calculated at successively larger (or successively smaller) values of time or displacement. The accuracy of numerical methods increases as the interval for the time or displacement increase is made smaller.

Using Eq. (2–87), which is the equation of motion, and the Euler and the trapezoidal methods to solve the previously mentioned mechanical vibration problem, we have the following equations for the solution of this nonlinear problem:

$$\ddot{x}_t = \frac{g}{W}(F \sin \omega t_t - a|\dot{x}_t|^b \dot{x}_t - kx_t) \qquad (2\text{-}94)$$

$$\dot{x}_{t+\Delta t} = \dot{x}_t + (\Delta t)\ddot{x}_t \qquad (2\text{-}95)$$

$$x_{t+\Delta t} = x_t + \left(\frac{\Delta t}{2}\right)(\dot{x}_t + \dot{x}_{t+\Delta t}) \qquad (2\text{-}96)$$

where damping coefficient $c = a|\dot{x}|^b$. Because an ocsillatory motion (i.e., a vibration) will result, this problem will require a very small time interval Δt. Thus, it is preferable to approximate $\dot{x}_{t+\Delta t}$ by a better method than the Euler method, and for this reason more accurate methods are discussed in Sec. 2.15. Section 7.15 contains a digital computer program that solves this nonlinear vibration problem using the Euler and trapezoidal numerical methods. Thus, this computer program can be used to have these tedious and repetive numerical calculations performed by a digital computer. Section 7.24 has a program that solves a similar nonlinear vibration problem that has input graphs.

Illustrative Example 2.21. Suppose that the acceleration $\ddot{y}$ of a mass was equal to $-4y^2$ ft²/sec² where t is the time in seconds and y is the displacement in feet of the mass from a reference position. If the initial velocity is zero and the initial displacement is 2 ft, solve for the displacement of this mass from $t = 0$ to $t = 0.200$ sec using the trapezoidal and Euler methods and an interval of 0.050 sec. Calculate the errors of these numerical approximations.

Solution. In mathematical form, this problem requires us to solve the differential equation $\ddot{y} + 4y^2 = 0$, where $\dot{y}(0) = 0$ and $y(0) = 2.000$. Thus, from the given differential equation, we have

$$\ddot{y}(t) = -4[y(t)]^2$$

Application of Eq. (2–89) gives

$$\dot{y}(t + \Delta t) = \dot{y}(t) + (\Delta t)\ddot{y}(t)$$

Applying Eq. (2-92), we obtain

$$y(t + \Delta t) = y(t) + \left(\frac{\Delta t}{2}\right)[\dot{y}(t) + \dot{y}(t + \Delta t)]$$

Using the three previous equations, we obtain the following numerical results:

$$\ddot{y}(0) = -(4.00)(2.000)^2 = -16.000$$
$$\dot{y}(0.050) = 0 + (0.050)(-16.000) = -0.800$$
$$y(0.050) = 2.000 + (0.025)(0 - 0.800) = 1.980$$
$$\ddot{y}(0.050) = -(4.00)(1.980)^2 = -15.680$$
$$\dot{y}(0.100) = -0.800 + (0.050)(-15.680) = -1.584$$
$$y(0.100) = 1.980 + (0.025)(-0.800 - 1.584) = 1.920$$
$$\ddot{y}(0.100) = -(4.00)(1.920)^2 = -14.746$$
$$\dot{y}(0.150) = -1.584 + (0.050)(-14.746) = -2.321$$
$$y(0.150) = 1.920 + (0.025)(-1.584 - 2.321) = 1.822$$
$$\ddot{y}(0.150) = -(4.00)(1.822)^2 = -13.279$$
$$\dot{y}(0.200) = -2.321 + (0.050)(-13.279) = -2.985$$
$$y(0.200) = 1.822 + (0.025)(-2.321 - 2.985) = 1.787$$

The solution for the given nonlinear differential equation is $2/(4t^2 + 1)$. Thus, substitution for time t shows that the exact values for $y(0.050)$, $y(0.100)$, $y(0.150)$, and $y(0.200)$ are 1.980, 1.923, 1.835, and 1.724 ft, respectively. Thus, the numerical approximation errors are equal to 0.000, -0.003, -0.013, and -0.035 ft at $t = 0.050, 0.100, 0.150$, and 0.200 sec, respectively. These errors could be reduced by using a smaller value for the interval Δt or by using a more accurate method than the Euler method to compute $\dot{y}(t + \Delta t)$.

2.14   Other Numerical Methods for Solving Ordinary Differential Equations

We have shown how to solve a nonlinear vibration problem by numerical means in Sec. 2.14. Let us now briefly discuss the numerical solution of a first-order differential equation in general fashion. As stated previously, it is advantageous to solve most nonlinear differential equations by numerical methods, since most cannot be solved by classical or Laplace transform means. Numerical means can also be useful when the differential equation is linear, but the coefficients or the forcing function vary with time (i.e., the independent variable), since this variation often greatly complicates an analytical solution procedure. As we shall see later in Chap. 7, numerical techniques are especially useful if this variation with time is so involved that it has to be furnished as a graph or a table. Unlike solutions by analytical methods, however, only a little more complexity will be added when numerical methods are used to solve such problems.

Numerical methods are approximate solution techniques; therefore a certain amount of error will result. There are three types of error involved: truncation, roundoff, and inherent. *Truncation error* is the error that is due to the use of an approximation method. For example, the unshaded area under the curve of Fig. 2–17 is the error due to the use of the Euler method. *Roundoff error* is the error that is due to the use of a finite number of digits in the calculations (e.g., three slide rule digits). *Inherent error* is the error that is due to the use of an approximately correct (i.e., approximated) term (due to previous truncation and roundoff errors) in a problem equation. For example, an approximate value of y was used to calculate $\ddot{y}(0.100)$ and $\ddot{y}(0.150)$ in Ill. Ex. 2.20. These errors generally decrease as the Δt-or Δx-interval is made smaller. We can use the *interval-halving method* (which is easy to apply for digital computer solutions) to determine the proper interval size. That is, if the problem results obtained by using an interval Δt agree with the results obtained by using the interval $(\Delta t/2)$, then the Δt-interval chosen is satisfactory.

As mentioned previously, the Euler method is a rather inaccurate method and better means are desirable, especially for vibration problems where the derivatives $\ddot{x}$ and $\dot{x}$ vary rapidly. The Euler method assumes that the slope is constant in each Δt-interval, which is not the case in most problems. Because the slope usually varies in an interval, it would be better to use the slope at the midpoint of an interval, instead of the slope at the forward end of the interval as the Euler method does. Let us consider an interval of size $2(\Delta t)$. If we use the slope at the midpoint of this interval as an approximate average-slope value for this interval, substitution in the calculus equation

$$f(t + \Delta t) - f(t - \Delta t) = \int_{t-\Delta t}^{t+\Delta t} f'(\xi)\, d\xi \qquad (2\text{–}97)$$

results in the following equation after the integral is evaluated

$$f(t + \Delta t) = f(t - \Delta t) + 2(\Delta t)\, f'(t) \qquad (2\text{–}98)$$

where the constant term $f'(t)$ is treated as the average-slope value for $f'(\xi)$. This would be an exact approximation if $f'(\xi)$ is linear in this $2(\Delta t)$-interval. We shall arbitrarily call the technique given by Eq. (2–98) the *midpoint-slope method*. Equation (2–98) can be rewritten in subscript form as follows:

$$f_{t+\Delta t} = f_{t-\Delta t} + 2(\Delta t)\, f'_t \qquad (2\text{–}99)$$

The *Adams method* approximates the slope of $f(t)$ by a polynomial of degree n over $(n + 1)$ equally spaced points. For the simplest case, this method assumes that $f'(t)$ has a linear straight-line variation in each Δt-interval, instead of being constant as the Euler method assumes. A straight line is drawn between $f'(t - \Delta t)$ and $f'(t)$, and is extrapolated to $f'(t + \Delta t)$. Thus, the slope m of this straight line is $[f'(t) - f'(t - \Delta t)]/\Delta t$. Using this value of m in the following straight-line equation,

$$f'(\xi) = f'(t) + m(\xi - t)$$

and substituting the above equation into Eq. (2–88) in Sec. 2.14, we obtain the following result after the integral is evaluated:

$$f(t + \Delta t) = f(t) + \left(\frac{\Delta t}{2}\right)[3f'(t) - f'(t - \Delta t)] \qquad \textbf{(2–100)}$$

This result approximates $f(\xi)$ by a quadratic polynomial in the interval from $(t - \Delta t)$ to $(t + \Delta t)$ because the derivation integrated a straight-line approximation for $f'(\xi)$. Equation (2–100) can be written in subscript form as follows:

$$f_{t+\Delta t} = f_t + \left(\frac{\Delta t}{2}\right)(3f_t' - f_{t-\Delta t}') \qquad \textbf{(2–101)}$$

One disadvantage of both the midpoint-slope and Adams methods is that they need the values of both $f(t_0)$ and $f(t_0 + \Delta t)$ (i.e., two starting values), where t_0 is the initial value of time, to calculate $f(t_0 + 2\Delta t)$, but usually only one starting value, $f(t_0)$, is given in physical problems. Thus, we might want to approximate $f(t_0 + \Delta t)$ by the first three terms of a Taylor series as follows:

$$f(t_0 + \Delta t) = f(t_0) + (\Delta t) f'(t_0) + \frac{(\Delta t)^2}{2} f''(t_0) \qquad \textbf{(2–102)}$$

It should be noted that the Euler method is the first two terms of this Taylor series.

There are many numerical methods for solving ordinary differential equations, and we have covered three of the more easily applied methods (i.e., Euler, midpoint-slope, and Adams). The *Runge-Kutta method*, though more difficult to apply, is one of the more accurate numerical methods for solving ordinary differential equations. It has a big additional advantage in that it requires only one value of $f(t)$, which is usually given, in order to apply (i.e., start) the method.

The *Milne method*, which is given by the following equation and is easier than the Runge-Kutta method to apply, requires four starting values for $f(t)$:

$$f(t + \Delta t) = f(t - 3\Delta t) + \tfrac{4}{3}(\Delta t)[2f'(t) - f'(t - \Delta t) + 2f'(t - 2\Delta t)] \qquad \textbf{(2–103)}$$

Let us write the first-order, ordinary differential equation to be solved in the following generalized, functional-notation form:

$$\frac{dy}{dt} = F(t, y) \tag{2-104}$$

In other words, we want to solve the preceding differential equation for $y(t)$. The functional term $F(t, y)$ is the solution of the ordinary differential equation for the derivative dy/dt. We are using this functional form because it is easier to express the Runge-Kutta formulas in a functional form.

The *second-order accuracy* Runge-Kutta equations are:

$$a_1(t) = (\Delta t)F(t, y(t)) \tag{2-105a}$$

$$a_2(t) = (\Delta t)F(t + \Delta t, y(t) + a_1(t)) \tag{2-105b}$$

$$y(t + \Delta t) = y(t) + \frac{1}{2}[a_1(t) + a_2(t)] \tag{2-105c}$$

The *third-order accuracy* Runge-Kutta equations are:

$$a_1(t) = (\Delta t)F(t, y(t)) \tag{2-106a}$$

$$a_2(t) = (\Delta t)F\left(t + \frac{\Delta t}{2}, y(t) + \frac{a_1(t)}{2}\right) \tag{2-106b}$$

$$a_3(t) = (\Delta t)F(t + \Delta t, y(t) + 2a_2(t) - a_1(t)) \tag{2-106c}$$

$$y(t + \Delta t) = y(t) + \frac{1}{6}[a_1(t) + 4a_2(t) + a_3(t)] \tag{2-106d}$$

The *fourth-order accuracy* Runge-Kutta equations are:

$$a_1(t) = (\Delta t)F(t, y(t)) \tag{2-107a}$$

$$a_2(t) = (\Delta t)F\left(t + \frac{\Delta t}{2}, y(t) + \frac{a_1(t)}{2}\right) \tag{2-107b}$$

$$a_3(t) = (\Delta t)F\left(t + \frac{\Delta t}{2}, y(t) + \frac{a_2(t)}{2}\right) \tag{2-107c}$$

$$a_4(t) = (\Delta t)F(t + \Delta t, y(t) + a_3(t)) \tag{2-107d}$$

$$y(t + \Delta t) = y(t) + \frac{1}{6}[a_1(t) + 2a_2(t) + 2a_3(t) + a_4(t)] \tag{2-107e}$$

The Runge-Kutta method is also much more accurate than the trapezoidal method. Thus, we may wish to obtain a rather accurate solution of the nonlinear vibration problem in Sec. 2.14 by using the Runge-Kutta method to calculate both velocity $\dot{x}_{t+\Delta t}$ and displacement $x_{t+\Delta t}$. We could also use the Adams or midpoint-slope method to calculate velocity $\dot{x}_{t+\Delta t}$, the trapezoidal method to calculate displacement $x_{t+\Delta t}$, and the interval-halving method to obtain a proper Δt-interval. A first-order differential equation is used in Ill. Exs. 2.22 to 2.24 of this section in order to demonstrate to the reader the accuracies of the Euler, mid-point-slope, Adams, and Runge-Kutta methods. These results are briefly summarized in Table 2-1.

The *Newmark β method*, which is discussed in the Biggs and Newmark references in Appendix A-3, is a method that iterates for both $f'(t + \Delta t)$ and $f(t + \Delta t)$ within each Δt-time-interval. It is a numerical technique that has been widely applied to vibration systems, particularly to structural vibrations and other systems having several degrees of freedom. This method calculates $f'(t + \Delta t)$ by the trapezoidal method as follows:

$$f'(t + \Delta t) = f'(t) + \frac{\Delta t}{2}[f''(t) + f''(t + \Delta t)]$$

and $f(t + \Delta t)$ is calculated by

$$f(t + \Delta t) = f(t) + (\Delta t)f'(t) + [(\tfrac{1}{2} - \beta)f''(t) + \beta f''(t + \Delta t)](\Delta t)^2$$

The Newmark β method is an iteration method because a value for $f''(t + \Delta t)$ is needed in order to calculate both $f'(t + \Delta t)$ and $f(t + \Delta t)$. The last term in the preceding equation may be considered to be an acceleration-correction term for the Euler method, where the choice of the β-value predicts (or in effect assumes) the variation of acceleration $f''(t)$ within that time interval. Best results are usually obtained if β is chosen to be between $\tfrac{1}{4}$ and $\tfrac{1}{6}$. Note that the first three terms of a Taylor series for $f(t + \Delta t)$ result if $\beta = 0$. For $\beta = \tfrac{1}{6}$, the result will correspond to a linear-acceleration method that is not discussed in this text. For $\beta = \tfrac{1}{4}$, the $f(t + \Delta t)$ equation becomes

$$f(t + \Delta t) = f(t) + (\Delta t)f'(t) + \frac{(\Delta t)^2}{4}[f''(t) + f''(t + \Delta t)]$$

Illustrative Example 2.22.   Suppose that the velocity dy/dt of a particle is equal to $-4y^2t$ ft/sec where t is the time in seconds and y is the displacement of the particle from a reference position in feet. If the displacement equals 2 ft at time zero, solve for the displacement of this particle from $t = 0$ to $t = 0.200$ sec using the Adams method and an interval of 0.050 sec. Use the Euler method to approximate the second starting value. Calculate the errors of these numerical approximations.

Solution.   Since $dy/dt = -4y^2t$, we are essentially solving the first-order differential equation $dy/dt + 4y^2t = 0$, where $y(0) = 2.000$ ft. To calculate the second starting value, which is $y(0.050)$, we must first substitute in the given differential equation $\dot{y}(t) = -4[y(t)]^2t$ to compute $\dot{y}(0)$ and then apply Eq. (2-89) to approximate $y(0.050)$. These calculations are as follows:

$$\dot{y}(0) = -(4.00)(2.000)^2(0) = 0$$
$$y(0.050) = 2.000 + (0.050)(0) = 2.000 \text{ ft}$$

where the Euler-method approximation for this point was computed from the equation $y(\Delta t) = y(0) + (\Delta t)\dot{y}(0)$. Note that this method produced a rather poor approximation since $\dot{y}(0) = 0$ is not a good average velocity for the first interval.

To apply the Adams method (i.e., Eq. (2-100)) for the rest of this problem, we successively use the following equations:

$$\dot{y}(t) = -4[y(t)]^2 t$$

$$y(t+\Delta t) = y(t) + \left(\frac{\Delta t}{2}\right)[3\dot{y}(t) - \dot{y}(t - \Delta t)]$$

to obtain the following numerical results:

$$\dot{y}(0.050) = -(4.00)(2.000)^2(0.050) = -0.800 \text{ ft/sec}$$
$$y(0.100) = 2.000 + (0.025)(-2.400 - 0) = 1.940 \text{ ft}$$
$$\dot{y}(0.100) = -(4.00)(1.940)^2(0.100) = -1.506 \text{ ft/sec}$$
$$y(0.150) = 1.940 + (0.025)(-4.518 - 0.800) = 1.847 \text{ ft}$$
$$\dot{y}(0.150) = -(4.00)(1.847)^2(0.150) = -2.047 \text{ ft/sec}$$
$$y(0.200) = 1.847 + (0.025)(-6.141 + 1.506) = 1.731 \text{ ft}$$

In order to determine the errors in our computed results, we must first obtain the exact results. This is why we have chosen an example problem whose analytical solution may be easily obtained. Since the given differential equation can be rewritten in the following form:

$$\frac{dy}{y^2} = -4t \, dt$$

integration of both sides of the previous equation and evaluation of the integration constant from the initial condition $y(0) = 2$ gives us the following exact solution for displacement y:

$$y = \frac{2}{4t^2 + 1}$$

Substitution for time t in the previous equation shows that the exact values for $y(0.050)$, $y(0.100)$, $y(0.150)$, and $y(0.200)$ are 1.980, 1.923, 1.835, and 1.724 ft, respectively. Thus, the numerical approximation errors are equal to -0.020, -0.017, -0.012, and -0.007 ft at $t = 0.050, 0.100, 0.150$, and 0.200 sec. Note that the largest of these errors resulted from the use of the Euler method to approximate $y(0.050)$. These errors can be reduced by using a smaller value for Δt and by using a better method (e.g., the first three terms of a Taylor series) to approximate the second starting value. If the Euler method had been used throughout, the errors would be -0.020, -0.037, -0.048, and -0.053 ft at $t = 0.050, 0.100, 0.150$ and 0.200 sec. If an interval of 0.00625 sec had been used for applying the Euler method throughout this problem, the error would be only -0.006 ft at $t = 2.000$ sec.

Illustrative Example 2.23.   Solve Ill. Ex. 2.22 using the midpoint-slope method and an interval of 0.050 sec. Use the Euler method to approximate the second starting value. Calculate the errors of these numerical approximations.

Solution. We are given that $y(0) = 2.000$ ft. From Ill. Ex. 2.22, we find that $\dot{y}(0) = 0$, and that use of the Euler method approximates the value of $y(0.050)$ to be 2.000 ft. To apply the midpoint-slope method (i.e., Eq. (2-98)) for the rest of this problem, we successively use the following two equations:

$$\dot{y}(t) = -4[y(t)]^2 t$$
$$y(t + \Delta t) = y(t - \Delta t) + 2(\Delta t)\dot{y}(t)$$

to obtain the following numerical results:

$$\dot{y}(0.050) = -(4.00)(2.000)^2(0.050) = -0.800$$
$$y(0.100) = 2.000 + (0.100)(-0.800) = 1.920$$
$$\dot{y}(0.100) = -(4.00)(1.920)^2(0.100) = -1.475$$
$$y(0.150) = 2.000 + (0.100)(-1.475) = 1.852$$
$$\dot{y}(0.150) = -(4.00)(1.852)^2(0.150) = -2.058$$
$$y(0.200) = 1.920 + (0.100)(-2.058) = 1.714$$

The numerical approximation errors are equal to -0.020, 0.003, -0.017, and 0.010 ft at $t = 0.050$, 0.100, 0.150, and 0.200 sec, respectively. That is, the error in the computed value for $y(0.200)$ is $(1.724 - 1.714)$ or 0.010 ft. These errors would be reduced, of course, if we used a smaller interval (i.e., a smaller value for Δt). If we had obtained our second starting value by using the first three terms of a Taylor series as follows:

$$y(\Delta t) = y(0) + (\Delta t)\dot{y}(0) + \frac{(\Delta t)^2}{2} \ddot{y}(0)$$

we would greatly reduce the magnitude of our errors, since the largest error (in magnitude) in our previous calculations was the result of the calculation of $y(0.050)$ by the Euler method. To calculate the value of $\ddot{y}(0)$, we first differentiate the given differential equation to obtain $\ddot{y}(t) = -4[y(t)]^2$, and then substitute $y(0) = 2$ in this result to obtain $\ddot{y}(0) = -4(2)^2 = -16$. Use of this three-term Taylor series equation gives $y(0.050) = 1.980$, which is an exact result. Note that it was easy to compute $\ddot{y}(0)$ for this problem because the given differential equation is simple in form (e.g., the equation coefficients are not plotted or tabulated as they are for some professional engineering problems). Another improved way to calculate the second starting value would be to first approximate $y(\Delta t)$ by the Euler method and then to recalculate $y(\Delta t)$ by the trapezoidal method. This will also give us $y(0.050) = 1.980$.

Illustrative Example 2.24. Solve Ill. Ex. 2.22 using the second-order accuracy Runge-Kutta method. Calculate the errors of these numerical approximations when an interval of 0.050 is used.

Solution. We are given the first starting value which is $y(0) = 2.000$ ft. Now we wish to calculate the values $y(0.050)$, $y(0.100)$, $y(0.150)$, and $y(0.200)$ using the second-order accuracy Runge-Kutta method. Since $dy/dt = -4y^2 t$, then Eq. (2-104) states that function $F(t, y) = -4y^2 t$. Applying Eqs. (2-105a, b, c), we have

$$a_1(t) = (\Delta t)F[t, y(t)] = (\Delta t)[-4y(t)^2 t]$$
$$a_2(t) = (\Delta t)F[t + \Delta t, y(t) + a_1(t)] = (\Delta t)\{-4[y(t) + a_1(t)]^2(t + \Delta t)\}$$
$$y(t + \Delta t) = y(t) + \tfrac{1}{2}[a_1(t) + a_2(t)]$$

Using these three equations for each calculation of $y(t + \Delta t)$, we obtain the following numerical results:

$$a_1(0) = -(0.050)(4.00)(2.000)^2(0) = 0$$
$$a_2(0) = -(0.050)(4.00)(2.000 + 0)^2(0.050) = -0.040$$
$$y(0.050) = 2.000 + \tfrac{1}{2}(0 - 0.040) = 1.980 \text{ ft}$$
$$a_1(0.050) = -(0.050)(4.00)(1.980)^2(0.050) = -0.039$$
$$a_2(0.050) = -(0.050)(4.00)(1.980 - 0.039)^2(0.100) = -0.075$$
$$y(0.100) = 1.980 + \tfrac{1}{2}(-0.039 - 0.075) = 1.923 \text{ ft}$$
$$a_1(0.100) = -(0.050)(4.00)(1.923)^2(0.100) = -0.074$$
$$a_2(0.100) = -(0.050)(4.00)(1.923 - 0.074)^2(0.150) = -0.102$$
$$y(0.150) = 1.923 + \tfrac{1}{2}(-0.074 - 0.102) = 1.835 \text{ ft}$$
$$a_1(0.150) = -(0.050)(4.00)(1.835)^2(0.150) = -0.101$$
$$a_2(0.150) = -(0.050)(4.00)(1.835 - 0.101)^2(0.200) = -0.121$$
$$y(0.200) = 1.835 + \tfrac{1}{2}(-0.101 - 0.121) = 1.724 \text{ ft}$$

Comparison of these results with the analytic solution values furnished in Ill. Ex. 2.22 show that these results for $y(t)$ are exact, for the number of decimal places that were carried in the calculations. Thus, this illustrates, by example, the superior accuracy of the Runge-Kutta method when compared to the methods utilized in Ill. Exs. 2.22 and 2.23.

Table 2-1 tabulates, for comparison purposes, the errors that result when the Euler, Adams, midpoint-slope, and second-order accuracy Runge-Kutta methods are used to solve this problem when a time interval of 0.050 sec is used. It should be mentioned again that the accuracy for the Adams and midpoint-slope methods would improve if the second starting value $y(0.050)$ were more accurately approximated [e.g., by Eq. (2-102)] or if $y(0.050)$ were recalculated by the trapezoidal method.

TABLE 2-1

t (sec)	Exact y(t)	Errors			
		Euler	Adams	Midpoint-Slope	Runge-Kutta
0	2.000	0	0	0	0
0.050	1.980	−0.020	−0.020	−0.020	0.000
0.100	1.923	−0.037	−0.017	0.003	0.000
0.150	1.835	−0.048	−0.012	−0.017	0.000
0.200	1.724	−0.053	−0.007	0.010	0.000

2.15 Some Specific Analytical Methods for Solving Some Forced Vibrations of Nonlinear Systems

The reader may wonder why this text discusses the use of laborious numerical methods, which should be utilized on a digital computer, for solving nonlinear vibration problems. Analog and digital computer solution methods are emphasized in this text because there is no general analytical method for the exact solution of nonlinear differential equations. That is, a nonlinear vibration has a nonlinear differential equation of motion, and each problem type must be treated separately since solution methods vary according to the form of the nonlinear differential equation. This had been previously emphasized in Sec. 1.9, which discusses free vibrations of nonlinear systems. Exact solutions are known for only a few nonlinear problem types. Graphical solution methods are very laborious, and most special analytical solution methods give only approximate solutions that are valid only when the nonlinear terms in the equation of motion have a small magnitude. If the external force varies with time t, if the damping force varies with velocity $\dot{x}$, and if the spring force varies with displacement x, the equation of motion for the forced vibration of the damped spring-mass system in Fig. 2–3 can be expressed in the following functional form:

$$\ddot{x} = f(\dot{x}, x, t) \tag{2-108}$$

Let us assume that the acceleration $\ddot{x}$ has a nonlinear variation with $\dot{x}$ or x or with both terms. Illustrative Example 2.25 shows how Eq. (2–108) can be transformed into a mathematical form that permits a graphical analysis by the *phase-plane method* which was briefly discussed in Sec. 1.9. If Eq. (2–108) can be rewritten in the form $\ddot{x} + a^2 x + bF(\dot{x}, x, t) = 0$, where b has a very small value and $F(\dot{x}, x, t)$ represents a nonlinear function, we can then obtain an approximate analytical solution by use of the *perturbation method*. This analytical solution method is exemplified in Ill. Ex. 1.24 of Sec. 1.9, but is much more thoroughly discussed in the R. Bellman reference in Appendix A–3. If the nonlinear vibration is undamped, so that its equation of motion has the form $\ddot{x} = f(x, t)$, then an *iterative method* might be tried if the nonlinear term has a small magnitude. In this case, we *assume* a solution $x_1(t)$, and then substitute it in the equation of motion to obtain

$$\ddot{x}_2 = f[x_1(t), t] \tag{2-109}$$

Integrating the previous equation and integrating the result, where the arbitrary integration constants are evaluated from the two initial conditions, we obtain a new approximation $x_2(t)$ for the solution. We then substitute $x_2(t)$ in the equation of motion to obtain $\ddot{x}_3$. Two integrations give us a new approximation $x_3(t)$. We keep repeating this procedure, which is exemplified in the third W. T. Thomson reference in Appendix A-3, to see if a convergence occurs (as determined by some specific convergence criteria). In summary, the second derivative of the next approximation solution $x_{n+1}(t)$ is computed from the equation $\ddot{x}_{n+1} = f[x_n(t), t]$, where $x_n(t)$ is the previously assumed solution. Another method which requires that we *assume* the form of the solution is the *Ritz-averaging method* which is discussed in the L. S. Jacobsen and R. S. Ayre reference in Appendix A-3. This method can be applied to vibration problems whose equation of motion has the functional form

$$\ddot{x} = f(t) + g(\dot{x}) + h(x) \qquad (2\text{-}110)$$

where functions $g(\dot{x})$ and $h(x)$ must be odd functions [i.e., $h(-x) = -h(x)$]. The method determines an approximation solution, consisting of n terms, of the following form:

$$x = a_1 f_1(t) + a_2 f_2(t) + \ldots + a_n f_n(t) \qquad (2\text{-}111)$$

where the functions $f_1(t)$ to $f_n(t)$ are *assumed* and the coefficients a_1 to a_n are computed by an integral-averaging technique that is stipulated by this method. Since it is difficult to predict the approximating functions $f_i(t)$, this method is more useful when the forcing function $f(t)$ in the equation of motion is harmonic [e.g., $(F/m) \cos \omega t$] so that the following two-term approximation can be used:

$$x = a_1 \sin \omega t + a_2 \cos \omega t = X \cos (\omega t - \phi) \qquad (2\text{-}112)$$

It can be shown that $\tan \phi = 0$ (i.e., $\phi = 0$ or π), and that amplitude X can be computed from the following equation when the system has no damping, so that $g(\dot{x}) = 0$ in the equation of motion

$$\left[\frac{4}{\pi x} \int_0^{\pi/2} h(X \cos \gamma) \cos \gamma \, d\gamma \right] + \left(\frac{\omega}{\omega_n} \right)^2 = \pm \frac{F}{kX} \qquad (2\text{-}113)$$

Illustrative Example 2.25. Show that Eq. (2-108) can be transformed into the form of Eq. (1-52) in Sec. 1.9 so that the phase-plane method may be applied.

Solution. This graphical procedure for solving nonlinear differential equations of the form of Eq. (2-108) is called the *phase-plane delta method*, which is discussed and exemplified in the L. S. Jacobsen and R. S. Ayre reference in Appendix A-3. Adding and subtracting the term $\omega_n^2 x$ to Eq. (2-108) gives

$$\ddot{x} = [f(\dot{x}, x, t) - \omega_n^2 x] + \omega_n^2 x$$

Substitution of $\ddot{x} = \dot{x}(d\dot{x}/dx)$ and

$$\delta = \frac{1}{\omega_n^2} [f(\dot{x}, x, t) - \omega_n^2 x]$$

into the equation for $\ddot{x}$ gives

$$\dot{x} \left(\frac{d\dot{x}}{dx} \right) = \omega_n^2 (x + \delta)$$

Substitution of $v = \dot{x}/\omega_n$ into the previous equation gives the following equation (after some rearrangement):

$$\frac{dv}{dx} = \frac{x + \delta}{v}$$

for which we can obtain phase-plane plots of v versus x.

Illustrative Example 2.26.   Find the steady-state amplitude of vibration of the mass in Fig. 2-1 using the Ritz-averaging method.

Solution.   Since we know that the steady-state displacement will be harmonic for this linear spring-mass system, we can use Eq. (2-113) to calculate amplitude X. Since $h(x) = -kx$, this gives

$$\frac{4}{\pi X} \int_0^{\pi/2} X \cos^2 \gamma \, d\gamma = \left(\frac{\omega}{\omega_n} \right)^2 \mp \frac{F}{kX}$$

Since the left-hand term equals $(4/\pi)(\pi/4)$, or unity, the solution is $X = (F/k)/(1 - r^2)$, where $r = \omega/\omega_n$, which as expected agrees with the exact result given by Eq. (2-2) of Sec. 2.2.

PROBLEMS

2.1.   Obtain Eq. (2–3) from Eq. (2–2).

2.2.   For the undamped system in Fig. 2–1, suppose that the applied force is $20 \sin 15t$ lb when time t is in sec. Find the equation for the motion (steady-state plus transient) of the mass if $W = 38.6$ lb, $k = 60$ lb/in., $\dot{x}(0) = 2$ in., and $x(0) = 8$ in./sec. Also, find the equation for the velocity variation with time.

2.3.   For the undamped system in Fig. 2–1, suppose that the applied force is $30 \sin (8t - \pi/6)$ instead of $F \sin \omega t$. Find the equation for the motion (steady-state plus transient) of the mass if $W = 77.2$ lb, $k = 60$ lb/in., and the initial displacement and the initial velocity of the mass are both zero. Also, find the equation of motion if the applied force is $(20 \sin 15t + 8 \cos 30t + 14 \sin 2t)$.

2.4.   Verify the calculations of the coefficients C and D in Ill. Ex. 2.2.

2.5.   If the upper hinge point for a vertical, simple pendulum of weight W and length L (as shown in Fig. 1–13) is given a horizontal motion $x = e \sin \omega t$, write the equation of motion for this system and find the equation for the steady-state angular displacement $\theta(t)$ for weight W.

2.6. For the damped spring-mass system in Fig. 2–3, suppose that $k = 30$ lb/in., $c = 0.38$ lb-sec/in., $W = 38.6$ lb, and the applied force is $15 \sin 10t$ lb when time t is in seconds. Calculate the transient and steady-state displacement of the mass when the initial velocity is zero and the initial displacement is 3 in.

2.7. Repeat Prob. 2.6, if the initial displacement is zero and the initial velocity is 5 in./sec.

2.8. Repeat Prob. 2.6, if the applied force is $30 \cos 15t$ lb, for t in seconds, and the initial velocity is 8 in./sec.

2.9. For the spring-mass system in Prob. 2.2, suppose that we added a linear damper, as shown in Fig. 2–3, where $c = 1.3$ lb-sec/in. Calculate the transient and steady-state displacement of the mass when both the initial displacement and initial velocity equal zero.

2.10. Verify Eqs. (2–12) and (2–13).

2.11. Determine the resonant frequencies, the resonant phase angles, and the resonant amplitudes for the two damped spring-mass systems in Probs. 2.6 and 2.9.

2.12. Find the applied frequency at which the peak amplitude of Fig. 2–4 will occur for a specified damping ratio value ζ. Also find the corresponding phase angle and the peak amplitude.

2.13. For the damped spring-mass system in Fig. 2–3, determine the damping coefficient c for this system when an applied force $15 \sin 24t$ lb, for t in sec, results in a resonant amplitude of 7.2 in.

2.14. Repeat Ill. Ex. 2.4 for an applied force that is proportional to the displacement of the mass (i.e., $F_0 x$).

2.15. Suppose that the system in Fig. 2–3 is excited by a force that is proportional to the acceleration of the mass, rather than by the force $F \sin \omega t$. Discuss the stability of the system.

2.16. For the damped disk-shaft system in Fig. 2–6, suppose that $I = 24$ in.-lb-sec^2, $k_t = 10^4$ in.-lb/rad, and $c_t = 0.380$ lb-sec. Find the steady-state angular displacement of the disk when an oscillating applied torque of $60 \sin 18t$ in.-lb, for t in sec, is applied to this disk.

2.17. For the damped disk-shaft system in Fig. 2–6, suppose that $I = 32$ in.-lb-sec^2, $k_t = 7(10^5)$ in.-lb/rad, and the applied torque is $40 \sin 12t$ in.-lb for t in sec. If the resonant amplitude is measured to be $0.0163°$, determine the undamped natural frequency of the system and the viscous damping coefficient c_t.

2.18. For the damped system in Fig. 2–3, suppose that the applied force is $F \cos(\omega t + \alpha)$ instead of $F \sin \omega t$. Obtain the equation for the steady-state displacement using both the method of undetermined coefficients and the mechanical impedance method.

2.19. Find the steady-state displacements for Probs. 2.6 and 2.9 using the mechanical impedance method.

2.20. Use the mechanical impedance method to find the steady-state displacement of the damped spring-mass system in Fig. 2–3 when the applied force is

$$F(t) = F_1 \sin (\omega_1 t + \alpha_1) + F_2 \cos (\omega_2 t + \alpha_2)$$

2.21. Let force $F(t)$ be a periodic force of period T, where $F(t) = F_0$ for $0 < t < T/2$ and $F(t) = 0$ for $T/2 < t < T$. That is, this force is a rectified rectangular wave (i.e., the negative portions of Fig. 2–7 are removed). Express this periodic force in Fourier series form. Find the steady-state displacement $x(t)$ when this periodic force is applied to the damped spring-mass system in Fig. 2–7.

2.22. Express the periodic force in Fig. 2–19 as a Fourier series. Find the steady-state displacement $x(t)$ when this periodic force is applied to the damped spring-mass system in Fig. 2–3.

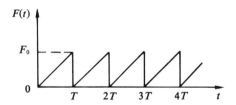

Fig. 2-19

2.23. Let force $F(t)$ be a periodic force of period T, where $F(t) = F_0 \sin (2\pi/T)$ for $0 \leq t \leq T/2$ and $F(t) = 0$ for $T/2 \leq t \leq T$. That is, this force is a rectified sine wave. Express this periodic force in Fourier series form. Find the steady-state displacement $x(t)$ when this periodic force is applied to the damped spring-mass system in Fig. 2–7.

2.24. If the machine in Fig. 2–8 weighs 300 lb, is supported on eight springs where $k = 400$ lb/in. for each spring, and has an applied force of $120 \sin 30t$; find the amplitude of the machine oscillation, the transmissibility, and the maximum force transmitted to the foundation if the damping ratio is 0.14.

2.25. Suppose that the machine in Fig. 2–8 weighs 500 lb and has a vertical applied force of 300 sin 40t lb, for t in sec. If the damping ratio for the supporting dampers is 0.18, specify the spring constant for the four supporting springs (of the same type) so that only 20% of the applied force is transmitted to the foundation (i.e., the maximum force transmitted to the foundation is reduced by 80%).

2.26. Delicate instruments are mounted on an isolating material that has negligible damping. If an experiment shows that this isolating material deflects $\frac{1}{4}$ in. under a 60-lb weight, find the amplitude of the oscillatory motion transmitted for a 3000–cycles/sec oscillation that has an amplitude of 0.002 in.

2.27. Repeat Prob. 2.25 if the isolating material has a damping ratio of 0.12.

2.28. An 800–lb machine is mounted on springs whose total spring constant is 6000 lb/in. A vertical 40-lb piston, whose motion may be considered to be harmonic, moves up and down in this machine with a frequency of 300 cycles/min and has a 12–in. stroke. Find the amplitude of the machine oscillation and the amplitude of the force transmitted to the foundation if the machine is also mounted on viscous dampers that have a damping ratio of 0.23.

2.29. A vibrometer whose natural frequency is 1.4 rad/sec is attached to a machine undergoing a vibration of 2 cycles/sec. If the vibrometer mass has a 0.5–in. amplitude relative to its frame, determine the vibration amplitude for the machine.

2.30. A vibrometer, whose damping is negligible, has a natural frequency of 1.5 rad/sec. What is the lower frequency limit of this measuring instrument to keep the error below 3%?

3.21. Suppose that the vibrometer in Prob. 2.30 is used to determine the magnitude of vibration for a machine structure. If the machine is running at 120 rpm, and if the vibrometer gives a relatives displacement reading of 0.007 in., determine the magnitudes of the displacement, velocity, and acceleration for the machine part.

2.32. A 5-lb mass is suspended inside a box by a vertical spring that is attached to the top inside surface of this box. This box is placed on top of a shake table whose displacement variation is 0.14 sin 7t in., for t in sec. If the spring constant is 30 lb/in., find the oscillation amplitude of the mass.

2.33. An accelerometer, whose natural frequency is 9 rad/sec and whose damping ratio is 0.65, is attached to a vibrating system. If the frequency of this vibrating system is 2 cycles/sec, and if the recorded amplitude that is read on the dial is 0.06 in., determine the maximum amplitude of this vibrating system.

2.34. An accelerometer is made by mounting a 3-lb mass on rubberized material that has a damping ratio of 0.73 and which deflects 0.002 in. under a 100-lb load. If the amplitude that is read on the dial indicator for a 4–cycle/sec vibration is 0.007 in., determine the maximum acceleration of the member to which this instrument is attached.

2.35. A large motor operating at 2000 rpm has an unbalance that causes its support to vibrate. If the unbalanced force is 100 lb, and if the support has a spring constant of 40,000 lb/in. and a damping constant of 150 lb-sec/in., find the amplitude of the forced vibration of this motor if its weight is 700 lb. Also, find the natural frequency of the support and the vibration amplitude for the motor if the support had no damping.

2.36. A machine that weighs 300 lb has a 60-lb rotor whose eccentricity is 0.35 in. This machine, which operates at 400 rpm, is mounted on springs whose total spring constant is 600 lb/in. and the damping ratio is 0.10. If the machine is constrained to a vertical vibration, determine the vibration amplitude of this machine.

2.37. For the machine in Prob. 2.37, calculate the vibration amplitude if the support has no damping and specify the damping ratio that is required to reduce this amplitude by 60%.

2.38. A piston oscillates in a vertical cylinder with the harmonic motion $8 \cos 12t$ in. for t in sec. The cylinder weighs 4 lb and is supported by a spring whose spring constant is 30 lb/in. If the damping between the piston and the cylinder wall can be approximated by a viscous damping constant of 1.2 lb–sec/in., determine the amplitude of the resultant cylinder vibration and the corresponding phase angle with respect to the excitation force.

2.39. A 200-lb machine is mounted on pads on top of a support whose damping ratio is 0.15 and which deflects 2 in. under a 9000-lb load. A 5-lb piston within this machine has a 60-cycles/sec reciprocating motion whose stroke is 6 in. Determine the amplitude of the resultant machine vibration and the phase angle with respect to the excitation force.

2.40. A 30-lb rotor of 0.00103-in. eccentricity is firmly mounted at the center of a metal shaft, whose EI–value is $2.5(10^7)$ lb-in^2, of 1.5-in diameter. If the shaft length between the bearing supports is 22 in., find the critical speed and the vibration amplitude of the rotor and the force transmitted to the bearing supports when rotating at 2400 rpm.

2.41. Verify the results in Ill. Ex. 2.10.

2.42. A 1-ton vehicle, whose total spring constant for its suspension system is 200 lb/in., travels on a sinusoidally-shaped road whose wave length, or period, is 14 ft and whose amplitude is 0.200 ft. That is, this vehicle travels over a rough road whose profile may be approximated by a sine wave. If the damping ratio of the shock-absorber is 0.4, determine the critical speed of this vehicle and the amplitude of the steady-state vibration when this vehicle travels at 35 mph. Repeat these calculations if this vehicle carries an 800-lb load.

2.43. Repeat Prob. 2.42 if another vehicle, whose springs are compressed 3 in. due to its own weight, travels over the same road where the shock-absorber damping ratio is 0.170.

2.44. A small instrument is mounted on a 300-lb table that is isolated from the foundation by springs, whose total spring constant is 180 lb/in., and by dampers, whose damping ratio is 0.15. If the foundation has the vibratory motion $0.2 \sin 15t$ inch for t in sec, determine the motion of this small instrument.

2.45. If the spring-mass system in Fig. 2–10 is dropped from an elevation d to a cement foundation, find the resulting motion of mass m.

2.46. If the base of the spring-mass system in Fig. 2–10 lies on top of a cam which causes the motion of the base to be a sawtooth wave of amplitude X and period T (see Fig. 2–19), determine the steady-state displacement of mass m if the system is initially at rest.

2.47. If the base of the spring-mass system in Fig. 2–10 is given a rectangular-wave motion (see Fig. 2–7) of amplitude X and period T, determine the steady-state displacement of mass m.

2.48. Repeat Prob. 2.47 if the base motion is a rectified sine wave of amplitude X and period T.

2.49. A horizontal bar of length L is supported by a vertical spring (that is mounted between its left end and the floor) and by a fulcrum that is located a ft from its left end. A concentrated weight W is mounted at the right end of the bar. Assuming the bar to be rigid and of negligible weight, determine the vertical displacement equation for weight W when the fulcrum is given a vertical oscillating displacement $b \sin \omega t$.

2.50. For the spring-mass system in Fig. 2–20, write the equation of motion and use the mechanical impedance method to find the steady-state displacement and phase angle of mass m when $x_2 = X_2 \sin \omega t$.

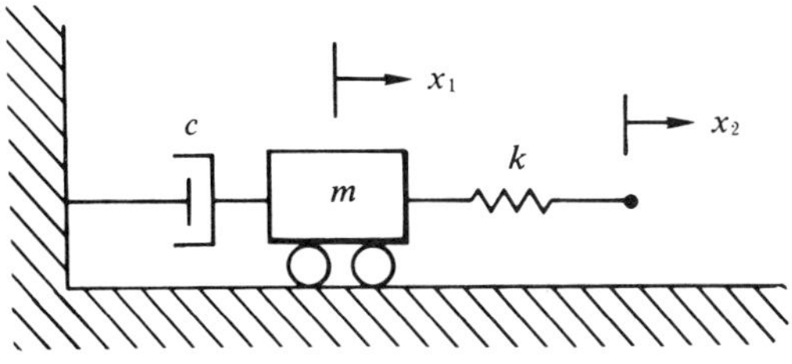

Fig. 2-20

2.51. Repeat Prob. 2.50, first interchanging the locations of the spring and damper in Fig. 2–20 to form a different spring-mass system.

2.52. Determine the two critical speeds for the bearing-support system in Ill. Ex. 2.11.

2.53. Obtain the inverse transforms of the following:

(a) $F(s) = \dfrac{s^4 + 5s^2 - 3}{s(s^2 - 4)(s^2 + 3s - 2)}$

(b) $F(s) = \dfrac{s^4 + 2s^3 + 3s - 1}{s^3(s^2 + 4)}$

(c) $F(s) = \dfrac{s^2 + 1}{s(s + 1)^2(s + 2)^3}$

(d) $F(s) = \dfrac{3s + 2}{(s^2 + 2)(s^2 + 4)}$

2.54. Solve the following two differential equations using Laplace transforms, where the initial values of y and d^2y/dt^2 are zero and the initial value of dy/dt is 50 for both equations.

(a) $\dfrac{d^3y}{dt^3} - 3\dfrac{dy}{dt} + 2y = 2 \sin 100t - 4 \cos 100t$

(b) $\dfrac{d^3y}{dt^3} - 3\dfrac{d^2y}{dt^2} - 2\dfrac{dy}{dt} = 4e^{50t} + 100$

2.55. Verify Eq. (2–71). Use Eq. (2–71) to find the Laplace transforms of $t \sin at$, $t \cos at$, $t \sinh at$, and $t \cosh at$. Use Eq. (2–72) to find the Laplace transforms of $\sin at/t$ and $\cos at/t$.

2.56. Use the two shifting theorems to find the Laplace transforms of $te^{-at}[u(t - b)]$ and $at^2[u(t - b)]$.

2.57. Obtain the Laplace transform of a symmetrical triangular wave of amplitude M and period T by first using the result of Ill. Ex. 2.15 to obtain the Laplace transform of a rectangular wave of height $2M/T$ and then applying Eq. (2–70). That is, note that the integral of a rectangular wave is a symmetrical triangular wave.

2.58. Use Eq. (2–75) and the Laplace transform of a sawtooth pulse (see Table A–2) to obtain the Laplace transform of a sawtooth wave of amplitude M and period T.

2.59. Use the convolution theorem to determine the Laplace transform of $s/[(s^2 + a^2)(s + b)]$.

2.60. What is the value of the function $f(t)$ at $t = 0$ if its Laplace transform is given by

$$F(s) = K \frac{(s + a) \cos \alpha + b \sin \alpha}{(s + a)^2 + b^2}$$

2.61. Derive the Laplace transforms of the four pulse functions in Fig. 2–15 that are given in Table A–2 in the Appendix.

2.62. Repeat Prob. 2.45 using Laplace transforms.

2.63. For the spring-mass system in Fig. 2–3, find the equation for the displacement of the mass when a sawtooth pulse force of magnitude F_0 and duration T is suddenly applied to the mass without an initial displacement or velocity.

2.64. Repeat Prob. 2.63 when the applied force is a symmetrical triangular pulse of magnitude F and duration T.

2.65 Repeat Prob. 2.63 when the applied force is a sinusoidal pulse of magnitude F and duration T.

2.66. For the damped spring-mass system in Fig. 2–3, suppose that $W = 77.2$ lb, $k = 30$ lb/in., $c = 1.2$ lb-sec/in., and the applied force is $15 \sin (20t - \pi/6)$ lb for t in sec. Find the displacement equation for the mass if the initial displacement and velocity are both zero.

2.67. For the system in Ill. Ex. 2.6, use Laplace transforms to obtain the total displacement of the mass.

2.68. Repeat Prob. 2.67, if the applied force is a sawtooth wave of magnitude F and duration T.

2.69. Use Laplace transforms to obtain the total displacement of the mass in Prob. 2.46.

2.70. Repeat Prob. 2.47, using Laplace transforms to obtain the total displacement of the mass, instead of by classical means.

2.71. If the base of the spring-mass system in Fig. 2–10 is given a rectangular pulse motion (see Fig. 2–15a) of amplitude X and duration T, determine the displacement of mass m.

2.72. Repeat Prob. 2.71, if the applied displacement is a sawtooth pulse of magnitude X and duration T.

2.73. Repeat Prob. 2.71, if the applied displacement is a symmetrical triangular pulse of magnitude X and duration T.

2.74. If the system in Fig. 2–10 has negligible damping and may be dropped a height h, use Laplace transforms to specify the spring constant k to keep the maximum acceleration of the mass below a value b.

2.75. For the spring-mass system in Fig. 2–20, use Laplace transforms to find the displacement of the mass when displacement x_2 is a sawtooth pulse of amplitude X and duration T.

2.76. Repeat Prob. 2.75, if displacement x_2 is a sawtooth wave of amplitude X and period T.

2.77. Repeat Prob. 2.75, if displacement x_2 is rectangular pulse. Repeat if displacement x_2 is a symmetrical triangular pulse.

2.78. Repeat Prob. 2.75, using the system of Prob. 2.51.

2.79. Calculate the mean-square value for a displacement $x(t)$ that is represented by:
(a) $X \cos \omega t$;
(b) a rectangular wave of amplitude X and period T;
(c) a sawtooth wave of amplitude X and period T;
(d) a symmetrical triangular wave of amplitude X and period T;
(e) a rectified sine wave of amplitude X and period T.

2.80. A random vibration has a constant spectral-density value of 0.012 in.^{2}/(cycles/sec) between 40 and 800 cycles/sec. If this value is zero outside this frequency range, calculate the standard deviation, the mean-square value, and the root-mean-square value if the mean value is 4.8 inches.

2.81. For the damped spring-mass system in Fig. 2–3, suppose that $W = 77.2$ lb, $k = 30$ lb/in., and the damping ratio is 0.16. Determine the mean-square response of the mass when the applied force $F(t)$ is

$$F(t) = 50 (\cos 10t + \cos 40t + \cos 3t + \cos 2t)$$

2.82. Repeat Ill. Ex. 2.21 using a Δt-value of 0.025 sec. Compare your two sets of errors.

2.83. For the nonlinear spring-mass system described by Eq. (2–87), suppose that $W = 77.2$ lb, $a = 0.34$, $b = 0.4$, $k = 30$ lb/in., and that the applied force is $60 \sin 12t$ lb for t in sec. If the initial displacement is 3 in. and the initial velocity is 2 in./sec, solve this problem numerically using the Euler and trapezoidal methods, as discussed in Sec. 2.13, from $t = 0$ to $t = 0.200$ sec, using a Δt-value of 0.02 sec.

2.84. Repeat Prob. 2.83 using a Δt-value of 0.010 sec, and compare your results.

2.85. Repeat Ill. Ex. 2.22 using a Δt-value of 0.025 sec. Compare your two sets of errors.

2.86. Repeat Ill. Ex. 2.23 using Δt-value of 0.025 sec. Compare your two sets of errors.

2.87. Repeat Prob. 2.83, using the midpoint-slope method instead of the Euler method to calculate $\dot{x}_{t+\Delta t}$. Use the Euler method to compute $\dot{x}(\Delta t)$ as the second starting value for the midpoint-slope method.

2.88. Repeat Prob. 2.86, using the Adams method instead of the midpoint-slope method.

2.89. Solve the nonlinear vibration in Prob. 2.83 numerically by using the second-order-accuracy Runge-Kutta method to compute both $\dot{x}_{t+\Delta t}$ and $x_{t+\Delta t}$.

2.90. Use the Ritz-averaging method to compute the amplitude X and phase angle ϕ for an approximation solution of the form $X \cos (\omega t - \phi)$ for the Duffing equation of motion which is

$$m\ddot{x} + k(x \pm ax^3) = F \cos \omega t$$

3 Multiple-Degree-of-Freedom Vibrations

3.1 Introduction

The first two chapters dealt with systems that had only one degree of freedom. This chapter will analyze the free and forced vibrations of mechanical systems that have several degrees of freedom. In the previous two chapters, we saw that many practical vibration problems can be represented by a one-degree-of-freedom system. Most actual systems, however, have several masses and several degrees of freedom, and their vibrating motions are more complex than those for a single-degree-of-freedom system. This chapter will show how to obtain the equations of motion by Newton's second law, by the Lagrange equations, and by the use of influence coefficients. We shall see that the addition of more degrees of freedom to a two-degree-of-freedom system increases the labor of the solution procedure, but does not introduce any new physical principles. Chapter 4 will cover various methods for computing the natural frequencies of a multiple-degree-of-freedom system. Each flexibly-connected mass in a multiple-degree-of-freedom system can move independently of the other masses, and only under certain conditions will all of the masses undergo a harmonic motion at the same frequency. Since all of the masses have the same frequency, they all attain their amplitudes at the same time, even if two of them move in opposite directions. When such a harmonic motion occurs, that frequency at which it occurs is a *natural frequency* for the system. We call such a motion a *principal mode of vibration*, and the number of principal modes (and also natural frequencies) for a system is equal to the number of degrees of freedom (i.e., independent coordinates) for that system. The oscillation at the lowest natural frequency is called the *first mode*, the oscillation at the next higher natural frequency is called the *second mode*, etc. This chapter will show how we can represent a free vibration (including those that are periodic, but not harmonic) of a multiple-degree-of-freedom system by the superposition of principal-mode vibrations of unit amplitude. For example, the free-vibration motion of an undamped, two-degree-of-freedom system can be expressed as a linear combination of its two principal modes of vibration.

3.2   Undamped Free Vibrations

Before we discuss a specific example, let us first discuss how to use Newton's second law to write the equations of motion for a multiple-degree-of-freedom, mechanical system. For a translational, spring-mass system, we first choose the direction for positive displacement (i.e., up, down, right, or left). This will also be the direction for positive velocity and acceleration, since they are timewise derivatives of displacement. From Newton's second law, it is seen that this will also be the direction for positive force. Second, we assume a displacement for each mass in the system. To minimize the confusion that can be caused by manipulating negative signs, it is probably easiest to move each mass in the positive direction. Also, we assume the relative magnitudes of the displacements (i.e., assume which displacement is the largest, next largest, etc.). The result will be correct for any chosen set of relative displacements, because the equations of motion must hold for all possible conditions that do not violate physical constraints. The third and final step is to isolate each mass, draw its free-body diagram, and write the Newton's second law equation for each mass in the system.

For rotational, disk-shaft systems, the procedure for obtaining the equations of motion is very similar. The direction (clockwise or counter-clockwise) for positive angular displacement, angular velocity, angular acceleration, and torque must be chosen. Each disk or other type of inertia in the system is rotated, preferably in a positive direction, and the relative magnitudes of these angular deflections are assumed. Then, each disk or inertia is isolated and a free-body diagram is drawn. The $\sum M = I\ddot{\theta}$ equation is written for each isolated rigid body. To go further, suppose that we had a two-shaft system, where these two shafts (whose torsional spring constants are k_{t1} and k_{t2}) are connected by a pair of gears. We can represent this geared system by a non-geared system, where the two shafts are rigidly connected at their ends. If we let n be the gear ratio (i.e., the number of driver teeth divided by the number of driven teeth), the second shaft must now have a torsional spring constant of $n^2 k_{t2}$, in order to make the two systems equivalent. Note that the second shaft rotates n times as fast as the first shaft in the geared system, while both shafts rotate at the same speed in the non-geared system because they are rigidly connected. If we wish to also include the inertia of the pair of gears, we can put a disk, whose moment of inertia equals that of the gear-pair, at the point where the two shafts are connected.

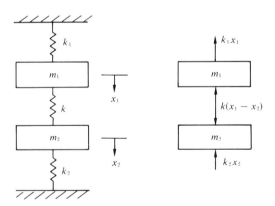

Fig. 3-1

Let us write the equations of motion for the spring- mass system shown in Fig. 3–1. First, let the positive direction be downward for both coordinates x_1 and x_2. Second, move both masses in the positive downward direction and let displacement x_1 be larger than displacement x_2, thus compressing the middle spring. It should be noted that the same equations of motion would result if $x_2 > x_1$ (in which case the middle spring is stretched). Application of Newton's second law for each mass gives

$$m_1 \ddot{x}_1 = -k_1 x_1 - k(x_1 - x_2)$$
$$m_2 \ddot{x}_2 = k(x_1 - x_2) - k_2 x_2$$

The signs of the spring forces in the first equation of motion are both negative because these two forces both act in the upward (i.e., negative) direction. The middle spring is compressed an amount $(x_1 - x_2)$; it exerts a negative upward force on mass m_1 and it exerts a positive downward force on mass m_2. These equations of motion may be rewritten as follows:

$$m_1 \ddot{x}_1 + (k_1 + k)x_1 - kx_2 = 0 \tag{3-1a}$$
$$m_2 \ddot{x}_2 + (k_2 + k)x_2 - kx_1 = 0 \tag{3-1b}$$

These equations are a pair of simultaneous, linear homogeneous differential equations that we learned how to solve in our differential equations course. To show how such simultaneous differential equations are solved in general fashion for review purposes, Ill. Exs. 3.5 and 3.6 solve a numerical version of this two-degree-of-freedom problem by classical means and by Laplace transforms. Since the only derivatives in these simultaneous differential equations are of the second order (i.e., $\ddot{x}_1$ and $\ddot{x}_2$), a simpler solution procedure may be employed. That is, because of the form of these two differential equations, we know that the solution will be periodic and can be expressed as a superposition of harmonic (i.e., sine and cosine) terms. Thus, we can solve this pair of differential equations by letting

$$x_1 = A \sin(\omega t + \alpha) \tag{3-2a}$$
$$x_2 = B \sin(\omega t + \alpha) \tag{3-2b}$$

where ω is a natural frequency, and by substituting these values of x_1 and x_2 into Eqs. (3-1a, b). This substitution, followed by a division of each equation by $\sin(\omega t + \alpha)$, gives

$$(k_1 + k - m_1 \omega^2)A - kB = 0 \tag{3-3a}$$
$$(k_2 + k - m_2 \omega^2)B - kA = 0 \tag{3-3b}$$

These linear algebraic equations in A and B are satisfied if $A = B = 0$, but this solution does not result in a vibration. The other possible solution can be obtained by setting the determinant of the coefficients of A and B equal to zero as follows:

$$\begin{vmatrix} k_1 + k - m_1 \omega^2 & -k \\ -k & k_2 + k - m_2 \omega^2 \end{vmatrix} = 0 \tag{3-4}$$

Expansion of this determinant, followed by a division by $m_1 m_2$, gives

$$\omega^4 - \left[\frac{k_1 + k}{m_1} + \frac{k_2 + k}{m_2}\right]\omega^2 + \frac{k_1 k_2 + k_1 k + k_2 k}{m_1 m_2} = 0 \qquad (3\text{-}5)$$

Equation (3-4), and hence also Eq. (3-5), is called the *characteristic equation* (and sometimes the *frequency equation*) of the system. Since Eq. (3-5) is a quadratic in ω^2, we can use the quadratic formula to obtain ω_1^2 and ω_2^2. The values for ω_1^2 and ω_2^2 will always be real and positive, so that the four mathematical solutions for natural frequency ω are $\pm\omega_1$ and $\pm\omega_2$. Since $\sin(-\omega_1)t = \sin \omega_1 t$ or $-\sin \omega_1 t$, we only have to utilize the $+\omega_1$-and $+\omega_2$-values in our solution. Since Eqs. (3-1a, b) are linear differential equations, their complete solution is a linear combination of the two possible solutions. Thus, the form of the complete solution is

$$x_1 = A_1 \sin(\omega_1 t + \alpha_1) + A_2 \sin(\omega_2 t + \alpha_2) \qquad (3\text{-}6a)$$
$$x_2 = B_1 \sin(\omega_1 t + \alpha_1) + B_2 \sin(\omega_2 t + \alpha_2) \qquad (3\text{-}6b)$$

Using Eqs. (3-3a, b), we find that the coefficient ratios are given by

$$\frac{A_1}{B_1} = \frac{k}{k_1 + k - m_1\omega_1^2} = \frac{k_2 + k - m_2\omega_1^2}{k} \qquad (3\text{-}7a)$$

$$\frac{A_2}{B_2} = \frac{k}{k_1 + k - m_1\omega_2^2} = \frac{k_2 + k - m_2\omega_2^2}{k} \qquad (3\text{-}7b)$$

which means that we can compute amplitudes B_1 and B_2 after A_1 and A_2 have been determined. We can compute the values of coefficients A_1 and A_2 and phase angles α_1 and α_2 from the four initial conditions $x_1(0)$, $x_2(0)$, $\dot{x}_1(0)$, and $\dot{x}_2(0)$ (i.e., the two initial displacements and the two initial velocities). Note from Eqs. (3-6a, b) that the solutions for x_1 and x_2 are not harmonic (if the A_1-, A_2-, B_1-, and B_2-coefficients are non-zero), but both of these two solutions are the sum of two harmonic motions at natural frequencies ω_1 and ω_2. Certain initial conditions can cause A_1 (and hence B_1) to equal zero, and Eqs. (3-6a, b) show that the resultant motion for both masses will be harmonic at the same natural frequency ω_2. Thus, from our discussion in Sec. 3.1, we have a principal mode of vibration. Likewise, the other principal mode of vibration occurs when A_2 (and hence B_2) equals zero. It can be shown that a principal-mode vibration results for a system with equal masses and equal spring constants when the initial velocities are zero and $x_1(0) = \pm x_2(0)$. Thus, we have found the two principal-mode vibrations for this two-degree-of-freedom system whose amplitude ratios are given by Eqs. (3-7a, b) and which vibrate at natural frequencies ω_1 and ω_2.

For the first mode (i.e., at the lower natural frequency ω_1), it can be shown that the two masses move in the same direction. That is, the amplitude ratio A_1/B_1 is positive in Eq. (3–7a) if ω_1 is the lower natural frequency. These two masses will always move in the opposite direction for the second mode because substitution of the solution for ω_2 [i.e., the higher ω-solution from Eq. (3–5)] into Eq. (3–7b) gives a negative value for the second-mode amplitude ratio A_2/B_2. Figure 3–2a can represent the first mode of vibration since the two masses move in the same direction, while Fig. 3–2b can represent the second vibration mode since the two masses move in opposite directions. For the same reason, Figs. 3–2a, b can also represent the two principal modes for the vertical vibration of the system in Fig. 3–2c, which consists of two masses and a horizontal elastic string.

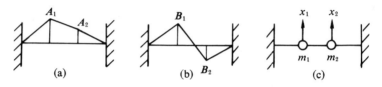

Fig. 3-2

Applying Newton's second law, we find that the equations of motion for the three-degree-of-freedom spring-mass system in Fig. 3–3 are

$$m_1\ddot{x}_1 = -k_1x_1 - k_2(x_1 - x_2)$$
$$m_2\ddot{x}_2 = k_2(x_1 - x_2) - k_3(x_2 - x_3)$$
$$m_3\ddot{x}_3 = k_3(x_2 - x_3) - k_4x_3$$

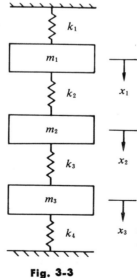

Fig. 3-3

We shall use the same solution procedure that we used before. Substitution of $x_1 = A \sin(\omega t + \alpha)$, $x_2 = B(\sin \omega t + \alpha)$, and $x_3 = C(\sin \omega t + \alpha)$ into these equations of motion, followed by a division by sin $(\omega t + \alpha)$, gives

$$(k_1 + k_2 - m_1 \omega^2)A - k_2 B = 0 \tag{3-8a}$$
$$(k_2 + k_3 - m_2 \omega^2)B - k_2 A - k_3 C = 0 \tag{3-8b}$$
$$(k_3 + k_4 - m_3 \omega^2)C - k_3 B = 0 \tag{3-8c}$$

Equating the determinant of the A-, B-, and C-coefficients to zero, we obtain the characteristic equation as follows:

$$\begin{vmatrix} k_1 + k_2 - m_1 \omega^2 & -k_2 & 0 \\ -k_2 & k_2 + k_3 - m_2 \omega^2 & -k_3 \\ 0 & -k_3 & k_3 + k_4 - m_3 \omega^2 \end{vmatrix} = 0 \tag{3-9}$$

which is a polynomial equation of degree 3 in ω^2. Similarly, for a free and undamped, four-degree-of-freedom system, we can use the same solution procedure to obtain a characteristic polynomial equation of degree 4 in ω^2. Thus, for higher-degree-of-freedom systems, we need to know how to solve for the roots of polynomial equations (which is discussed in Sec. 3.3) in order to determine the natural freqencies of the principal vibration modes, as a part of this type of solution procedure. Chapter 4 furnishes other methods for calculating natural frequencies. The solution for this three-degree-of-freedom problem is

$$x_1 = A_1 \sin(\omega_1 t + \alpha_1) + A_2 \sin(\omega_2 t + \alpha_2) + A_3 \sin(\omega_3 t + \alpha_3) \tag{3-10a}$$

$$x_2 = B_1 \sin(\omega_1 t + \alpha_1) + B_2 \sin(\omega_2 t + \alpha_2) + B_3 \sin(\omega_3 t + \alpha_3) \tag{3-10b}$$

$$x_3 = C_1 \sin(\omega_1 t + \alpha_1) + C_2 \sin(\omega_2 t + \alpha_2) + C_3 \sin(\omega_3 t + \alpha_3) \tag{3-10c}$$

where natural frequencies ω_1, ω_2, and ω_3 can be obtained from Eq. (3-9), the characteristic equation, and where the coefficient ratios $A_1/B_1, A_2/B_2, A_3/B_3, A_1/C_1, A_2/C_2$, and A_3/C_3 can be determined from Eqs. (3-8a, b, c). The value of $A_1, A_2, A_3, \alpha_1, \alpha_2$, and α_3 can be obtained from the three initial displacements and the three initial velocities.

Illustrative Example 3.1.   Suppose that in Fig. 3-1 each mass weighs 38.6 lb, the middle spring has a spring constant of 3.50 lb per in., and the other two springs have a spring constant of 2.50 lb per in. Find the variation of the displacement of each mass with time, if the initial displacement of the upper mass is two in. and the initial displacement of the lower mass and the initial velocities of both masses all equal zero.

Solution.   Since $g = 386$ in/sec^2, $m_1 = m_2 = 38.6/386 = 0.100$, and it is also given that $k_1 = k_2 = 2.50$ and $k = 3.50$. Substitution of these values in Eq. (3-5) gives

$$\omega^4 - \left(\frac{6.00}{0.100} + \frac{6.00}{0.100}\right)\omega^2 + \frac{6.25 + 8.75 + 8.75}{0.0100} = 0$$

Since $\omega^4 - 120\omega^2 + 2375 = 0$, use of the quadratic formula gives $\omega^2 = (120 \pm \sqrt{4900})/2 = 25.0$ and 95.0. Thus, $\omega_1 = 5.00$ rad/sec and $\omega_2 = 9.76$ rad/sec. Substitution in Eqs. (3-7a, b) gives

$$\frac{A_1}{B_1} = \frac{3.50}{2.50 + 3.50 - 2.50} = 1.00$$

$$\frac{A_2}{B_2} = \frac{3.50}{2.50 + 3.50 - 9.50} = -1.00$$

Thus, substitution in Eqs. (3-6a, b) gives

$$x_1 = A_1 \sin(5t + \alpha_1) + A_2 \sin(9.76t + \alpha_2)$$
$$x_2 = A_1 \sin(5t + \alpha_1) - A_2 \sin(9.76t + \alpha_2)$$

Use of the initial conditions $x_1(0) = 2$ and $x_2(0) = 0$ gives

$$2 = A_1 \sin \alpha_1 + A_2 \sin \alpha_2$$
$$0 = A_1 \sin \alpha_1 - A_2 \sin \alpha_2$$

from which we find that $A_1 \sin \alpha_1 = A_2 \sin \alpha_2 = 1$. Use of the initial conditions $\dot{x}_1(0) = \dot{x}_2(0) = 0$ gives

$$0 = 5A_1 \cos \alpha_1 + 9.76A_2 \cos \alpha_2$$
$$0 = 5A_1 \cos \alpha_1 - 9.76A_2 \cos \alpha_2$$

which are satisfied by $\cos \alpha_1 = \cos \alpha_2 = 0$ (i.e., by $\alpha_1 = \alpha_2 = 90°$). Thus, $A_1 = A_2 = 1$. Since $\sin(\omega t + 90°) = \cos \omega t$, substitution in Eqs. (3-6a, b) gives the solutions

$$x_1 = \cos 5t + \cos 9.76t$$
$$x_2 = \cos 5t - \cos 9.76t$$

Illustrative Example 3.2.   Write the characteristic equation for the disk-shaft system shown in Fig. 3-4.

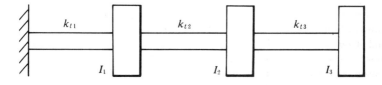

Fig. 3-4

Solution.   To ease our solution, we shall make use of the analogy between spring-mass and disk-shaft systems, as discussed in Sec. 2.4 and shown in Table 2-1. Thus, we can substitute I_i for m_i, k_{t1} for k_1, k_{t2} for k_2, k_{t3} for k_3, and $k_4 = 0$ into Eq. (3-9) to obtain the characteristic equation as follows:

$$\begin{vmatrix} k_{t1} + k_{t2} - I_1\omega^2 & -k_{t2} & 0 \\ -k_{t2} & k_{t2} + k_{t3} - I_2\omega^2 & -k_{t3} \\ 0 & -k_{t3} & k_{t3} - I_3\omega^2 \end{vmatrix} = 0$$

Illustrative Example 3.3.   Determine the characteristic equation and the principal-mode amplitude ratios for the disk-shaft system shown in Fig. 3-5.

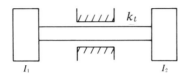

Fig. 3-5

Solution.   We shall make use of the analogy between spring-mass and disk-shaft systems, as discussed in Sec. 2.4. From Eqs. (3-6a, b), the form of the solution is

$$\theta_1 = A_1 \sin(\omega_1 t + \alpha_1) + A_2 \sin(\omega_2 t + \alpha_2)$$
$$\theta_2 = B_1 \sin(\omega_1 t + \alpha_1) + B_2 \sin(\omega_2 t + \alpha_2)$$

where θ_1 is the angular displacement of the first disk and θ_2 is the angular displacement of the second disk. The values of A_1, A_2, α_1, and α_2 can be obtained from the four initial values $\theta_1(0)$, $\theta_2(0)$, $\dot{\theta}_1(0)$, and $\dot{\theta}_2(0)$. To obtain our characteristic equation and principal-mode amplitude ratios, we substitute k_t for k, $k_1 = k_2 = 0$, I_1 for m_1, and I_2 for m_2 into Eqs. (3-5) and (3-7a, b) to obtain

$$\omega^4 - \left(\frac{k_t}{I_1} + \frac{k_t}{I_2}\right)\omega^2 = \omega^2\left[\omega^2 - \left(\frac{k_t}{I_1} + \frac{k_t}{I_2}\right)\right] = 0$$

$$\frac{A_1}{B_1} = \frac{k_t}{k_t - I_1\omega_1^2}$$

$$\frac{A_2}{B_2} = \frac{k_t}{k_t - I_2\omega_2^2}$$

From the characteristic equation, we see that $\omega_1^2 = 0$ and $\omega_2^2 = k_t/I_1 + k_t/I_2$. Substitution of these values in the previous two equations gives $A_1/B_1 = 1$ and $A_2/B_2 = -I_2/I_1$. Since natural frequency $\omega_1 = 0$, the first mode motion moves as a rigid body without any distortion of the elastic shaft. For this reason, any multiple-degree-of-freedom system that has one of its natural frequencies equal to zero is called a *semi-definite system*.

Illustrative Example 3.4.   In Sec. 1.2, it was shown that the body motion of an automobile, with regard to vehicle suspension, can be represented by the two-degree-of-freedom system shown in Fig. 1-6, which consists of two springs and a bar of weight $W = mg$ and centroidal moment of inertia I_g. Furnish the equations for the motions of this automobile body.

Solution. Using x and θ, as defined in Sec. 1.2, as the coordinates for this two-degree-of-freedom system, the equations of motion are

$$m\ddot{x} = \Sigma F = -k_1(x - L_1\theta) - k_2(x + L_2\theta)$$
$$I_g\ddot{\theta} = \Sigma M_g = k_1(x - L_1\theta)L_1 - k_2(x + L_2\theta)L_2$$

which can be rewritten as

$$m\ddot{x} + (k_1 + k_2)x - (k_1L_1 - k_2L_2)\theta = 0$$
$$I_g\ddot{\theta} + (k_1L_1^2 + k_2L_2^2)\theta - (k_1L_1 - k_2L_2)x = 0$$

which have the same mathematical form as Eqs. (3-1a, b) if the coefficients are equated. From Eqs. (3-6a, b), the form of the solution is

$$x = A_1 \sin(\omega_1 t + \alpha_1) + A_2 \sin(\omega_2 t + \alpha_2)$$
$$\theta = B_1 \sin(\omega_1 t + \alpha_1) + B_2 \sin(\omega_2 t + \alpha_2)$$

where A_1, A_2, α_1, and α_2 can be obtained from the four initial conditions $x(0)$, $\dot{x}(0)$, $\theta(0)$, and $\dot{\theta}(0)$. As for the two principal-mode motions, the center of rotation can be a point on the bar or a point outside the bar for either mode motion. Using the analogy for Eq. (3-4), natural frequencies ω_1 and ω_2 can be computed from

$$\begin{vmatrix} k_1 + k_2 - m\omega^2 & -k_1L_1 + k_2L_2 \\ -k_1L_1 + k_2L_2 & k_1^2L_1^2 + k_2^2L_2^2 - I_g\omega^2 \end{vmatrix} = 0$$

and the ratios A_1/B_1 and A_2/B_2, using the analogy for Eqs. (3-7a, b), can be computed from

$$\frac{A_1}{B_1} = \frac{-k_1L_1 + k_2L_2}{k_1 + k_2 - m\omega_1^2}$$
$$\frac{A_2}{B_2} = \frac{-k_1L_1 + k_2L_2}{k_1 + k_2 - m\omega_2^2}$$

Illustrative Example 3.5. Solve Ill. Ex. 3.1 using the classical method for solving simultaneous, ordinary differential equations.

Solution. The equations of motion are given by Eqs. (3-1a, b). Substituting numerical values, we have

$$0.100\ddot{x}_1 + 6.00x_1 - 3.50x_2 = 0$$
$$0.100\ddot{x}_2 + 6.00x_2 - 3.50x_1 = 0$$

Writing the above equations in differential operator form (where $D \equiv d/dt$) after multiplying each differential equation by ten, we obtain

$$(D^2 + 60)x_1 - 35x_2 = 0$$
$$-35x_1 + (D^2 + 60)x_2 = 0$$

We can obtain an equation in x_1 by operating on (in effect, multiplying) the first equation with $(D^2 + 60)$, multiplying the second equation by 35, and adding to obtain

$$(D^2 + 60)^2 x_1 - (35)^2 x_1 = 0$$

The same result could have been obtained by Cramer's rule, since application of this rule gives

$$\begin{vmatrix} D^2 + 60 & -35 \\ -35 & D^2 + 60 \end{vmatrix} x_1 = \begin{vmatrix} 0 & -35 \\ 0 & D^2 + 60 \end{vmatrix} = 0$$

Rewriting the equation for x_1, we obtain

$$(D^4 + 120D^2 + 2375)x_1 = 0$$

Using the quadratic formula, the roots for D^2 in the above equation are -95.0 and -25.0. Thus, the four roots are $\pm i9.76$ and $\pm i5.00$ so that the solution for x_1 is

$$x_1 = A \cos 9.76t + B \sin 9.76t + C \cos 5.00t + D \sin 5.00t$$

We can obtain the solution for x_2 by the same method as we did for x_1, by using Cramer's rule, or by substituting the solution for x_1 into one of the equations of motion and solving the resulting differential equation for x_2. The result when using one of these methods is

$$x_2 = E \cos 9.76t + F \sin 9.76t + G \cos 5.00t + H \sin 5.00t$$

Substitution of the above equations for x_1 and x_2 into one of the original differential equations shows that $E = -A$, $F = -B$, $G = C$, and $H = D$. The four specified initial conditions give us four algebraic equations for solving for the constants A, B, C, and D. Since $x_1(0) = 2$ and $x_2(0) = 0$, the initial displacement conditions give

$$2 = A + C$$
$$0 = E + G = -A + C$$

so that $A = C = 1$. Since $\dot{x}_1(0) = \dot{x}_2(0) = 0$, the initial velocity conditions give

$$0 = 9.76B + 5.00D$$
$$0 = 9.76F + 5.00H = -9.76B + 5.00D$$

which shows that $B = D = 0$. Thus, the solution for this problem is

$$x_1 = \cos 9.76t + \cos 5.00t$$
$$x_2 = -\cos 9.76t + \cos 5.00t$$

which is the same as the result in Ill. Ex. 3.1.

Illustrative Example 3.6.   Solve Ill. Ex. 3.1 using the Laplace transform method for solving simultaneous differential equations.

Solution.   The two differential equations are given in Ill. Ex. 3.5. Let $X \equiv X(s) \equiv L\{x_1(t)\}$ and $Y \equiv Y(s) \equiv L\{x_2(t)\}$. Since $x_1(0) = 2$ and $x_2(0) = \dot{x}_1(0) = \dot{x}_2(0) = 0$, the Laplace transforms of these two differential equations, after multiplication by ten, are

$$(s^2 X - 2) + 60X - 35Y = 0$$
$$s^2 Y + 60 Y - 35X = 0$$

which may be rewritten as

$$(s^2 + 60)X - 35Y = 2$$
$$-35X + (s^2 + 60)Y = 0$$

Use of Cramer's rule gives

$$X \equiv L\{x_1(t)\} = \frac{\begin{vmatrix} 2 & -35 \\ 0 & s^2 + 60 \end{vmatrix}}{\begin{vmatrix} s^2 + 60 & -35 \\ -35 & s^2 + 60 \end{vmatrix}} = \frac{2s^2 + 120}{s^4 + 120s^2 + 2375}$$

$$Y \equiv L\{x_2(t)\} = \frac{\begin{vmatrix} s^2 + 60 & 2 \\ -35 & 0 \end{vmatrix}}{\begin{vmatrix} s^2 + 60 & -35 \\ -35 & s^2 + 60 \end{vmatrix}} = \frac{-70}{s^4 + 120s^2 + 2375}$$

Since $s^4 + 120s^2 + 2375 = (s^2 + 25)(s^2 + 95)$, we have

$$X = \frac{As + B}{s^2 + 25} + \frac{Cs + D}{s^2 + 95}$$
$$Y = \frac{Es + F}{s^2 + 25} + \frac{Gs + H}{s^2 + 95}$$

Since $[(s^2 + 25)X(s)]_{s = i5} = [(2s^2 + 120)/(s^2 + 95)]_{s=i5} = 1$, $A = 1$ and $B = 0$. Similar calculations show that $C = 1$, $D = 0$, $E = 1$, $F = 0$, $G = -1$, and $H = 0$. Thus, the inverse transforms of $X(s)$ and $Y(s)$ are

$$x_1(t) = \cos 5t + \cos \sqrt{95}t$$
$$x_2(t) = \cos 5t - \cos \sqrt{95}t$$

which agrees with the results of Ill. Ex. 3.1.

3.3   Methods for Obtaining Roots of Polynomial Equations

Section 3.2 gave us a method for determining the timewise displacement variation for an undamped free vibration of a multiple-degree-of-freedom system. In this method, we had to obtain the n roots of a polynomial equation (i.e., the ω^2 soultion values for the characteristic equation) in order to obtain the n natural frequencies for an n-degree-of-freedom system. A *polynomial equation*, $f(x) = 0$, *of degree n* has the following form

$$f(x) = a_0 x^n + a_1 x^{n-1} + \cdots + a_{n-1}x + a_n = 0 \qquad (3\text{--}11)$$

If the n roots of this polynomial equation are r_1 to r_n, then this polynomial equation can be rewritten in the following factored form:

$$f(x) = a_0(x - r_1)(x - r_2) \cdots (x - r_n) = 0 \qquad (3\text{-}12)$$

One familiar means for determining the roots r_1 to r_n of Eq. (3-11) is to use the Newton-Raphson method, which the reader learned in calculus, and the quadratic formula. The iteration equation for the *Newton-Raphson method* is

$$x_{n+1} = x_n - \frac{f(x_n)}{f'(x_n)} \qquad (3\text{-}13)$$

A value x_1 is assumed and Eq. (3-13) is iteratively applied until a convergence to $x_{n+1} = x_n$ is attained. It should be warned that the Newton-Raphson method may not converge to a solution if $f'(x) = 0$ near a root. After the first root of the polynomial equation, r_1, is obtained, we divide the polynomial equation by $(x - r_1)$ to obtain a second polynomial of degree $(n - 1)$. This division can be performed by use of synthetic division procedures, which is exemplified in Ill. Ex. 3.7. The next root, r_2, is again obtained by the Newton-Raphson method, and another division is performed to obtain a polynomial of degree $(n - 2)$. This procedure is repeated until we are left with a second-degree polynomial of the form

$$cx^2 + dx + e = 0$$

whose roots can be computed from the *quadratic formula*

$$r_{n-1}, r_n = \frac{-d \pm \sqrt{d^2 - 4ce}}{2c}$$

The Newton-Raphson method can also be used to solve for the complex roots of a polynomial equation. A complex starting value must be assumed, and the rest of the procedure is similar to that for finding a real root. The calculations are longer and more laborious, because of the complex arithmetic operations that are required. The solution of a polynomial by the Newton-Raphson method is exemplified in Ill. Ex. 3.8.

There are many available methods for solving polynomial equations, and some have specialized applications. Some of the more important ones, besides the Newton-Raphson method, are the Bairstow, Bernoulli, Birge-Vieta, Graeffe, Lehmer, and Lin methods. The details for applying these methods can be found in texts on numerical analysis. Also of interest is a method devised by H. Rutishauser, based on the quotient-difference algorithm, which simultaneously approximates *all* of the roots of a polynomial, instead of calculating them one at a time, as most methods do. The quotient-difference algorithm is discussed in in the P. Henrici reference in Appendix A-3. The *Birge-Vieta method* is a procedure for applying the Newton-Raphson method.

The *Bairstow* and the *Lin* methods are advantageous for finding the the complex roots of a polynomial equation with real coefficients. Since we know that these roots will occur in complex conjugate pairs if all of the polynomial coefficients are real, then $(a - bi)$ is a root if $(a + bi)$ is a root. Thus, the given polynomial equation will have the following *quadratic factor*:

$$(x - a - bi)(x - a + bi) = x^2 + mx + n$$

Both the Bairstow and Lin methods solve for coefficients m and n. Once these two coefficients have been determined, the pair of roots can be obtained by use of the quadratic formula.

Lehmer's method is very advantageous for finding the roots of a polynomial that has complex coefficients, even though this method does not converge as rapidly as the Newton-Raphson method. It has the advantage that little modification is required when special cases are encountered. Most methods encounter a difficulty when two or more roots of a polynomial equation are close together; that is, it is not easy to find them to a high degree of accuracy. *Graeffe's method* alleviates this difficulty by transforming the original polynomial equation into an equation whose roots are squares of the original roots. This transformation is exemplified in Ill. Ex. 3.9. Thus, the roots of the transformed equation are more widely separated.

Because the *Bernoulli method* is self-starting and because it considers only two types of cases, this method is well suited for use in a digital computer computation. For these reasons, it is sometimes used to calculate starting values for a Newton-Raphson iteration on a digital computer. This method computes the real root r_1, or a pair of complex roots, of the given polynomial that has the largest absolute value. If the polynomial is divided by $(x - r_1)$, the method can be reapplied to this second polynomial to obtain the root that has the second largest absolute value. The use of this method is exemplified and explained in more detail in Ill. Ex. 3.10.

Unfortunately, no single method can be used to satisfactorily find, with complete certainty, all of the roots of a polynomial equation, for all of the possibilities that may be encountered. The Bairstow, Birge-Vieta, Lin, and Newton-Raphson methods will not always converge for every initial guess for a root. The Bernoulli, Graeffe, and Lehmer methods will converge to a solution with certainty (though sometimes it may be slow for a pair of nearly equal roots), but the size of the numbers involved grows very large during each successive iteration cycle. For this reason, it is best to find the roots of a polynomial equation (especially when using a digital computer) by using a procedure that combines two or more solution techniques.

Illnstrative Example 3.7.   Use synthetic division to verify that the roots of the following polynomial equation are 1, 2, and 3.

$$f(x) = x^3 - 6x^2 + 11x - 6 = 0$$

Solution.   As the reader has learned previously in an algebra course, the synthetic division procedure for the division of $f(x)$ in Eq. (3-11) by $(x - r)$ may be summarized by the following tabular form. The fact that synthetic division involves only additions and multiplications makes this method relatively easy to apply.

a_0	a_1	a_2	a_3	$\cdots$ a_{n-1}	a_n
	rb_0	rb_1	rb_2	$\cdots$ rb_{n-2}	rb_{n-1}
b_0	b_1	b_2	b_3	$\cdots$ b_{n-1}	b_n

Note that the first row consists of the n coefficients for the polynomial $f(x)$. The b_i terms in the third row are obtained by addition of the two terms above it; that is, b_1 to b_n are computed from the equation $b_i = a_i + rb_{i-1}$. The second row terms are obtained by the multiplication of b_{i-1} by r, as shown. The answer or quotient for this division is the polynomial $b_0 x^{n-1} + b_1 x^{n-2} + \cdots + b_{n-1}$ with a remainder of b_n. This remainder b_n is equal to $f(r)$, so that synthetic division can also be used to calculate the value of $f(x)$ for a specific value of x. Such a calculation is required when the Newton-Raphson method is applied. Use of synthetic division to divide the given polynomial by $(x - 3)$ gives

1	-6	11	-6
	3	-9	6
1	-3	2	0

Thus, since the remainder equals zero, we know that $x = 3$ is a root of this polynomial. Since the quotient is $x^2 - 3x + 2$, we can now use synthetic division as follows to divide this quotient by $(x - 2)$.

1	-3	2
	2	-2
1	-1	0

Since the remainder equals zero, we know that $x = 2$ is another root of the given polynomial equation. Since the quotient is $x - 1$, we know that $x = 1$ is the third root of the given polynomial equation. It can be seen from this example how a synthetic division procedure may be easily programmed for a digital computer solution, where the inputs are r and the a_i coefficients for polynomial $f(x)$, by using the equations $b_0 = a_0$ and $b_i = a_i + rb_{i-1}$, for $i \geq 1$, to obtain the b_0 to b_n output terms.

Illustrative Example 3.8.　　Use the Newton-Raphson method to obtain the roots of the following polynomial equation:

$$f(x) = x^4 - x^3 - 8x^2 - 48 = 0$$

Solution.　　The Newton-Raphson method employs the iterative equation

$$x_{n+1} = x_n - \frac{f(x_n)}{f'(x_n)}$$

Since $f'(x) = 4x^3 - 3x^2 - 16x - 4$, the previous equation may be rewritten as follows, when we substitute for $f(x_n)$ and $f'(x_n)$.

$$x_{n+1} = x_n - \frac{x_n^4 - x_n^3 - 8x_n^2 - 4x_n - 48}{4x_n^3 - 3x_n^2 - 16x_n - 4}$$

Assumption of $x_1 = 5.00$ gives

$$x_2 = 5.00 - \frac{(5)^4 - (5)^3 - 8(5)^2 - 4(5) - 48}{4(5)^3 - 3(5)^2 - 16(5) - 4} = 5.00 - \frac{232}{341} = 4.32$$

If we keep applying the iteration equation, we will come to a step where $x_{n+1} = x_n = 4.00$. Thus, one root is $x = 4.00$. To obtain another root, we first divide the given polynomial $f(x)$ by $(x - 4)$ to obtain the following quotient $g(x)$

$$g(x) = \frac{f(x)}{x - 4} = x^3 + 3x^2 + 4x + 12$$

where this operation can be performed by synthetic division as shown in Ill. Ex. 3.7. To obtain the next root by the Newton-Raphson method, we employ the iteration equation

$$x_{n+1} = x_n - \frac{g(x_n)}{g'(x_n)}$$

which can be rewritten as follows, after we substitute for $g(x_n)$ and $g'(x_n)$.

$$x_{n+1} = x_n - \frac{x_n^3 + 3x_n^2 + 4x_n + 12}{3x_n^2 + 6x_n + 4}$$

Assumption of $x_1 = -4.00$ gives

$$x_2 = -4.00 - \frac{(-4)^3 + 3(-4)^2 + 4(-4) + 12}{3(-4)^2 + 6(-4) + 4} = -4.00 - \left(\frac{20}{28}\right) = -3.28$$

If we keep applying the previous iteration equation, we will come to a step where $x_{n+1} = x_n = -3.00$. Thus, a second root is $x = -3$. If we divide $g(x)$ by $(x + 3)$, we obtain $g(x)/(x + 3) = x^2 + 4$. Thus, the last two roots are the solutions of the equation $x^2 + 4 = 0$, which are $x = 2i$ and $x = -2i$. Thus, the four roots are $4, -3, 2i$, and $-2i$, and the given polynomial equation, $f(x) = 0$, may be expressed in the factored form $(x - 4)(x + 3)(x^2 + 4) = 0$.

The previous procedure for applying the Newton-Raphson method can be simplified by using synthetic division. That is, we shall use synthetic division to calculate both $f(x_n)$ and $f'(x_n)$. Let us now use synthetic division to divide $f(x)$ by $(x - 5)$ and also the quotient by $(x - 5)$, as follows, where synthetic division details have already been discussed in Ill. Ex. 3.7.

1	-1	-8	-4	-48	
	5	20	60	280	
1	4	12	56	$232 = f(5)$	
	5	45	285		
1	9	57	$341 = f'(5)$		

The values of $f(5)$ and $f'(5)$ are located at the denoted places in the previous calculation. That is, the remainder in the first division by $(x - x_n)$ equals $f(x_n)$ or $f(5)$, and the remainder in the second division equals $f'(x_n)$ or $f'(5)$. Thus, it is seen that synthetic division can be effectively used to calculate $f(x_n)$, $f'(x_n)$, and the coefficients of the next polynomial for a Newton-Raphson iteration. In like manner, $g(-4)$ and $g'(-4)$ may be computed as follows:

1	3	4	12	
	-4	4	-32	
1	-1	8	$-20 = g(-4)$	
	-4	20		
1	-5	$28 = g'(-4)$		

Illustrative Example 3.9.   Show how Eq. (3-12) may be transformed into an equation $F(x)$ whose roots are the squares of the roots $r_1, r_2, \ldots,$ and r_n for Eq. (3-11).

Solution.   This can be done by multiplying Eq. (3-12) by $(-1)^n f(-x)$, where it can be shown that

$$(-1)^n f(-x) = a_0(x + r_1)(x + r_2) \cdots (x + r_n)$$

Denoting the result of the multiplication $[f(x)][(-1)^n f(-x)]$ as $F(x)$, we obtain the following polynomial equation in x^2:

$$F(x) = a_0^2(x^2 - r_1^2)(x^2 - r_2^2) \cdots (x - r_n^2) = 0$$

If we let z equal x^2, the previous equation can be rewritten as follows, where $F(x)$ is now a polynomial of degree n in z.

$$F(x) = a_0^2(z - r_1^2)(z - r_2^2) \cdots (z - r_n^2)$$
$$= d_0 z^n + d_1 z^{n-1} + \cdots + d_{n-1} z + d_n = 0$$

where it can be seen that the roots of the previous polynomial equation in z are $r_1^2, r_2^2, \ldots,$ and r_n^2, which are the squares of the roots for Eqs. (3-11) and (3-12). Thus, we only have to solve the previous polynomial equation for its n roots, and then take the square roots of these results to obtain the roots for Eq. (3-11). Included in Graeffe's method is a procedure and equation for computing the d_i-coefficients, for the previous polynomial equation of degree n in z, directly from the a_i-coefficients for Eq. (3-11). For example, note the pattern of the following equations, which calculate the first four d_i-values.

$$d_0 = a_0^2$$
$$d_1 = -a_1^2 + 2a_0 a_2$$
$$d_2 = a_2^2 - 2a_1 a_3 + 2a_0 a_4$$
$$d_3 = -a_3^2 + 2a_2 a_4 - 2a_1 a_5 + 2a_0 a_6$$

Illustrative Example 3.10.   Use the Bernoulli method to find the root r_1 of the following polynomial equation which has the largest absolute value.

$$x^3 - 6x^2 + 11x - 6 = 0$$

Solution.   Since the given polynomial equation can be written in the factored form $(x - 3)(x - 2)(x - 1) = 0$, which is verified in Ill. Ex. 3.7, we know that our answer should be $r_1 = 3$. Essentially the Bernoulli method considers the following two basic cases, where reference is made to Eq. (3-12), where the a_i-coefficients are all real, and where $|r_1| \geq |r_2| \geq |r_3| \geq \cdots \geq |r_n|$. For the first case, $|r_1| > |r_2|$ and this method will compute root r_1, which is called the *dominant root* because it is the largest in absolute value. For the second case, roots r_1 and r_2 are complex conjugates and $|r_2| > |r_3|$. The second case calculates the complex roots r_1 and r_2. To handle other cases, the given polynomial can be transformed to a form that satisfies one of these two basic cases. This method can perform poorly if one or more roots almost equals the dominant root r_1 in value. The problem for this example falls within the category of the first case.

The Bernoulli method for case one calculates a sequence G_i. The ratio G_{i+1}/G_i converges to the value of the dominant root for large values of i. The procedure for case two is similar except that different sequences T_i and V_i, calculated by different equations, are formed. For a cubic polynomial of the form $x_3 + a_1 x^2 + a_2 x + a_3$, $G_1 = -a_1$, $G_2 = -a_1 G_1 - 2a_2$, and $G_3 = -a_1 G_2 - a_2 G_1 - 3a_3$. Other values of G_i can be computed from the following recursion equation:

$$G_i = -(a_1 G_{i-1} + a_2 G_{i-2} + a_3 G_{i-3})$$

Since $a_1 = -6$, $a_2 = 11$, and $a_3 = -6$ for this problem, we obtain $G_1 = 6$, $G_2 = 44$, $G_3 = 216$, and other values of G_i can be computed from the recursion formula

$$G_i = 6G_{i-1} - 11G_{i-2} + 6G_{i-3} = 6(G_{i-1} + G_{i-3}) - 11G_{i-2}$$

Use of this recursion formula gives $G_4 = 848$ and $G_5 = 2976$. Since $G_2/G_1 = 7.33$, $G_3/G_2 = 4.92$, $G_4/G_3 = 3.92$, and $G_5/G_4 = 3.52$, it is seen that the sequence G_{i+1}/G_i is converging toward the value of the dominant root, which is 3.00. As was mentioned previously, the Bernoulli method can be used for a few iterations to obtain starting values for an application of the Newton-Raphson method, which can be used to calculate the root in more accurate fashion (that is, to a greater number of significant digits). Thus, we would be combining two solution methods, as previously recommended, in order to find the roots of a polynomial equation.

3.4   Damped Free Vibrations

Suppose that the two-degree-of-freedom system in Fig. 3–1 in Sec. 3.2 had a linear damper, whose damping coefficient is c, connected between masses m_1 and m_2. Since this adds a damping force of magnitude $c(\dot{x}_1 - \dot{x}_2)$, application of Newton's second law for each mass gives

$$m_1\ddot{x}_1 = -k_1x_1 - k(x_1 - x_2) - c(\dot{x}_1 - \dot{x}_2)$$
$$m_2\ddot{x}_2 = k(x_1 - x_2) - k_2x_2 + c(\dot{x}_1 - \dot{x}_2)$$

which can be rewritten as

$$m_1\ddot{x}_1 + c\dot{x}_1 + (k_1 + k)x_1 - c\dot{x}_2 - kx_2 = 0 \qquad \text{(3–14a)}$$
$$m_2\ddot{x}_2 + c\dot{x}_2 + (k_2 + k)x_2 - c\dot{x}_1 - kx_1 = 0 \qquad \text{(3–14b)}$$

Because of the presence of the damping-force terms $c\dot{x}_1$ and $c\dot{x}_2$ in Eqs. (3–14a, b), we cannot use the solution technique in Sec. 3.2 which assumes a superposition of harmonic components. That is, the solution for this damped system can be a superposition of both aperiodic and decreasing-amplitude oscillatory components, so we shall use the classical-solution-method assumption that $x_1 = Ae^{st}$ and $x_2 = Be^{st}$. Substituting these equations for x_1 and x_2 into Eqs. (3–14a, b) and dividing the result by e^{st}, we have

$$(m_1s^2 + cs + k_1 + k)A - (cs + k)B = 0 \qquad \text{(3–15a)}$$
$$(m_2s^2 + cs + k_2 + k)B - (cs + k)A = 0 \qquad \text{(3–15b)}$$

This set of linear algebraic equations has a solution (other than the trivial one $A = B = 0$) only if the following determinant for coefficients A and B equals zero, as shown.

$$\begin{vmatrix} m_1s^2 + cs + k_1 + k & -cs - k \\ -cs - k & m_2s^2 + cs + k_2 + k \end{vmatrix} = 0 \quad . \quad \text{(3–16)}$$

Expansion of the previous determinant gives the following characteristic equation of the system, which must be solved to determine the four values for s.

$$[m_1s^2 + cs + k_1 + k][m_2s^2 + cs + k_2 + k] - (cs + k)^2 = 0 \qquad \text{(3–17)}$$

The roots (i.e., the values of s) of this characteristic equation, which is a fourth-degree polynomial, may be real or complex. Methods for obtaining roots of polynomial equations were discussed in Sec. 3.3. Since the coefficients of Eq. (3–17) are real, we know that all complex roots will occur as complex conjugates. From Sec. 3.3, we know that the Bairstow and Lin methods are advantageous for finding such roots. The Lin method is exemplified in Ill. Ex. 3.12 in this section. Using superposition, the general solution can be expressed as

$$x_1 = A_1e^{s_1t} + A_2e^{s_2t} + A_3e^{s_3t} + A_4e^{s_4t} \qquad \text{(3–18a)}$$
$$x_2 = B_1e^{s_1t} + B_2e^{s_2t} + B_3e^{s_3t} + B_4e^{s_4t} \qquad \text{(3–18b)}$$

where the coefficients A_1, A_2, A_3, and A_4 can be found from the four initial conditions (i.e., the two initial velocities and the two initial displacements). Use of Eqs. (3–15a, b) shows that the amplitude ratios, which can be real or complex, are given by

$$\frac{A_i}{B_i} = \frac{cs_i + k}{m_1 s_i^2 + cs_i + k_1 + k} = \frac{m_2 s_i^2 + cs_i + k_2 + k}{cs_i + k} \qquad (3\text{–}19)$$

where $i = 1, 2, 3$, or 4. The linear damper causes the displacements x_1 and x_2 to decrease with time, as was similarly shown in Sec. 1.7 for a linearly-damped single-degree-of-freedom system. Thus, a root s_1, s_2, s_3, or s_4 must be negative if it is real, or must have a negative real part if it is complex. If two of the roots are real and the other two roots are complex conjugates, then the displacements x_1 and x_2 can be expressed in the following form, which consists of both aperiodic and damped oscillatory components.

$$x_1 = A_1 e^{s_1 t} + A_2 e^{s_2 t} + C e^{-bt} \sin(\omega_d t + \alpha_1)$$
$$x_2 = B_1 e^{s_1 t} + B_2 e^{s_2 t} + D e^{-bt} \sin(\omega_d t + \alpha_2)$$

The last term in each of the previous two equations can be obtained by a derivation that is similar to that used in Sec. 1.7 to obtain Eq. (1–44) from Eq. (1–39) when roots n_1 and n_2 are complex conjugates. If roots s_1 to s_4 are all complex, then the displacements x_1 and x_2 consist of a superposition of only decreasing-amplitude, oscillatory components. If the roots s_1 to s_4 are all real, then the motions x_1 and x_2 are not oscillatory, since Eqs. (3–18a, b) show that all of the components are aperiodic.

Illustrative Example 3.11. Suppose, for the damped vibration system discussed in this section, that we added a linear damper of damping coefficient c_1 atop mass m_1 (which is parallel to spring k_1) and a linear damper of damping coefficient c_2 below mass m_2. Discuss the effect on the solutions for displacements x_1 and x_2.

Solution. Since these two new dampers add a damping force $-c_1 \dot{x}_1$ to the first Newton's second law equation and a damping force $-c_2 \dot{x}_2$ to the second Newton equation, the equations of motion can be obtained by substituting $(c_1 + c)\dot{x}_1$ for $c\dot{x}_1$ in Eq. (3–14a) and $(c_2 + c)\dot{x}_2$ for $c\dot{x}_2$ in Eq. (3–14b). A similar substitution in Eq. (3–17) gives the following new characteristic equation:

$$[m_1 s^2 + (c_1 + c)s + k_1 + k][m_2 s^2 + (c_2 + c)s + k_2 + k] - (cs + k)^2 = 0$$
$$(3\text{–}20)$$

Since the only difference between Eqs. (3–17) and (3–20) are the polynomial coefficients, the rest of the solution details have already been covered in this section.

Illustrative Example 3.12.   Suppose that the characteristic equation for a damped, two-degree-of-freedom system is given by

$$s^4 - s^3 - 8s^2 - 4s - 48 = 0$$

Find the four roots, s_1 to s_4, of this characteristic equation using Lin's method.

Solution.   Lin's method consists of repeated divisions by trial quadratic divisors until the remainder approaches zero. It is a controlled, trial-and-error division technique that factors a polynomial of even degree into a product of quadratic expressions. We can use the last three terms of the given polynomial to obtain a first trial divisor as follows:

$$\frac{-8s^2 - 4s - 48}{-8} = s^2 + 0.50s + 6$$

Division of the characteristic equation by this trial divisor gives

$$
\begin{array}{r}
s^2 - \quad 1.5s - 13.25 \\
s^2 + 0.50s + 6 \overline{\smash{\big)}\ s^4 - \quad s^3 + \quad 8s^2 - \quad 4s - 48} \\
\underline{s^4 + 0.5s^3 + \quad 6s^2} \\
-1.5s^3 - \quad 14s^2 - \quad 4s \\
\underline{-1.5s^3 - 0.75s^2 - \quad 9s} \\
-13.25s^2 + \quad 5s - 48 \qquad * \\
\underline{-13.25s^2 - 6.625s - 79.5} \\
11.625s + 31.5
\end{array}
$$

Since the remainder is not close to zero, we are not close to a solution. A second quadratic trial divisor is obtained from the expression in the preceding division that is denoted by an asterisk, as follows:

$$\frac{-13.25s^2 + 5s - 48}{-13.25} = s^2 - 0.377s + 3.62$$

If we keep applying this procedure, the trial divisor will approach $(s^2 + 4)$, the quotient will approach $(s^2 - s - 12)$, and the remainder will approach zero. Thus, the given characteristic equation can now be expressed in the factored form $(s^2 + 4)(s^2 - s - 12) = 0$. At this point, two of the roots may be obtained from the quadratic equation $s^2 + 4 = 0$. This gives us the roots $s = 2i$ and $s = -2i$. The other two roots can be obtained from the quadratic equation $s^2 - s - 12 = 0$. Use of the quadratic formula gives us the roots $s = 4$ and $s = -3$.

3.5   Forced Vibrations

Suppose that the two-degree-of-freedom system in Fig. 3–1 of Sec. 3.2 had the harmonic external force $F \sin \omega t$ applied to mass m_1. The application of Newton's second law for each mass gives

$$m_1 \ddot{x}_1 = -k_1 x_1 - k(x_1 - x_2) + F \sin \omega t$$
$$m_2 \ddot{x}_2 = k(x_1 - x_2) - k_2 x_2$$

which can be rewritten as

$$m_1 \ddot{x}_1 + (k_1 + k)x_1 - kx_2 = F \sin \omega t \qquad \text{(3–21a)}$$
$$m_2 \ddot{x}_2 + (k_2 + k)x_2 - kx_1 = 0 \qquad \text{(3–21b)}$$

The transient solutions are given in Sec. 3.2 since they are the solutions of Eqs. (3–1a, b). Since we know that the resulting steady-state motion will have the same frequency as the exciting force when a harmonic force acts on the system, we can use the mechanical impedance method to obtain the steady-state solutions. Also note that we can superpose steady-state solutions for individual forces to get the total steady-state solution when several harmonic, external forces act on this linear mechanical system. Substituting $Fe^{i\omega t}$ for $F \sin \omega t$, $\bar{X}_1 e^{i\omega t}$ for x_1, and $\bar{X}_2 e^{i\omega t}$ for x_2, where amplitudes $\bar{X}_1$ and $\bar{X}_2$ can be complex numbers, into Eqs. (3–21a, b) and dividing the result by $e^{i\omega t}$, we obtain

$$(-m_1 \omega^2 + k_1 + k)\bar{X}_1 - k\bar{X}_2 = F \qquad \text{(3–22a)}$$
$$(-m_2 \omega^2 + k_2 + k)\bar{X}_2 - k\bar{X}_1 = 0 \qquad \text{(3–22b)}$$

where the relation $\ddot{x}_j = -\omega^2 \bar{X}_j e^{i\omega t}$ was employed. Solving these two algebraic equations for amplitudes $\bar{X}_1$ and $\bar{X}_2$ by Cramer's rule gives

$$\bar{X}_1 \equiv X_1 e^{-i\phi_1} = \frac{\begin{vmatrix} F & -k \\ 0 & -m_2 \omega^2 + k_2 + k \end{vmatrix}}{|\Delta|} = \frac{(k_2 + k - m_2 \omega^2)F}{|\Delta|} \qquad \text{(3–23a)}$$

$$\bar{X}_2 \equiv X_2 e^{-i\phi_2} = \frac{\begin{vmatrix} -m_1 \omega^2 + k_1 + k & F \\ -k & 0 \end{vmatrix}}{|\Delta|} = \frac{kF}{|\Delta|} \qquad \text{(3–23b)}$$

where the denominator determinant $|\Delta|$ is given by

$$|\Delta| = \begin{vmatrix} -m_1 \omega^2 + k_1 + k & -k \\ -k & -m_2 \omega^2 + k_2 + k \end{vmatrix} \qquad \text{(3–23c)}$$

Since Eq. (3–23c) is the same as Eq. (3–4), and also Eq. (3–5), in Sec. 3.2, we have shown that the characteristic equation for the system can also be obtained by using the mechanical impedance method. Thus, the ω-roots of Eq. (3–23c) are the natural frequencies of the system. Since the excitation force $F \sin \omega t$ equals $Im\ [Fe^{i\omega t}]$ and since X_1, X_2, ϕ_1, and ϕ_2 are the displacement amplitudes and phase angles, the steady-state displacement motions are given by

$$x_1 = Im[\bar{X}_1 e^{i\omega t}] = Im[X_1 e^{-i\phi_1} e^{i\omega t}] = X_1 \sin(\omega t - \phi_1) \qquad \text{(3–24a)}$$
$$x_2 = Im[\bar{X}_2 e^{i\omega t}] = Im[X_2 e^{-i\phi_2} e^{i\omega t}] = X_2 \sin(\omega t - \phi_2) \qquad \text{(3–24b)}$$

Since Eqs. (3–23a, b, c) show that the numerators and denominators for $\bar{X}_1$ and $\bar{X}_2$ are all real numbers, then $\bar{X}_1$ and $\bar{X}_2$ are both real numbers, which are equal to amplitudes X_1 and X_2, and phase angles ϕ_1 and ϕ_2 must equal either zero or $180°$. Thus, displacement x_1 (and also x_2) is either always in the same direction or always in the opposite direction as the applied force, as was also shown in Sec. 2.2 for an undamped, single-degree-of-freedom system for displacement x. Let ω_1 and ω_2 denote the natural frequencies of the system, where $\omega_2 > \omega_1$. Calculations show that both masses move in the same direction as the applied force (i.e., $\phi_1 = \phi_2 = 0°$) if $\omega^2 < \omega_1^2$ and that both masses move in the opposite direction as the applied force (i.e., $\phi_1 = \phi_2 = 180°$) if $\omega_1^2 < \omega^2 < (k_2 + k)/m_2$. If $(k_2 + k)/m_2 < \omega^2 < \omega_2^2$, then $\phi_1 = 0°$ and $\phi_2 = 180°$; and $\phi_1 = 180°$ and $\phi_2 = 0°$ when $\omega^2 > \omega_2^2$. Since ϕ_1 and ϕ_2 must equal $0°$ or $180°$, the displacement equations are given by

$$x_1 = \pm X_1 \sin \omega t \tag{3-25a}$$
$$x_2 = \pm X_2 \sin \omega t \tag{3-25b}$$

where the magnitudes of X_1 and X_2 may be found from Eqs. (3–23a, b, c), since $X_i = |\bar{X}_i|$. The sign of displacement x_j is positive if $\phi_j = 0°$ and negative if $\phi_j = 180°$. Note in Eq. (3–23a) that amplitude X_1 equals zero when ω^2 equals $(k_2 + k)/m_2$, and note in Eq. (3–23b) that amplitude X_2 never equals zero. If the excitation frequency ω equals one of the two natural frequencies of this system, Eqs. (3–23a, b, c) show that resonance occurs (as also happens for a single-degree-of-freedom system) since amplitudes X_1 and X_2 become infinite.

Now suppose that this two-degree-of-freedom system with the applied force also had a linear damper, whose damping coefficient is c, connected between masses m_1 and m_2. Since this adds a damping force of magnitude $c(\dot{x}_1 - \dot{x}_2)$, the equations of motion now become

$$m_1\ddot{x}_1 + c\dot{x}_1 + (k_1 + k)x_1 - c\dot{x}_2 - kx_2 = F \sin \omega t \tag{3-26a}$$
$$m_2\ddot{x}_2 + c\dot{x}_2 + (k_2 + k)x_2 - c\dot{x}_1 - kx_1 = 0 \tag{3-26b}$$

The transient solutions, which die out with time, were discussed in Sec. 3.4, and the steady-state solutions can be obtained by the mechanical impedance method. Substituting $Fe^{i\omega t}$ for $F \sin \omega t$, $\bar{X}_1 e^{i\omega t}$ for x_1, and $\bar{X}_2 e^{i\omega t}$ for x_2, where $\bar{X}_1$ and $\bar{X}_2$ are complex amplitudes, into Eqs. (3–26a, b), and dividing the result by $e^{i\omega t}$, we obtain

$$(k_1 + k - m_1\omega^2 + ic\omega)\bar{X}_1 - (k + ic\omega)\bar{X}_2 = F \tag{3-27a}$$
$$(k_2 + k - m_2\omega^2 + ic\omega)\bar{X}_2 - (k + ic\omega)\bar{X}_1 = 0 \tag{3-27b}$$

Use of Cramer's rule gives

$$\bar{X}_1 \equiv X_1 e^{-i\phi_1} = \frac{\begin{vmatrix} F & -k - ic\omega \\ 0 & k_2 + k - m_2\omega^2 + ic\omega \end{vmatrix}}{|\Delta|} \tag{3-28a}$$

$$\bar{X}_2 \equiv X_2 e^{-i\phi_2} = \frac{\begin{vmatrix} k_1 + k - m_1\omega^2 + ic\omega & F \\ -k - ic\omega & 0 \end{vmatrix}}{|\Delta|} \qquad \text{(3-28b)}$$

where the denominator determinant is given by

$$|\Delta| = \begin{vmatrix} k_1 + k - m_1\omega^2 + ic\omega & -k - ic\omega \\ -k - ic\omega & k_2 + k - m_2\omega^2 + ic\omega \end{vmatrix} = 0$$

(3-28c)

Since the applied force is $F \sin \omega t$, the steady-state displacements are given by

$$x_1 = X_1 \sin(\omega t - \phi_1) \qquad \text{(3-29a)}$$
$$x_2 = X_2 \sin(\omega t - \phi_2) \qquad \text{(3-29b)}$$

where amplitudes X_1 and X_2 and phase angles ϕ_1 and ϕ_2 may be found from Eqs. (3-28a, b, c). Since Eqs. (3-28a, b, c) show that both $\bar{X}_1$ and $\bar{X}_2$ are complex numbers of the form $D + iE$, the phase angles ϕ_1 and ϕ_2 will not equal $0°$ or $180°$ unless $c = 0$. If the applied force is $F \cos \omega t$, the steady-state displacements will be of the form $x_j = X_j \cos(\omega t - \phi_j)$ where $X_1, X_2, \phi_1,$ and ϕ_2 may be found from Eqs. (3-28a, b, c). If the applied force is $Fe^{i\omega t}$, the steady-state displacements are $x_1 = X_1 e^{i(\omega t - \varphi_1)}$ and $x_2 = X_2 e^{i(\omega t - \varphi_2)}$. It can be seen that the algebraic calculations become more tedious when the mechanical impedance method is applied to find the steady-state motions for higher-degree-of-freedom systems, especially those that have damping.

3.6   Dynamic Absorber

A *dynamic absorber* is a single-degree-of-freedom system (e.g., a mass supported by a spring) that is added to another single-degree-of-freedom system, thus causing the new system to have two degrees of freedom and two natural frequencies. Suppose that we had a machine of mass m_1 that is supported by springs, whose total spring constant is k_1, on a foundation. If the excitation frequency ω and the natural frequency, which equals $\sqrt{k_1/m_1}$, were nearly equal, an excessive force would be transmitted to the foundation. This excess force could be corrected by attaching a dynamic absorber, consisting of a mass m_2 and a spring as shown in Fig. 3-6, to this machine.

Since the system in Fig. 3-6 can be represented by the system in Fig. 3-1 of Sec. 3.2 with spring k_2 removed and with an external force $F \sin \omega t$ applied to mass m_1, the equations of motion are given by Eqs. (3-21a, b) in Sec. 3.5 where $k_2 = 0$ (since spring k_2 is removed). Thus,

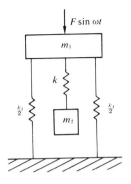

Fig. 3-6

use of the results in Sec. 3.5 shows that the steady-state displacements are given by

$$x_1 = X_1 \sin(\omega t - \phi_1) \qquad \text{(3-30a)}$$
$$x_2 = X_2 \sin(\omega t - \phi_2) \qquad \text{(3-30b)}$$

where phase angles ϕ_1 and ϕ_2 equal $0°$ or $180°$. Use of Eqs. (3-23a, b, c), where $k_2 = 0$ (since spring k_2 is removed from the system in Fig. 3-1), show that amplitudes X_1 and X_2 are given by

$$X_1 = |\bar{X}_1| = \left| \frac{(k - m_2\omega^2)F}{|\Delta|} \right| \qquad \text{(3-31a)}$$

$$X_2 = |\bar{X}_2| = \left| \frac{kF}{|\Delta|} \right| \qquad \text{(3-31b)}$$

where the denominator $|\Delta|$ is given by

$$|\Delta| = (k_1 + k - m_1\omega^2)(k - m_2\omega^2) - k^2 = 0 \qquad \text{(3-31c)}$$

Note in Eqs. (3-31a, b, c) that $X_1 = \bar{X}_1 = 0$ and $\bar{X}_2 = -F/k$ when $(k - m_2\omega^2)$ equals zero. Thus, from Eqs. (3-30a, b) we see that $x_1 = 0$, $x_2 = -(F/k)\sin \omega t$, and $\phi_2 = 180°$ (i.e., mass m_2 moves always in the opposite direction to that of the applied force) when $\omega = \sqrt{k/m_2}$. The two natural frequencies for this two-degree-of-freedom system can be found from Eq. (3-31c), which is the characteristic equation.

A dynamic absorber is used to reduce the amplitude of the machine or original system (i.e., mass m_1) when the excitation frequency of the applied force (or equivalent force) nearly equals $\sqrt{k_1/m_1}$, which is the natural frequency of the original system. The dynamic absorber is much more useful for devices whose excitation frequency is constant, and is not as practical for variable-speed devices (whose excitation frequency changes) since an absorber is effective only over a narrow frequency range. That is, the k_2 and m_2 values for a dynamic absorber are chosen so that $x_1 = 0$ for a specific value of ω, the excitation frequency. Since $x_1 = 0$ when $\omega = \sqrt{k/m_2}$, and since the natural frequency of the original system is $\sqrt{k_1/m_1}$, an undamped dynamic absorber is usually designed so that $k/m_2 = k_1/m_1$. That is, it is designed so that its natural frequency $\sqrt{k_2/m_2}$ equals the excitation frequency ω (to cause both displacement x_1 and the force $k_1 x_1$ transmitted to the foundation to equal zero) when this excitation frequency ω nearly equals the original natural frequency $\sqrt{k_1/m_1}$. Thus, the excessive vibration and forces of constant-speed machines can be reduced by the addition of a relatively small mass and spring to this machine. If we had a damped dynamic absorber, where a linear damper is placed between masses m_1 and m_2, we could use Eqs. (3–28a, b, c) and (3–29a, b) in Sec. 3.5 (after setting $k_2 = 0$) in order to analyze the effect of damping in a vibration absorber.

3.7   The Lagrange Equations

So far in this chapter, we have been using Newton's second law to obtain the equations of motion for vibrating systems. The reader has learned from dynamics that it is possible to specify some systems by more than one set of independent coordinates, and that the number of independent coordinates in each set is equal to the number of degrees of freedom for the given system. This is further discussed in Sec. 3.9; and any such set of independent coordinates is called a *generalized coordinates set*. Generalized coordinates are a set of independent parameters which completely specify the system location and which are independent of any constraints. The Lagrange method is an energy method that allows us to write the equations of motion in terms of any set of generalized coordinates (which may avoid some tedious coordinate transformation calculations), and it is a very powerful analysis method for certain complex physical systems (as will be shown later in this section). The fundamental form of Lagrange's equations can be expressed in terms of generalized coordinates q_i as follows:

$$\frac{d}{dt}\frac{\partial(\text{KE})}{\partial \dot{q}_i} - \frac{\partial(\text{KE})}{\partial q_i} + \frac{\partial(\text{PE})}{\partial q_i} + \frac{\partial(\text{DE})}{\partial \dot{q}_i} = Q_i \qquad (3\text{–}32)$$

where KE is the total kinetic energy of the system, PE is the total potential energy of the system, DE is the total energy dissipated in the system when the damping is linear, Q_i is a generalized external force (or a nonlinear damping force) acting on the system that is associated with q_i, and q_i is a generalized coordinate that describes a position of the system. The subscript i denotes that there are n such equations for an n-degree-of-freedom system since there must be n generalized coordinates, q_1 to q_n, to describe this system. That is, the Lagrange equations yield as many equations of motion as there are degrees of freedom. Equation (3–32) is derived in Ill. Ex. 3.15. For a conservative (i.e., frictionless) system, the Q_i-and DE-terms in Eq. (3–32) (which contains a $\partial(\mathrm{PE})/\partial q_i$-term) equal zero. Thus, the Lagrange equations for a *conservative system* are

$$\frac{d}{dt}\frac{\partial(\mathrm{KE})}{\partial \dot{q}_i} - \frac{\partial(\mathrm{KE})}{\partial q_i} + \frac{\partial(\mathrm{PE})}{\partial q_i} = 0 \qquad (3\text{–}33)$$

where $i = 1$ to n. If we let $L = (\mathrm{KE} - \mathrm{PE})$, where L is called the *Lagrangian*, Eq. (3–33) can be rewritten as follows since $\partial(\mathrm{PE})/\partial \dot{q}_i = 0$:

$$\frac{d}{dt}\left(\frac{\partial L}{\partial \dot{q}_i}\right) - \frac{\partial L}{\partial q_i} = 0 \qquad (3\text{–}34)$$

It must be emphasized that Eqs. (3–33) and (3–34) hold only for conservative systems, while Eq. (3–32) can also be applied to linearly damped, non-conservative systems. A form of Eq. (3–32) is derived in Ill. Ex. 3.15 to show how the Lagrange equations can be derived.

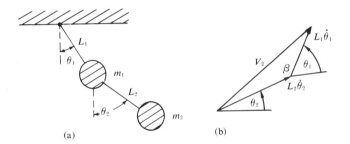

Fig. 3-7

It is easiest to first use a conservative system to illustrate the use of Lagrange's equations for obtaining the n equations of motion, because all that is needed is to write the total kinetic and potential energies for the system and then substitute these results in Eq. (3–33), the frictionless Lagrange equation. A classic example is the double pendulum shown in Fig. 3–7a, whose equations of motion are very tedious to obtain by Newton's second law. Since each mass can move independently of the other mass, the system has two degrees of freedom, and two generalized coordinates are needed to describe this system. Let the generalized coordinates be angles θ_1 and θ_2, as shown in Fig. 3–7a, and assume that these two angles have small maximum values. The kinetic energy of the system is given by

$$KE = \tfrac{1}{2}m_1 v_1^2 + \tfrac{1}{2}m_2 v_2^2$$

where v_1 and v_2 are the velocities of masses m_1 and m_2. Since $v_1 = L_1\dot{\theta}_1$ and $v_2 = v_1 \leftrightarrow L_2\dot{\theta}_2$, where $\leftrightarrow$ denotes a vector sum,

$$v_2 = L_1\dot{\theta}_1 \leftrightarrow L_2\dot{\theta}_2$$

The velocity component $L_1\dot{\theta}_1$ acts perpendicular to rod L_1 and the velocity component $L_2\dot{\theta}_2$ acts perpendicular to rod L_2. The vector equation for velocity v_2 is geometrically shown in Fig. 3–7b. Since $\beta = 180° - (\theta_1 - \theta_2) = 180° + (\theta_2 - \theta_1)$ and since $-\cos\beta = \cos(\theta_2 - \theta_1)$, use of the law of cosines gives

$$v_2^2 = L_1^2\dot{\theta}_1^2 + L_2^2\dot{\theta}_2^2 + 2(L_1\dot{\theta}_1)(L_2\dot{\theta}_2)\cos(\theta_2 - \theta_1)$$

Substitution of this result for v_2^2 into the KE-equation, after using the approximation that $\cos(\theta_2 - \theta_1) \approx 1$ for small values of θ_2 and θ_1, gives

$$KE = \tfrac{1}{2}m_1 L_1^2\dot{\theta}_1^2 + \tfrac{1}{2}m_2(L_1\dot{\theta}_1^2 + L_2\dot{\theta}_2^2 + 2L_1 L_2\dot{\theta}_1\dot{\theta}_2)$$
$$= \tfrac{1}{2}(m_1 + m_2)L_1^2\dot{\theta}_1^2 + \tfrac{1}{2}m_2 L_2^2\dot{\theta}_2^2 + m_2 L_1 L_2\dot{\theta}_1\dot{\theta}_2$$

Since $\partial(KE)/\partial\dot{\theta}_1 = (m_1 + m_2)L_1^2\dot{\theta}_1 + m_2 L_1 L_2\dot{\theta}_2$,

$$\frac{d}{dt}\left(\frac{\partial(KE)}{\partial\dot{\theta}_1}\right) = (m_1 + m_2)L_1^2\ddot{\theta}_1 + m_2 L_1 L_2\ddot{\theta}_2$$

Since $\partial(KE)/\partial\dot{\theta}_2 = m_2 L_2^2\dot{\theta}_2 + m_2 L_1 L_2\dot{\theta}_1$,

$$\frac{d}{dt}\left(\frac{\partial(KE)}{\partial\dot{\theta}_2}\right) = m_2 L_2^2\ddot{\theta}_2 + m_2 L_1 L_2\ddot{\theta}_1$$

Also note that $\partial(KE)/\partial\theta_1 = \partial(KE)/\partial\theta_2 = 0$. If mass m_1 is moved an angle θ_1, masses m_1 and m_2 are both raised a height $h_1 = L_1(1 - \cos\theta_1)$, and the potential energy of mass m_1 is $m_1 g L_1(1 - \cos\theta_1)$. If we also move mass m_2 and angle θ_2, then mass m_2 is raised a total height $h_2 = h_1 + L_2(1 - \cos\theta_2)$, and the total potential energy PE of the system is

$$PE = m_1 g L_1(1 - \cos\theta_1) + m_2 g[L_1(1 - \cos\theta_1) + L_2(1 - \cos\theta_2)]$$

Since $(1 - \cos\theta) \approx \theta^2/2$ when θ is small, we can use the approximation

$$PE = \tfrac{1}{2}(m_1 + m_2)g L_1\theta_1^2 + \tfrac{1}{2}m_2 g L_2\theta_2^2$$

Use of the previous equation gives

$$\frac{\partial(PE)}{\partial\theta_1} = (m_1 + m_2)g L_1\theta_1$$

$$\frac{\partial(PE)}{\partial\theta_2} = m_2 g L_2\theta_2$$

Since this system is frictionless, we can use Eq. (3–33) for the Lagrange equations. Since the generalized coordinates are θ_1 and θ_2 (i.e., $q_1 = \theta_1$ and $q_2 = \theta_2$), these equations can be written as

$$\frac{d}{dt}\left(\frac{\partial(\mathrm{KE})}{\partial\dot{\theta}_1}\right) - \frac{\partial(\mathrm{KE})}{\partial\theta_1} + \frac{\partial(\mathrm{PE})}{\partial\theta_1} = 0$$

$$\frac{d}{dt}\left(\frac{\partial(\mathrm{KE})}{\partial\dot{\theta}_2}\right) - \frac{\partial(\mathrm{KE})}{\partial\theta_2} + \frac{\partial(\mathrm{PE})}{\partial\theta_2} = 0$$

Substitution of the computed values of the terms in the previous two equations gives the following equations of motion for this double pendulum:

$$(m_1 + m_2)L_1\ddot{\theta}_1 + m_2L_2\ddot{\theta}_2 + (m_1 + m_2)g\theta_1 = 0 \qquad (3\text{-}35)$$
$$L_1\ddot{\theta}_1 + L_2\ddot{\theta}_2 + g\theta_2 = 0 \qquad (3\text{-}36)$$

where L_1 was factored out of the first Lagrange equation of motion to obtain Eq. (3-35), and m_2L_2 was factored out of the second Lagrange equation to obtain Eq. (3-36).

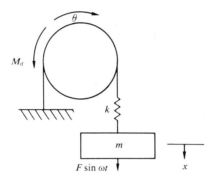

Fig. 3-8

Let us now use the two-degree-of-freedom system shown in Fig. 3-8, that has a linear damping torque M_d, for a relatively simple example of how Eq. (3-32), the non-conservative Lagrange equation, is applied. For this system, the equations of motion are more easily obtained by applying the $\Sigma F = m\ddot{x}$ (i.e., Newton's second law) and the $\Sigma M_g = I_g\ddot{\theta}$ equations to the mass and disk, respectively, which gives

$$m\ddot{x} = -k(x - x_1) + F\sin\omega t = k(r\theta - x) + F\sin\omega t$$
$$I_g\ddot{\theta} = k(x - x_1)r - c_t\dot{\theta} = krx - kr^2\theta - c_t\dot{\theta}$$

where r is the disk radius, $x_1 = r\theta$, and $c_t\dot{\theta}$ is the linear damping torque M_d on the disk that resists its rotational motion, and where it is assumed that there is no slipping between the rope and the disk. For the Lagrange equations, we shall use coordinates x and θ, rather than

x and x_1. Substitution of $q_1 = x$ and $q_2 = \theta$ into Eq. (3–32) gives us the following pair of Lagrange equations:

$$\frac{d}{dt}\frac{\partial(\text{KE})}{\partial\dot\theta} - \frac{\partial(\text{KE})}{\partial\theta} + \frac{\partial(\text{PE})}{\partial\theta} + \frac{\partial(\text{DE})}{\partial\dot\theta} = Q_1$$

$$\frac{d}{dt}\frac{\partial(\text{KE})}{\partial\dot x} - \frac{\partial(\text{KE})}{\partial x} + \frac{\partial(\text{PE})}{\partial x} + \frac{\partial(\text{DE})}{\partial\dot x} = Q_2$$

Since the total kinetic energy is that due to disk rotation plus that due to the translation of mass m,

$$\text{KE} = \tfrac{1}{2}I_g\dot\theta^2 + \tfrac{1}{2}m\dot x^2$$

Since $\partial(\text{KE})/\partial\dot\theta = I_g\dot\theta$ and $\partial(\text{KE})/\partial\dot x = m\dot x$, $d/dt[\partial(\text{KE})/\partial\dot\theta] = I_g\ddot\theta$ and $d/dt[\partial(\text{KE})/\partial\dot x] = m\ddot x$. Since KE does not vary with θ or x, $\partial(\text{KE})/\partial\theta = \partial(\text{KE})/\partial x = 0$. The potential energy of the system is due to the deformation $(x - r\theta)$ of the spring, and is given by

$$\text{PE} = \tfrac{1}{2}k(x - r\theta)^2$$

Thus, $\partial(\text{PE})/\partial\theta = -kr(x - r\theta)$ and $\partial(\text{PE})/\partial x = k(x - r\theta)$. Since the damping torque on the disk equals $c_t\dot\theta$, and since mass m has no damping force, the dissipated energy DE equals $c_t\dot\theta^2/2$. Thus, $\partial(\text{DE})/\partial\dot\theta = c_t\dot\theta$ and $\partial(\text{DE})/\partial\dot x = 0$. Since there is no external force or moment acting in the direction of θ, $Q_1 = 0$. Since the external force $F\sin\omega t$ acts in the direction of x, $Q_2 = F\sin\omega t$. Note that this force performs the work $dW = (F\sin\omega t)dx$ for an infinitesimal displacement dx. Substitution of our calculated terms into the two Lagrange equations gives the following equations of motion, which are equivalent to those derived by Newton's second law:

$$I_g\ddot\theta - kr(x - r\theta) + c_t\dot\theta = 0$$
$$m\ddot x + k(x - r\theta) = F\sin\omega t$$

Illustrative Example 3.13. Use the Lagrange equations to obtain the equations of motion for the spherical pendulum of length L that is shown in Fig. 3-9.

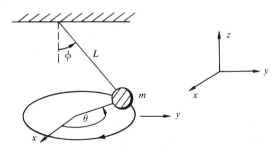

Fig. 3-9

Solution.   The total kinetic energy KE of this system is

$$KE = \tfrac{1}{2} m(\dot{x}^2 + \dot{y}^2 + \dot{z}^2)$$

This system has two degrees of freedom, and we shall let the two generalized coordinates be angles ϕ and θ, which are shown in Fig. 3-9. Note that another pair of generalized coordinates could be x and y, since z is related to x and y by the equation $x^2 + y^2 + z^2 = L^2$. Since Fig. 3-9 shows that the component of length L in the x-y-plane is $L \sin \phi$, $x = L \sin \phi \cos \theta$ and $y = L \sin \phi \sin \theta$. Figure 3-9 also shows that $z = L \cos \phi$. Computing $\dot{x}$, $\dot{y}$, and $\dot{z}$, and substituting the results in the KE-equation gives

$$KE = \tfrac{1}{2} m[L^2 \sin^2 \phi(\sin^2 \theta + \cos^2 \theta)\dot{\theta}^2 + L^2 \cos^2 \phi(\cos^2 \theta + \sin^2 \theta)\dot{\phi}^2$$
$$+ L^2 \dot{\phi}^2 \sin^2 \phi - 2\dot{\phi}\dot{\theta} \sin \phi \cos \phi \sin \theta \cos \theta + 2\dot{\phi}\dot{\theta} \sin \phi \cos \phi \sin \theta \cos \theta]$$
$$= \tfrac{1}{2} mL^2(\dot{\theta}^2 \sin^2 \phi + \dot{\phi}^2)$$

where $\dot{x} = -(L \sin \phi \sin \theta)\dot{\theta} + (L \cos \theta \cos \phi)\dot{\phi}$, $\cos^2 \theta + \sin^2 \theta = 1$, etc. The total potential energy PE of the system is given by

$$PE = mgL(1 - \cos \phi)$$

Since this system is frictionless, Eq. (3-33) may be used. Since $q_1 = \theta$ and $q_2 = \phi$, the Lagrange equations are

$$\frac{d}{dt} \frac{\partial(KE)}{\partial \dot{\theta}} - \frac{\partial(KE)}{\partial \theta} + \frac{\partial(PE)}{\partial \theta} = 0$$
$$\frac{d}{dt} \frac{\partial(KE)}{\partial \dot{\phi}} - \frac{\partial(KE)}{\partial \phi} + \frac{\partial(PE)}{\partial \phi} = 0$$

Substituting our equations for KE and PE into the previous two Lagrange equations, we obtain the following equations of motion for this spherical pendulum:

$$\ddot{\theta} \sin \phi + 2\dot{\theta}\dot{\phi} \cos \phi = 0$$
$$\ddot{\phi} - \dot{\theta}^2 \sin \phi \cos \phi + g \sin \phi/L = 0$$

Illustrative Example 3.14.   Use the Lagrange equations to obtain the equations of motion for the coupled-pendulum system that is shown in Fig. 3-10. Each pendulum is of length L.

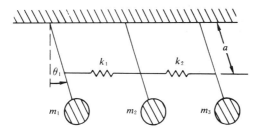

Fig. 3-10

Solution.   This is a three-degree-of-freedom system, and we shall let the three generalized coordinates be the oscillation angles, θ_1, θ_2, and θ_3, of the three masses. The kinetic energy KE of the system is

$$\text{KE} = \tfrac{1}{2} m_1 L^2 \dot{\theta}_1^2 + \tfrac{1}{2} m_2 L^2 \dot{\theta}_2^2 + \tfrac{1}{2} m_3 L^2 \dot{\theta}_3^2$$

Since the potential energy of the system is due to both the spring deformations and the change of the weight elevations, we have

$$\text{PE} = m_1 gL(1 - \cos \theta_1) + m_2 gL(1 - \cos \theta_2) + m_3 gL(1 - \cos \theta_3)$$
$$+ \tfrac{1}{2} k_1(a \sin \theta_2 - a \sin \theta_1)^2 + \tfrac{1}{2} k_2(a \sin \theta_3 - a \sin \theta_2)^2$$

Since this system is frictionless, substitution of $q_1 = \theta_1$, $q_2 = \theta_2$, and $q_3 = \theta_3$ into Eq. (3-33) gives the following three Lagrange equations:

$$\frac{d}{dt} \frac{\partial(\text{KE})}{\partial \dot{\theta}_1} - \frac{\partial(\text{KE})}{\partial \theta_1} + \frac{\partial(\text{PE})}{\partial \theta_1} = 0$$

$$\frac{d}{dt} \frac{\partial(\text{KE})}{\partial \dot{\theta}_2} - \frac{\partial(\text{KE})}{\partial \theta_2} + \frac{\partial(\text{PE})}{\partial \theta_2} = 0$$

$$\frac{d}{dt} \frac{\partial(\text{KE})}{\partial \dot{\theta}_3} - \frac{\partial(\text{KE})}{\partial \theta_3} + \frac{\partial(\text{PE})}{\partial \theta_3} = 0$$

Since $d/dt[\partial(\text{KE})/\partial \dot{\theta}_1] = m_1 L^2 \ddot{\theta}_1$, $\partial(\text{KE})/\partial \theta_1 = 0$, and $\partial(\text{PE})/\partial \theta_1 = m_1 gL \sin \theta_1 - k_1 a^2(\sin \theta_2 - \sin \theta_1)\cos \theta_1$, the first equation of motion is

$$m_1 L^2 \ddot{\theta}_1 + m_1 gL \sin \theta_1 - k_1 a^2(\sin \theta_2 - \sin \theta_1)\cos \theta_1 = 0$$

In like manner, the substitution of our equations for KE and PE in the last two Lagrange equations gives us the last two equations of motion, which are

$$m_2 L^2 \ddot{\theta}_2 + m_2 gL \sin \theta_2 + k_1 a^2(\sin \theta_2 - \sin \theta_1)\cos \theta_2$$
$$- k_2 a^2(\sin \theta_3 - \sin \theta_2)\cos \theta_2 = 0$$
$$m_3 L^2 \ddot{\theta}_3 + m_3 gL \sin \theta_3 + k_2 a^2(\sin \theta_3 - \sin \theta_2)\cos \theta_3 = 0$$

If the maximum values of θ_1, θ_2, and θ_3 are all small, we could simplify our equations of motion by use of the approximations $\sin \theta_i \approx \theta_i$ and $\cos \theta_i \approx \theta_i$.

Illustrative Example 3.15.   Derive the Lagrange equations.

Solution.   Let us assume a k-degree-of-freedom system that has n particles. The kinetic energy KE of this system is

$$\text{KE} = \sum_{j=1}^{n} \frac{m_j}{2} (\dot{x}_j^2 + \dot{y}_j^2 + \dot{z}_j^2)$$

Let the k generalized coordinates for this system be denoted by $q_1, q_2, \ldots, q_k$, and let us represent the relationship between the generalized coordinates and the (x, y, z)-coordinates by the following functional forms, where $f_{ji}(u, v)$ denotes a function of u and v.

$$x_j = f_{j1}(q_1, q_2, \ldots, q_k)$$
$$y_j = f_{j2}(q_1, q_2, \ldots, q_k)$$
$$z_j = f_{j3}(q_1, q_2, \ldots, q_k)$$

Differentiation of the x_j-equation with respect to time gives

$$\dot{x}_j = \frac{\partial f_{j1}}{\partial q_1}\dot{q}_1 + \frac{\partial f_{j1}}{\partial q_2}\dot{q}_2 + \cdots + \frac{\partial f_{j1}}{\partial q_n}\dot{q}_n = \sum_{i=1}^{k}\frac{\partial f_{j1}}{\partial q_i}\dot{q}_i$$

In like manner, $\dot{y}_j = \sum_{i=1}^{k}(\partial f_{j2}/\partial q_i)\dot{q}_i$ and $\dot{z}_j = \sum_{i=1}^{k}(\partial f_{j3}/\partial q_i)\dot{q}_i$. Substitution of these three equations for $\dot{x}_j$, $\dot{y}_j$, and $\dot{z}_j$ into the KE-equation gives

$$\text{KE} = \frac{1}{2}\sum_{j=1}^{n}m_j\left[\left(\sum_{i=1}^{k}\frac{\partial f_{j1}}{\partial q_i}\dot{q}_i\right)^2 + \left(\sum_{i=1}^{k}\frac{\partial f_{j2}}{\partial q_i}\dot{q}_i\right)^2 + \left(\sum_{i=1}^{k}\frac{\partial f_{j3}}{\partial q_i}\dot{q}_i\right)^2\right] \quad (3\text{-}37)$$

Since $\partial(uv)^2/\partial u = 2(uv)v$,

$$\frac{\partial(\text{KE})}{\partial \dot{q}_i}\dot{q}_i = \sum_{j=1}^{n}m_j\left[\frac{\partial f_{j1}}{\partial q_i}\dot{q}_i \sum_{i=1}^{k}\frac{\partial f_{ji}}{\partial q_i}\dot{q}_i\right.$$
$$\left. + \frac{\partial f_{j2}}{\partial q_i}\dot{q}_i \sum_{i=1}^{k}\frac{\partial f_{j2}}{\partial q_i}\dot{q}_i + \frac{\partial f_{j3}}{\partial q_i}\dot{q}_i \sum_{i=1}^{k}\frac{\partial f_{j3}}{\partial q_i}\dot{q}_i\right]$$

Since the sum of all these $[\partial(\text{KE})/\partial q_i]\dot{q}_i$-terms is equal to twice the right-hand side of Eq. (3-37), we have the relation

$$2(\text{KE}) = \sum_{i=1}^{k}\frac{\partial(\text{KE})}{\partial \dot{q}_i}\dot{q}_i$$

Differentiation of the previous equation with respect to time gives

$$2\frac{d(\text{KE})}{dt} = \sum_{i=1}^{k}\left\{\frac{d}{dt}\left[\frac{\partial(\text{KE})}{\partial \dot{q}_i}\right]\dot{q}_i + \frac{\partial(\text{KE})}{\partial \dot{q}_1}\frac{d\dot{q}_i}{dt}\right\}$$

Since KE is a function of q_i and $\dot{q}_i$, where $i = 1$ to k, we can use the chain rule to obtain

$$\frac{d(\text{KE})}{dt} = \sum_{i=1}^{k}\left[\frac{\partial(\text{KE})}{\partial q_i}\frac{dq_i}{dt} + \frac{\partial(\text{KE})}{\partial \dot{q}_i}\frac{d\dot{q}_i}{dt}\right]$$

Subtraction of the two previous equations, and a multiplication by dt, gives

$$d(\text{KE}) = \sum_{i=1}^{k}\left\{\frac{d}{dt}\left[\frac{\partial(\text{KE})}{\partial \dot{q}_i}\right]dq_i - \frac{\partial(\text{KE})}{\partial q_i}dq_i\right\}$$

From virtual work principles, the infinitesimal work dW is as follows, where Q_i is a generalized force that has the same direction as displacement q_i:

$$dW = Q_1 dq_1 + Q_2 dq_2 + \cdots + Q_k dq_k = \sum_{i=1}^{k}Q_i dq_i$$

From our dynamics course, we have the work-energy equation $dW = d(\text{KE})$. Thus, we can equate the two previous equations for $d(\text{KE})$ and dW to obtain

$$\sum_{i=1}^{k}\left[\frac{d}{dt}\frac{\partial(\text{KE})}{\partial \dot{q}_i} - \frac{\partial(\text{KE})}{\partial q_i} - Q_i\right]dq_i = 0$$

Since the term within the brackets must equal zero, we have

$$\frac{d}{dt}\frac{\partial(\text{KE})}{\partial \dot{q}_i} - \frac{\partial(\text{KE})}{\partial q_i} = Q_i \quad (3\text{-}38)$$

Since $i = 1$ to k, there are k Lagrange equations of motion. The generalized force Q_i in Eq. (3-38) is the sum of the spring, damping, and external forces.

If Q_{i1} denotes an elastic or potential force (e.g., a spring force), then $Q_{i1} = -\partial(\text{PE})/\partial q_i$. If Q_{i2} denotes a linear damping force (i.e., the damping force is directly proportional to velocity $\dot{q}_i$), then $Q_{i2} = -\partial(\text{DE})/\partial \dot{q}_i$ where DE is the dissipated energy. Let Q_{i3} denote the sum of the external and nonlinear damping forces in the direction of q_i. Since $Q_i = Q_{i1} + Q_{i2} + Q_{i3}$, (Eq. 3–38), with its general definition of the Q_i-forces, is equivalent to Eq. (3–32), the non-conservative Lagrange equation.

3.8   Influence Coefficients

So far in this chapter, we have used Newton's second law and the Lagrange equations to obtain the equations of motion for multiple-degree-of-freedom systems. A third method is to use influence coefficients. An *influence coefficient* α_{ij} is defined to be the static deflection of the system at point i due to a unit force applied at point j when this unit force is the only force applied to the system. For rotational systems, α_{ij} is the angular displacement at location i due to a unit torque applied at location j. Since the values of the influence coefficients are easy to obtain experimentally by applying a unit load at point j and measuring the deflections at several i-locations, the influence-coefficient method is widely used in the analysis of complex elastic structures, especially those which are discontinuous. Since an influence coefficient is a static elastic property, we must include the inertia forces, as well as the externally applied forces, when influence coefficients are used to obtain the equations of motion of a dynamic system. As an example, suppose that we had a linear, three-degree-of-freedom system that is undergoing a free vibration at a principal mode whose frequency is ω. Because the inertia force $-m_i \ddot{x}_i$ for mass m_i equals $m_i \omega^2 x_i$ for a principal-mode vibration (where x_i is the deflection of mass m_i), and using the definition for influence coefficient α_{ij}, the deflections of the three masses are given by

$$x_1 = (m_1 \omega^2 x_1)\alpha_{11} + (m_2 \omega^2 x_2)\alpha_{12} + (m_3 \omega^2 x_3)\alpha_{13} \qquad \text{(3–39a)}$$
$$x_2 = (m_1 \omega^2 x_1)\alpha_{21} + (m_2 \omega^2 x_2)\alpha_{22} + (m_3 \omega^2 x_3)\alpha_{23} \qquad \text{(3–39b)}$$
$$x_3 = (m_1 \omega^2 x_1)\alpha_{31} + (m_2 \omega^2 x_2)\alpha_{32} + (m_3 \omega^2 x_3)\alpha_{33} \qquad \text{(3–39c)}$$

which are a new form of the equations of motion. After a division by ω^2, the previous equations can be rearranged to the following form:

$$\left(\alpha_{11} m_1 - \frac{1}{\omega^2}\right) x_1 + (\alpha_{12} m_2) x_2 + (\alpha_{13} m_3) x_3 = 0$$

$$(\alpha_{21} m_1) x_1 + \left(\alpha_{22} m_2 - \frac{1}{\omega^2}\right) x_2 + (\alpha_{23} m_3) x_3 = 0$$

$$(\alpha_{31} m_1) x_1 + (\alpha_{32} m_2) x_2 + \left(\alpha_{33} m_3 - \frac{1}{\omega^2}\right) x_3 = 0$$

The previous three algebraic equations have a non-trivial solution for displacements x_1, x_2, and x_3 only if the determinant of the coefficients of x_1, x_2, and x_3 vanishes, as follows.

$$\begin{vmatrix} \alpha_{11}m_1 - \dfrac{1}{\omega^2} & \alpha_{12}m_2 & \alpha_{13}m_3 \\[2mm] \alpha_{21}m_1 & \alpha_{22}m_2 - \dfrac{1}{\omega^2} & \alpha_{23}m_3 \\[2mm] \alpha_{31}m_1 & \alpha_{32}m_2 & \alpha_{33}m_3 - \dfrac{1}{\omega_2} \end{vmatrix} = 0 \qquad (3\text{-}40)$$

The previous determinant equation is a characteristic equation of the system since its expansion results in a polynomial of degree three in $1/\omega^2$, from which we can solve for the three natural frequencies of the system. It should be rather easy for the reader to rewrite Eqs. (3–39a, b, c) and (3–40) for an n-degree-of-freedom system.

A very useful influence-coefficient property is *Maxwell's reciprocity theorem* which states that $\alpha_{ij} = \alpha_{ji}$ for an elastic system, and which can be proved by strain energy (i.e., stored work due to elastic deformation) concepts. There are n^2 influence coefficients for an n-degree-of-freedom-system, but there are only $(n^2 + n)/2$ different values since $\alpha_{ji} = \alpha_{ij}$. It can also be shown that $\alpha_{ii} = 1/k_i$, where k_i is the elastic stiffness (i.e., force per unit deformation) at point i.

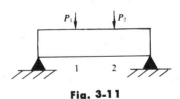

Fig. 3-11

Let us use strain energy methods to prove Maxwell's reciprocity theorem for the system shown in Fig. 3–11, which consists of a simply-supported beam with two concentrated loads P_1 and P_2. Since the beam can be loaded in two steps, assume that load P_1 is applied first at station 1 and then load P_2 is applied later at station 2. When load P_1 is applied alone at station 1, the influence coefficients are α_{11} and α_{21}, and the strain potential energy equals $(P_1^2 \alpha_{11})/2$. When load P_2 is applied after P_1, there will be an additional deflection $P_2\alpha_{12}$ at station 1 due to load P_2, and the work done by P_1 during this additional deflection is $P_1(P_2\alpha_{12})$. Since the strain potential energy for the elastic deformation at station 2 is $(P_2^2 \alpha_{22})/2$, the total potential energy PE in the system is

$$\text{PE} = \tfrac{1}{2}P_1^2 \alpha_{11} + P_1(P_2\alpha_{12}) + \tfrac{1}{2}P_2^2 \alpha_{22}$$

If we now apply load P_2 first at station 2 and then load P_1 at station 1, it can be shown that the total potential energy of the system is given by

$$\text{PE} = \tfrac{1}{2}P_2^2\alpha_{22} + P_2(P_1\alpha_{21}) + \tfrac{1}{2}P_1^2\alpha_{11}$$

where the $(P_2^2\alpha_{22})/2$ term is due to the application of load P_2, and the last two terms are due to the application of load P_1. Since the final states of this beam are the same for both loading procedures, the two total potential energies must be equal. Equating these two potential energy equations, we obtain $\alpha_{21} = \alpha_{12}$. Since we can obtain a similar relationship when several loads act on this system,

$$\alpha_{ij} = \alpha_{ji} \qquad (3\text{--}41)$$

which holds for any linear elastic system. We could also derive this result for this beam system by using the deflection equation for a simply-supported beam from our strength of materials course.

Illustrative Example 3.16. Determine the equation for the influence coefficients for a beam of length L that is simply supported at both ends.

Solution. From our strength of materials course, the deflection equation for this simply-supported beam is as follows if $x < a$:

$$y = \frac{Px(L-a)}{6EIL}[L^2 - x^2 - (L-a)^2]$$

where y is the vertical deflection at location x of this simply-supported beam due to a concentrated load P at location a. Both x and a are measured from the left end of the beam. Letting $P = 1.0$ lb, letting station i denote location x, and letting station j denote location a, the following equation gives us the influence coefficients for any pair of (x_i, x_j)-locations on this beam if $x_i < x_j$. We can use other formulas from our strength of materials course to obtain the influence coefficients for other simple elastic structures.

$$\alpha_{ij} = \frac{x_i(L-x_j)}{6EIL}[L^2 - x_i^2 - (L-x_j)^2] \qquad (3\text{--}42)$$

Illustrative Example 3.17. Suppose that a vertical external force $F\sin\omega t$ is applied to mass m_2 in the system shown in Fig. 3-2c, which consists of two masses attached to an elastic string. Write the equations of motion for this system using influence coefficients.

Solution. Since the force on mass m_1 is the inertia force $-m_1\ddot{x}_1$, and since the force on mass m_2 is $(F\sin\omega t - m_2\ddot{x}_2)$, the equations of motion are

$$x_1 = (-m_1\ddot{x}_1)\alpha_{11} + (F\sin\omega t - m_2\ddot{x}_2)\alpha_{12}$$
$$x_2 = (-m_1\ddot{x}_1)\alpha_{21} + (F\sin\omega t - m_2\ddot{x}_2)\alpha_{22}$$

where it should be noted that the $m_i g$ weight forces are balanced by the initial string deflection and that the x_i-deflections are measured from the static equilibrium position.

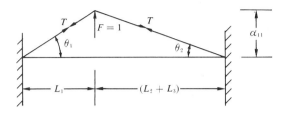

Fig. 3-12

We shall calculate the α_{ij}-influence-coefficients assuming that string tension T is constant, and that the maximum deflections are small. Figure 3-12 shows the free-body diagram when a unit force is applied to mass m_1 at station 1. We know that the deflection at station 1 is α_{11} and that the deflection at station 2 is α_{21} by virtue of the definition for α_{ij}. The lengths of the three string segments in Fig. 3-2c are denoted by L_1, L_2, and L_3. Since the vertical components of tension T must equal the unit force F, a vertical force balance gives

$$F = 1 = T\left(\frac{\alpha_{11}}{L_1}\right) + T\left(\frac{\alpha_{11}}{L_2 + L_3}\right)$$

where $\alpha_{11}/L_1 = \tan\theta_1 \approx \sin\theta_1$ and $\alpha_{11}/(L_2 + L_3) = \tan\theta_2 \approx \sin\theta_2$. Solving this equation for α_{11}, we have

$$\alpha_{11} = \frac{L_1}{T}\left(\frac{L_2 + L_3}{L_1 + L_2 + L_3}\right) = \frac{L_1}{LT}(L_2 + L_3)$$

where $L = (L_1 + L_2 + L_3)$ is the total string length. Use of similar triangles in Fig. 3-12 gives

$$\alpha_{21} = \frac{L_1}{T}\left(\frac{L_3}{L_1 + L_2 + L_3}\right) = \frac{L_1 L_3}{LT}$$

By Maxwell's reciprocity theorem, $\alpha_{12} = \alpha_{21} = (L_1 L_3)/(LT)$. In a similar manner, we can apply a unit vertical force on mass m_2 and use the same vertical-force-balance procedure to obtain $\alpha_{22} = L_3(L_1 + L_2)/(LT)$. The last step is to substitute these calculated α_{ij}-values in the previous equations of motion.

Illustrative Example 3.18.   Determine the influence coefficients for the three-degree-of-freedom, spring-mass system shown in Fig. 3-3 when spring k_4 is removed.

Solution.   When a downward unit force is applied to mass m_1, the spring of stiffness k_1 will stretch a distance $1/k_1$. Thus, $\alpha_{11} = 1/k_1$. Since the other two springs are not stretched, masses m_2 and m_3 move in rigid fashion with mass m_1. Thus, $\alpha_{21} = \alpha_{31} = \alpha_{11}$ since masses m_2 and m_3 move the same distance as m_1. Using Maxwell's reciprocity theorem, $\alpha_{12} = \alpha_{21}$ and $\alpha_{13} = \alpha_{31}$. So far, we have the following results:

$$\alpha_{11} = \alpha_{12} = a_{13} = \alpha_{21} = \alpha_{31} = \frac{1}{k_1} \qquad (3\text{-}43)$$

To calculate α_{22} and α_{32}, we apply a unit downward force on mass m_2. The springs k_1 and k_2, which are in series, will deflect a total distance $(1/k_1 + 1/k_2)$. Thus, $\alpha_{22} = (1/k_1 + 1/k_2)$. Because mass m_3 moves the same distance as mass m_2 for this loading, and because of Maxwell's reciprocity theorem, $\alpha_{32} = \alpha_{23} = \alpha_{22}$, and

$$\alpha_{22} = \alpha_{23} = \alpha_{32} = \frac{1}{k_1} + \frac{1}{k_2} = \frac{k_1 + k_2}{k_1 k_2} \qquad (3\text{-}44)$$

When a unit force is applied to mass m_3, all three springs, which are in series, move a total distance α_{33} which is

$$\alpha_{33} = \frac{1}{k_1} + \frac{1}{k_2} + \frac{1}{k_3} = \frac{k_1 k_2 + k_1 k_3 + k_2 k_3}{k_1 k_2 k_3} \qquad (3\text{-}45)$$

3.9   Generalized Coordinates, Coordinate Coupling, and Principal Coordinates

As stated several times before, an n-degree-of-freedom system can be defined mathematically by n independent coordinates, which can be lengths, angles, etc. If $z = f(x, y)$, as shown in Ill. Ex. 3.13 for a spherical pendulum, then only two of the three x, y, z-coordinates are independent. It is sometimes possible to spectify the configuration of a system by more than one set of independent coordinates, and any such set of coordinates is called *generalized coordinates*. As an example, in Ill. Ex. 3.13 of Sec. 3.7 we saw that the two-degree-of-freedom, spherical pendulum in Fig. 3–9 may be specified by coordinates (θ, ϕ), $(x, y), (x, z), (y, z)$, etc. As another example, consider the system in Fig. 1–6 of Sec. 1.2 which consists of a rigid bar that is supported by springs k_1 and k_2. Let x_i denote the vertical displacement of point i on the bar, and let there be m such points. Thus, the motion of this two-degree-of-freedom system can be specified by $(x_1, \theta), (x_2, \theta), \cdots, (x_m, \theta)$, $(x_1, x_2), (x_1, x_3), \cdots, (x_1, x_m), (x_2, x_3), (x_2, x_4)$, etc., where θ is the angle of bar rotation. The equations of motion when using coordinates (x, θ), where x is the vertical displacement at the center of gravity of the bar are given in Ill. Ex. 3.4 in Sec. 3.2.

It is seen that these two equations of motion are not independent of each other (i.e., they are coupled since each equation varies with x and θ), and that the coupling term is coefficient $(k_1 L_1 - k_2 L_2)$. Since two independent differential equations will result (i.e., each equation has only one unknown), the rectilinear and rotational motions are uncoupled only if we locate the center of gravity such that $k_1 L_1 = k_2 L_2$. The physical effect of coupling is that a force transmitted through a coupling spring or damper will cause a vibration in one part of a system to induce a vibration in another part of the system. There are two types of coupling. *Static coupling* is due to static displacements, while *dynamic coupling* is due to inertia forces. The equations of motion in Ill. Ex. 3.4 show that a (x, θ)-coordinate specification results in a static coupling because the first equation contains a $b\theta$-term and the second equation contains a bx-term, where $b = -(k_1L_1 - k_2L_2)$, but no dynamic coupling. If we use a (x_1, θ)-coordinate specification, where x_1 is located on the bar at a distance from the center of gravity, then dynamic coupling results because the equation of motion will contain a moment $m\ddot{x}_1e$ and a force $me\ddot{\theta}$. That is, the moment-balance equation of motion has the inertia-moment term $m\ddot{x}_1e$, and the force-balance equation of motion has the inertia-force term $me\ddot{\theta}$. The dynamic coupling term for this (x_1, θ)-coordinate-pair is coefficient me.

It is possible to choose a set of generalized coordinates for a system such that each equation of motion is independent of the other equations. Such a set of coordinates is called *principal coordinates*. That is, each differential equation of motion, when written in terms of its principal coordinates, has no static or dynamic coupling, and can be solved independently of the other equations since each equation has only one dependent or unknown variable. Principal coordinates are further exemplified in Ill. Ex. 3.19.

Illustrative Example 3.19.   Determine the principal coordinates of the spring-mass system in Fig. 3-1 of Sec. 3.2 when $k_1 = k = k_2 \equiv k$ and $m_1 = m_2 \equiv m$.

Solution.   Substitution in Eqs. (3-1a, b) in Sec. 3.2 gives the following equations of motion:

$$m\ddot{x}_1 + 2kx_1 - kx_2 = 0$$
$$m\ddot{x}_2 + 2kx_2 - kx_1 = 0$$

Since the addition of these two equations gives

$$m(\ddot{x}_1 + \ddot{x}_2) + k(x_1 + x_2) = 0$$

one principal coordinate is $q_1 = (x_1 + x_2)$. Since subtraction of the equations of motion gives

$$m(\ddot{x}_1 - \ddot{x}_2) + 3k(x_1 - x_2) = 0$$

the other principal coordinate is $q_2 = (x_1 - x_2)$. Using Eqs. (1-2) and (1-3) of Sec. 1.3, the solutions for q_1 and q_2 are

$$q_1 \equiv x_1 + x_2 = A \cos \sqrt{\frac{k}{m}}\, t + B \sin \sqrt{\frac{k}{m}}\, t$$

$$q_2 \equiv x_1 - x_2 = E \cos \sqrt{\frac{3k}{m}}\, t + F \sin \sqrt{\frac{3k}{m}}\, t$$

and the two natural frequencies of this system are $\omega_1 = \sqrt{k/m}$ and $\omega_2 = \sqrt{3k/m}$.

3.10   Orthogonality of the Principal Modes of Vibration

Consider a linear elastic system that has n degrees of freedom, and hence, n natural frequencies and n principal modes. The *orthogonality relation* between the principal modes of vibration (which is not proved in this text) is as follows for an n-degree-of-freedom system:

$$\sum_{i=1}^{n} m_i A_i^{(r)} A_i^{(s)} = 0 \qquad (3\text{--}46)$$

where $r \neq s$, $A_i^{(r)}$ are the amplitudes corresponding to the rth mode, and $A_i^{(s)}$ are the amplitudes corresponding to the sth mode. This orthogonality relation holds for oscillatory systems that possess static coupling, but *not dynamic coupling*. It was shown in Sec. 3.9 that it is possible to choose a set of n generalized coordinates such that the resulting equations of motion will not have dynamic coupling. Section 4.9 shows the usefulness of the orthogonality relation for calculating the higher natural frequencies of a system. For a two-degree-of-freedom system, the orthogonality relation is

$$m_1 A_1 A_2 + m_2 B_1 B_2 = 0 \qquad (3\text{--}47)$$

where A_1 and B_1 are the amplitudes of masses m_1 and m_2 during the first vibration mode, as shown by Eqs. (3–6a, b) in Sec. 3.2, and where A_2 and B_2 are the amplitudes of masses m_1 and m_2 during the second vibration mode. For a three-degree-of-freedom system, the orthogonality relations are as follows, where A_i, B_i, and C_i denote the amplitudes of the three masses during the ith principal mode of vibration.

$$m_1 A_1 A_2 + m_2 B_1 B_2 + m_3 C_1 C_2 = 0 \qquad (3\text{--}48\text{a})$$
$$m_1 A_1 A_3 + m_2 B_1 B_3 + m_3 C_1 C_3 = 0 \qquad (3\text{--}48\text{b})$$
$$m_1 A_2 A_3 + m_2 B_2 B_3 + m_3 C_2 C_3 = 0 \qquad (3\text{--}48\text{c})$$

Illustrative Example 3.20.   Show that the orthogonality relation holds for the spring-mass system in Ill. Ex. 3.1 in Sec. 3.2.

Solution.   In Ill. Ex. 3.1, we see that $m_1 = 0.100$, $m_2 = 0.100$, $A_1 = 1$, $A_2 = 1$, $B_1 = 1$, and $B_2 = -1$. Substitution in Eq. (3–47) gives

$$(0.100)(1)(1) + (0.100)(1)(-1) = 0$$

3.11   Numerical Solution of Nonlinear Vibrations

Simultaneous, nonlinear ordinary differential equations can be solved by a d-c electronic differential analyzer type of analog computer, as shown in Chap. 6, and by numerical methods on a digital computer.

In this section, we shall discuss how such nonlinear differential equations may be solved numerically. As a physical example of a two-degree-of-freedom nonlinear vibration, suppose that the spring-mass system in Fig. 3–1 in Sec. 3.2 also had a damper between mass m_1 and the ceiling and another damper between mass m_2 and the floor. Also suppose that each damper is nonlinear, where the magnitude of the damping force F_d equals $a|\dot{x}|^b \dot{x}$. Thus, the equations of motion are both nonlinear and are as follows:

$$m_1 \ddot{x}_1 = -k_1 x_1 - k(x_1 - x_2) - a_1|\dot{x}_1|^b \dot{x}_1$$
$$m_2 \ddot{x}_2 = k(x_1 - x_2) - k_2 x_2 - a_2|\dot{x}_2|^b \dot{x}_2$$

Let us solve these pair of nonlinear differential equations numerically using the Euler and trapezoidal methods, which are defined by Eqs. (2–90) and (2–93) of Sec. 2.13. Using the equations of motion and the Euler and trapezoidal methods, the sequence of calculations is as follows:

$$\ddot{x}_{1_t} = [-k_1 x_{1_t} - k(x_{1_t} - x_{2_t}) - a_1|\dot{x}_{1_t}|^b \dot{x}_{1_t}]/m_1$$
$$\ddot{x}_{2_t} = [k(x_{1_t} - x_{2_t}) - k_2 x_{2_t} - a_2|\dot{x}_{2_t}|^b \dot{x}_{2_t}]/m_2$$
$$\dot{x}_{1_{t+\Delta t}} = \dot{x}_{1_t} + \Delta t (\ddot{x}_{1_t})$$
$$\dot{x}_{2_{t+\Delta t}} = \dot{x}_{2_t} + \Delta t (\ddot{x}_{2_t})$$
$$x_{1_{t+\Delta t}} = x_{1_t} + \frac{\Delta t}{2} (\dot{x}_{1_t} + \dot{x}_{1_{t+\Delta t}})$$
$$x_{2_{t+\Delta t}} = x_{2_t} + \frac{\Delta t}{2} (\dot{x}_{2_t} + \dot{x}_{2_{t+\Delta t}})$$

where the value of time-interval Δt can be established using the interval-halving method, which is discussed in Sec. 2.14. As noted in Secs. 2.13 and 2.14, for a vibration problem such as this one, it is preferable to use the midpoint-slope or Adams method, which are defined by Eqs. (2–29) and (2–101), instead of the Euler method. A much more accurate numerical procedure would be to solve for $\dot{x}_{1_{t+\Delta t}}$, $\dot{x}_{2_{t+\Delta t}}$, $x_{1_{t+\Delta t}}$, and $x_{2_{t+\Delta t}}$ by the Runge-Kutta method.

As can be seen from the calculation sequence for this example, the solution of simultaneous, ordinary differential equations by most numerical methods (e.g., Euler, trapezoidal, midpoint-slope, and Adams) is rather straightforward. There is an exception when the Runge-Kutta method is used. Suppose that we wanted to solve the following pair of first-order ordinary differential equations written in functional form

$$\frac{dy_1}{dt} = F(t, y_1, y_2)$$
$$\frac{dy_2}{dt} = G(t, y_1, y_2)$$

If we solved this pair of ordinary differential equations by the second-order-accuracy *Runge-Kutta method*, the sequence of calculations would be

$$a_1(t) = (\Delta t)F[t, y_1(t), y_2(t)]$$
$$b_1(t) = (\Delta t)G[t, y_1(t), y_2(t)]$$
$$a_2(t) = (\Delta t)F[t + \Delta t, y_1(t) + a_1(t), y_2(t) + b_1(t)]$$
$$b_2(t) = (\Delta t)G[t + \Delta t, y_1(t) + a_1(t), y_2(t) + b_1(t)]$$
$$y_1(t + \Delta t) = y_1(t) + \tfrac{1}{2}[a_1(t) + a_2(t)]$$
$$y_2(t + \Delta t) = y_2(t) + \tfrac{1}{2}[b_1(t) + b_2(t)]$$

These previous equations should be compared with Eqs. (2–105a, b, c) which solve a single ordinary differential equation by this same method. From this comparison, it should not be difficult for the reader to see how simultaneous, ordinary differential equations are solved by the third- and fourth-order-accuracy Runge-Kutta methods.

PROBLEMS

3.1. Repeat Ill. Ex. 3.1 for the initial conditions $x_1(0) = \dot{x}_1(0) = \dot{x}_2(0) = 0$ and $x_2(0) = 2$ in.

3.2. Repeat Ill. Ex. 3.1 for the initial conditions $x_1(0) = x_2(0) = \dot{x}_2(0) = 0$ and $\dot{x}_1(0) = 3$ in./sec.

3.3. Solve Probs. 3.1 and 3.2 using Laplace transforms.

3.4. For the spring-mass system in Fig. 3–1, suppose that the initial velocities were both zero and that the initial displacements were both non-zero but equal in value. Determine the displacement equations for the two masses and state any conclusions that you may wish to make.

3.5. Determine the initial conditions so that the spring-mass system in Fig. 3–1 will vibrate at the first principal mode and at the second principal mode.

3.6. For the spring-mass system in Fig. 3–1, suppose that $m_1 = 7.72$ lb, $m_2 = 3.86$ lb, and that k_1, k_2, and $k = 2$, 7, and 3 lb/in., respectively. If $x_1(0) = 2$ in., and if $x_2(0) = \dot{x}_1(0) = \dot{x}_2(0) = 0$, determine the displacement equations for the two masses.

3.7. Solve Prob. 3.6 using the classical method for solving simultaneous, ordinary differential equations.

3.8. In Fig. 3–2c, suppose that $m_1 = m_2$ and that the three elastic string lengths all equal L. Write the equations of motion for this two-degree-of-freedom system. Use the analogy between these equations of motion and Eqs. (3–1a,b) to determine the characteristic equation from Eq. (3–4) and the principal-mode amplitude ratios from Eqs. (3–7a,b).

3.9. Write the equations of motion for the three, two-degree-of-freedom systems shown in Fig. 3–13. Determine the characteristic equation and the principal-mode amplitude ratios using the analogy method given in Prob. 3.8.

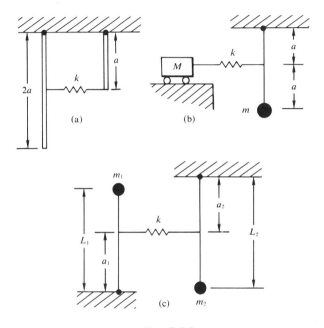

Fig. 3-13

3.10. Determine the charatceristic equation and the principal-mode amplitude ratios for the three-degree-of-freedom, disk-shaft system shown in Fig. 3–14. What is the lowest natural frequency? Draw the equivalent semi-definite, spring-mass system.

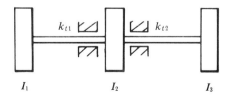

Fig. 3-14

3.11. For the disk-shaft system in Fig. 3–8, suppoe that all three disks had the same moment of inertia I and that all three shaft portions had the same spring constant k_t. Determine the three natural frequencies, the principal-mode amplitude ratios, and the angular displacement equations for the three disks.

3.12. For the semi-definite, disk-shaft system in Fig. 3–5, the moments of inertia of the two disks are 3000 and 6000 in.–lb–sec². Determine the shaft spring constant that would be necessary so that one natural frequency equals 4 cycles/sec.

3.13. A small heavy vehicle weighs 2 tons and has a 20–ft wheelbase. Its moment of inertia about the center of gravity is 5000 in.–lb–sec², and its center of gravity is located 8 ft from the front axle. Find the two natural frequencies and the $x(t)$- and $\theta(t)$-vehicle-motions if the front and rear springs have spring constants of 300 and 360 lb/in. respectively.

3.14. Determine the characteristic equation for the three-degree-of-freedom, spring-mass system shown in Fig. 3–15. Determine the three natural frequencies if all masses have the same weight and if all springs have the same spring constant.

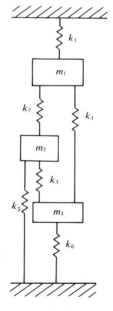

Fig. 3-15

3.15. Write the equations of motion and determine the characteristic equation for the branched spring-mass system shown in Fig. 3–16.

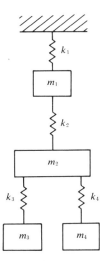

Fig. 3-16

3.16. Suppose that a mass M of cross-sectional area A is floating in water with a submergence depth h, where this floating mass supports a second mass m of volume V which is submerged in the water. The two masses, which are vertically aligned, are connected by a spring whose spring constant is k. Write the equations of motion and determine the characteristic equation and the principal-mode amplitude ratios for this two-degree-of-freedom system.

3.17. Two circular disks, whose masses are m_1 and m_2 and whose radii are r_2 and r_1 respectively, are connected by a spring of spring constant k, and roll without slipping on a horizontal floor. If this spring is connected to the centers of the two disks, determine the characteristic equation and the natural frequencies for this two-degree-of-freedom system.

3.18. Use the Newton-Raphson method and the quadratic formula to obtain the three roots of the following polynomial equation. To obtain your starting values for the Newton-Raphson method, use the Bernoulli method for four steps (i.e., calculate G_1 to G_4 and G_4/G_3).

$$x^3 - 6x^2 - 13x - 24 = 0$$

3.19. Use the Newton-Raphson method and the quadratic formula to obtain the four roots of the following polynomial equation.

$$x^4 + 2x^3 - 6x^2 - 18x - 27 = 0$$

3.20. Use the Lin method to obtain the four roots of the polynomial equation in Prob. 3.19.

3.21. Determine the equation for the amplitude ratios for the damped spring-mass system in Ill. Ex. 3.11.

3.22. Consider the two-degree-of-freedom, spring-mass system in Fig. 3–1, where the lower spring k_2 and the lower support are removed. Also suppose that the upper support is given the harmonic displacement A sin ωt. Use the mechanical impedance method to find the steady-state displacements of the two masses.

3.23. Use the mechanical impedance method to determine the steady-state displacements $x(t)$ and $x_1(t)$ for the system in Fig. 3–17, which has an elastically supported damper.

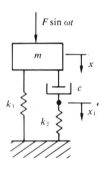

Fig. 3-17

3.24. In Fig. 3–3, suppose that an applied force F sin ωt is applied to mass m_1. Use the mechanical impedance method to find the steady-state displacements of the three masses.

3.25. In Fig. 3–4, suppose that a harmonic torque T sin ωt is applied to disk I_3. Use the mechanical impedance method to find the steady-state angular displacements of the three disks.

3.26. Suppose that the damped system discussed in Sec. 3.5 also had a damper c_1 between mass m_1 and the upper support and a damper c_2 between mass m_2 and the lower support. Rewrite Eqs. (3-26a, b), (3-27a,b), and (3-28a,b,c,) to hold for this new system.

3.27. The two-degree-of-freedom system in Fig. 3–18 consists of two simple pendulums, each of mass m and length L, where one pendulum has an exponential applied force $Fe^{i\omega t}$ and the motion of the other pendulum is resisted by a damper c. Use the mechanical impedance method to find the characteristic equation and the complex amplitudes of the two masses. Assume small angular amplitudes for the two pendulums.

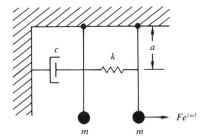

Fig. 3-18

3.28. Suppose that the machine-dynamic-absorber system in Fig. 3–6 had a damper c between masses m_1 and m_2. Use the mechanical impedance method to find the characteristic equation and the complex amplitudes of the two masses.

3.29. For the machine-dynamic-absorber system in Fig. 3–6, suppose that the machine weighed 77.2 lb and that the applied frequency were 600 cycles/sec. Specify the weight and the spring constant ratio k/k_1 for the dynamic absorber so that the nearest natural frequency is at least 25 per cent from the impressed frequency.

3.30. A concentrated mass m slides outward along a rod that is rotating clockwise at angular velocity ω about one end and in a vertical plane. Use the Lagrange equation to obtain the equation of motion for this system.

3.31. Use the Lagrange equations to obtain the equations of motion for small oscillations of the two systems shown in Fig. 3–19.

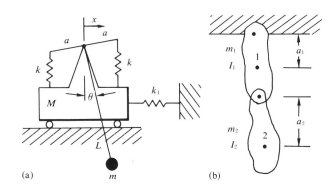

Fig. 3-19

3.32. For the double pendulum in Fig. 3–19b, suppose that the lower compound pendulum is replaced by a simple pendulum of length

L_2 and mass m_2 that is attached to the bottom of the upper compound pendulum. Use the Lagrange equations to obtain the equations of motion for the system for small oscillations.

3.33. Obtain the equations of motion of the two systems shown in Fig. 3–20 using the Lagrange equations.

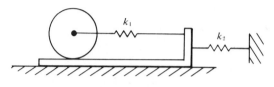

Fig. 3-20

3.34. Suppose that a pendulum of mass m and length L is attached to the edge of a circular disk of radius R that is rotating horizontally about a vertical axis with constant angular velocity ω. Obtain the equations of motion for this system using the Lagrange equations.

3.35. Suppose that we had four masses that were equally spaced on an elastic string of length L that was horizontal and tied to a support at both ends. Compute the influence-coefficient matrix for this system.

3.36. Use strength of materials theory to compute the influence coefficients for a cantilever beam that has a mass at the free end.

3.37. Compute the influence coefficients for the beam system in Fig. 3–11 if the beam can be considered to be massless, if the beam is of length L, load P_1 is $L/3$ ft from the left end, and load P_2 is $2L/3$ ft from the left end.

3.38. Compute the steady-state solutions for $x_1(t)$ and $x_2(t)$ for the elastic-string system in Ill. Ex. 3.17.

3.39. Compute the transient solution for $x_1(t)$ and $x_2(t)$ for the elastic-string system in Ill. Ex. 3.17.

3.40. Compute the influence coefficients for the two-degree-of-freedom, spring-mass system in Fig. 3–1.

3.41. Compute the influence coefficients for the three-degree-of-freedom, spring-mass system in Fig. 3–3.

3.42. Compute the influence coefficients for the three-degree-of-freedom, spring-mass system in Fig. 3–15 when all springs have the same spring constant value.

3.43. Determine the principal coordinates for the elastic-string system in Fig. 3–2 when the two masses are equally spaced a distance $L/3$ apart and when the two masses have the same weight.

3.44. Determine the principal coordinates for the two-degree-of-freedom, spring-mass system in Fig. 3–1 when all masses have the same weight and all springs have the same spring constant value.

3.45. Solve the system in Prob. 3.44 for displacements $x_1(t)$ and $x_2(t)$ and show that the principal vibration modes are orthogonal.

3.46. Use your solutions for Probs. 3.1, 3.2, and 3.6 to show that the principal vibration modes for these three spring-mass systems are orthogonal.

3.47. For the three-degree-of-freedom system in Fig. 3–3, suppose that all masses had the same weight and all springs had the same spring constant value. Find the displacements of the three masses when mass m_3 is given an initial displacement of one inch. All initial velocities and all other initial displacements equal zero. Use your results to show that the principal vibration modes for this spring-mass system are orthogonal.

3.48. Write the calculation sequence to solve the two-degree-of-freedom, nonlinear system in Sec. 3.11 using the midpoint-slope and trapezoidal methods. Use the Euler method to compute $\dot{x}_1(t_0 + \Delta t)$ and $\dot{x}_2(t_0 + \Delta t)$ as second starting values for the midpoint-slope method, where t_0 is the initial value of time.

3.49. Repeat Prob. 3.48, using the Adams method instead of the midpoint-slope method.

3.50. For the nonlinear spring-mass system in Sec. 3.11, let $W_1 = W_2 = 38.6$ lb, $k_1 = k_2 = k = 2$ lb/in., $a_1 = 0.030$, $b = 0.125$, $x_1(0) = 2$ in., and $\dot{x}_1(0) = x_1(0) = x_2(0) = 0$. Using a Δt-interval of 0.100 sec, calculate $x_1(t)$ and $x_2(t)$ numerically using the Euler and trapezoidal methods up to $t = 0.600$ sec.

3.51. Use the numerical methods of Prob. 3.48 to solve Prob. 3.50. Compare your two sets of results.

3.52. Use the numerical methods of Prob. 3.49 to solve Prob. 3.50. Compare your two sets of results.

4

Methods for Calculating Natural Frequencies

4.1   Introduction

This chapter presents several methods for calculating the natural frequencies of systems that have several degrees of freedom. Sections 3.2 and 3.4 in the previous chapter showed how we may determine the natural frequencies of a multiple-degree-of-freedom system. Because the presented procedure requires the evaluation of a determinant of size n and the determination of the roots of a polynomial of degree $2n$, the task of determining the natural frequencies by this procedure requires much mathematical work when n, the number of degrees of freedom, is large. It should be noted that this procedure requires that the characteristic equation be determined from the equation of motion and that the roots of this characteristic equation, which is a polynomial of degree $2n$, also be determined. Methods for calculating roots of polynomials (which are available to engineers as digital computer subroutines) are discussed in Sec. 3.3.

The methods covered in this chapter are special methods that calculate the natural frequencies directly from the equations of motion and do not require the calculation of the coefficients of the characteristic equation. It is advantageous to program many of these methods on digital computers, because their use is greatly eased when the digital computer performs the repetitive calculations that are involved. Because this is commonly done in professional practice, Chap. 7 contains digital computer programs that apply some of these methods. It should also be mentioned that the natural frequencies of complex structures are often determined by experimental means. An experimental determination can also be used to check the computed natural frequencies.

There are other analytical methods for calculating the natural frequencies of n-degree-of-freedom systems than those covered in this chapter. The advantages and limitations of the methods presented in this chapter are given in the descriptions of these methods. The methods presented in this chapter include most of the better known methods, and they are sufficient for most applications.

4.2 The Rayleigh Method

The *Rayleigh method* is a relatively simple method for calculating the approximate natural frequency of the first mode (i.e., the lowest or fundamental natural frequency). This method, which is commonly applied to beam and multi-mass systems, requires that the deflection curve for the mode (whose natural frequency is to be calculated) be estimated. Thus, the Rayleigh method is generally used only to calculate the fundamental natural frequency because it is difficult to estimate the deflection curves for the higher modes. Since this method is an energy method which assumes that the total mechanical energy (kinetic plus potential energy) is constant, an important limitation of this method is that it can only be applied to undamped (or conservative) systems.

We saw in Sec. 3.2 that all of the masses execute simple harmonic motion at one natural frequency when an undamped, multiple-degree-of-freedom system vibrates at a principal mode. The Rayleigh method, which was discussed in Sec. 1.4 for single-degree-of-freedom systems, assumes that the maximum kinetic energy equals the maximum potential energy at a principal mode. Because the deflection of the system during such a principal-mode vibration must be known in order to calculate these kinetic and potential energies, a reasonable dynamic deflection curve corresponding to the given mode must be assumed. If the correct dynamic deflection curve is assumed, the exact value of the natural frequency is obtained. If the assumed deflection curve is not the same as the actual dynamic deflection curve, the calculated natural frequency will be higher than the actual value. Thus, the accuracy of the Rayleigh method depends on how closely we can predict the dynamic deflection curve; and if the predicted curve is reasonably close, the calculated natural frequency will be slightly higher than the true value. The static deflection curve is often used to approximate the dynamic deflection curve for the first mode.

Let us first obtain the Rayleigh method equation for the lateral vibration of a multi-mass system, which consists of a beam or shaft of negligible mass that supports several lumped masses along its span. Suppose that there are n masses of weights $W_1, W_2, W_3, \cdots, W_n$ on this shaft. Let us approximate the dynamic deflection curve by the static deflection curve, and let $y_1, y_2, y_3, \cdots, y_n$ denote the static deflections of these n masses when treated as static loads. Since the kinetic energy of this system equals the sum of the kinetic energies of the n masses, and since $(\dot{y}_i)_{\max} = \omega(y_i)_{\max}$ at a principal mode, we have

$$\text{KE}_{\max} = \frac{\omega^2}{2g}(W_1 y_1^2 + W_2 y_2^2 + W_3 y_3^2 + \cdots + W_n y_n^2)$$

where ω is the natural frequency and where the maximum deflection of weight W_i is assumed to equal y_i, the static deflection due to weight W_i. Since the potential energy equals the beam strain energy, the total pontential energy of the system equals the work done by the three weights and is given by

$$\text{PE}_{\max} = \tfrac{1}{2}(W_1 y_1 + W_2 y_2 + W_3 y_3 + \cdots + W_n y_n)$$

After equating the equations for $\text{KE}_{\max}$ and $\text{PE}_{\max}$, we obtain the following equation for calculating the fundamental natural frequency ω for a multi-mass system. If this multi-mass system is rotating instead of undergoing a lateral (or vertical) vibration, then ω is the critical speed during whirl, as discussed in Ill. Ex. 2.9 of Sec. 2.9.

$$\omega^2 = \frac{g \sum\limits_{i=1}^{n} W_i y_i}{\sum\limits_{i=1}^{n} W_i y_i^2} \tag{4-1}$$

Now we shall obtain the Rayleigh method equation for application to beams and other systems with distributed mass and elasticity. Suppose that we have a beam of length L that weighs w lb per unit length.

Since potential energy PE equals the work done during beam bending,

$$PE = \frac{1}{2} \int M \, d\theta$$

where M is the bending moment and θ is the bending angle. From our strength of materials course, we know that $\theta \approx \tan \theta = dy/dx$ for small lateral deflections and that $M = EI(d^2y/dx^2)$, where y is the lateral or vertical deflection of the beam at horizontal location x and EI is the flexural rigidity of the beam. Substitution of $d\theta = (d^2y/dx^2)dx$ and $M = EI(d^2y/dx^2)$ into the PE equation gives

$$PE = \frac{1}{2} \int_0^L EI \left(\frac{d^2y}{dx^2}\right)^2 dx$$

where the previous equation gives PE_{max} if deflection y is the amplitude of the assumed deflection curve.

Since the kinetic energy of a beam segment of length dx and mass dm is $(\dot{y}^2/2)dm$ (where $dm = w \, dx/g$), and since a principal-mode motion is harmonic, the maximum kinetic energy of this beam (when vibrating laterally at a principal mode) is

$$KE_{max} = \frac{1}{2} \int \dot{y}^2 \, dm = \frac{\omega^2}{2} \int y^2 \, dm = \frac{\omega^2}{2g} \int_0^L wy^2 \, dx$$

where deflection y denotes the amplitude of the assumed deflection curve so that $\dot{y} = \omega y$. After equating PE_{max} and KE_{max}, we obtain

$$\omega^2 = \frac{g \int_0^L EI(d^2y/dx^2)^2 \, dx}{\int_0^L wy^2 \, dx} \qquad \text{(4-2)}$$

where the terms EI and w are constants if the beam is uniform (that is, if both the material and cross-sectional area are uniform). Thus, for uniform beams, the two integrations in Eq. (4-2) can usually be performed mathematically. It is important that the assumed deflection curve for $y(x)$ in Eq. (4-2) should satisfy, as far as possible, the four boundary conditions for the beam. For non-uniform beams (such as the discontinuous one shown in Fig. 7-8 of Sec. 7.25), both w and EI vary with x, and the two integrals in Eq. (4-2) are usually computed by graphical or numerical means. Since the $y(x)$-deflection curve for a non-uniform beam is difficult to obtain, and since $d^2y(x)/dx^2 = M(x)/EI(x)$, graphical-and numerical-integration techniques are usually employed to obtain this $y(x)$-curve. The use of numerical-integration techniques to calculate the natural frequencies of non-uniform beams by Rayleigh's method is discussed in Sec. 7.25.

Illustrative Example 4.1. Suppose that we had a massless beam that is 20 ft long, has an EI-value of 10^7 lb-in.^{2}, and is simply supported at both ends. Calculate the fundamental natural frequency for the lateral vibration of this beam system if it supports a 300-lb load that is located 7 ft from the left end and a 200-lb load that is located 12 ft from the left end.

Solution. From our strength of materials course, we find that the static deflection y at location x due to a concentrated load W, located a ft from the left end of a beam of length L, that is simply supported at both ends, is given by the following equation if $x \leq a$.

$$y = \frac{W(L-a)x}{6EIL} [L^2 - x^2 - (L-a)^2]$$

The deflections y_{12} and y_{22} at $x = 7$ ft and $x = 12$ ft due to a 200-lb load applied at $x = 12$ ft are

$$y_{12} = \frac{(200)(8)(7)}{(6)(20)(10^7)} (20^2 - 7^2 - 8^2)(12)^3 = 4.65 \text{ in.}$$

$$y_{22} = \frac{(200)(8)(12)}{(6)(20)(10^7)} (20^2 - 12^2 - 8^2)(12)^3 = 5.31 \text{ in.}$$

To use our deflection equation to calculate deflections y_{11} and y_{21}, let x be measured from the right end of the beam. Thus, the 300-lb load is applied at $a = 13$ ft and the deflections at $x = 8$ ft and 13 ft are

$$y_{21} = \frac{(300)(7)(8)}{(6)(20)(10^7)} (20^2 - 8^2 - 7^2)(12)^3 = 6.98 \text{ in.}$$

$$y_{11} = \frac{(300)(7)(13)}{(6)(20)(10^7)} (20^2 - 13^2 - 7^2)(12)^3 = 7.18 \text{ in.}$$

Using superposition, the total deflection due to both loads at $x = 7$ ft equals $(y_{11} + y_{12}) = (7.18 + 4.65)$ in., or 11.83 in. In like manner, use of superposition shows that the deflection due to both loads at $x = 12$ ft equals $(y_{12} + y_{22})$, or 12.29 in. Substitution of these results in Eq. (4-1) gives

$$\omega^2 = \frac{(386)[(300)(11.83) + (200)(12.29)]}{[(300)(11.83)^2 + (200)(12.29)^2]}$$

from which we find that our Rayleigh method approximation for natural frequency ω is 5.69 rad/sec. If this system were a rotating disk-shaft system (instead of a laterally vibrating beam system) that is supported on bearings, then this calculated value of ω would be the critical speed of this disk-shaft system.

Illustrative Example 4.2. Suppose that spring k_2 is removed from the spring-mass system in Fig. 3-1 of Sec. 3.2. Approximate the lowest natural frequency of this spring-mass system using Rayleigh's method.

Solution. For an undamped spring-mass system possessing n-degrees-of-freedom, equating the maximum kinetic and potential energies gives the following Rayleigh method equation:

$$\omega^2 = \frac{g \sum\limits_{i=1}^{n} W_i x_i}{\sum\limits_{i=1}^{n} W_i x_i^2} \tag{4-3}$$

where W_i and x_i are the weight and maximum deflection of mass i. Since spring k_2 is removed from the spring-mass system in Fig. 3-1, we can compute the static deflection values for x_1 and x_2, as follows:

$$x_1 = \frac{W_1 + W_2}{k_1}$$

$$x_2 = x_1 + \frac{W_2}{k} = \frac{W_1 + W_2}{k_1} + \frac{W_2}{k}$$

Since this system has two degrees of freedom, natural frequency ω can be approximated by substituting these x_1- and x_2-values in the following form of Eq. (4-3), where $W_i = m_i g$:

$$\omega = \sqrt{g \left(\frac{W_1 x_1 + W_2 x_2}{W_1 x_1^2 + W_2 x_2^2} \right)}$$

Illustrative Example 4.3. Use the Rayleigh method to approximate the lowest natural frequency of a uniform beam of length L that is simply supported at both ends and has no externally-applied loads.

Solution. From our strength of materials course, we know that the static deflection y of this beam at location x is given by

$$y = \left(\frac{w}{24EI} \right)(2Lx^3 - x^4 - L^3 x)$$

where w is the weight of this beam per unit length. Using two differentiations to compute $d^2 y / dx^2$, we obtain

$$\frac{d^2 y}{dx^2} = \left(\frac{w}{24EI} \right)(12Lx - 12x^2)$$

Substitution of these equations for y and $d^2 y / dx^2$ in Eq. (4-2) gives

$$\omega^2 = \frac{g(EI)(144) \int_0^L (Lx - x^2)^2 \, dx}{w \int_0^L (2Lx^3 - x^4 - L^3 x)^2 \, dx} = \frac{97.3g(EI)}{wL^4}$$

Thus, use of the static deflection curve gives us $\omega = 9.86\sqrt{g(EI)/wL}$ rad/sec, which also equals the exact natural-frequency value (up to three-digit significance) for this beam problem. The actual deflection curve for this problem is

$$y = \left(y_m \sin \frac{\pi x}{L} \right) \sin \omega t$$

where y_m is the maximum deflection of this beam (i.e., the maximum value of y at $x = L/2$). For a uniform, unloaded cantilever beam that weighs w lb per unit length, the fundamental natural frequency equals $3.52\sqrt{g(EI)/wL^4}$.

Substitution of the following static-deflection equation from our strength of materials course

$$y = \frac{w}{24EI}(x^4 - 4L^3x + 3L^4)$$

in Eq. (4-2) gives the slightly higher value, $\omega = 3.53\sqrt{g(EI)/wL^4}$.

4.3   The Holzer Method

The *Holzer method* is a tabular method for calculating natural frequencies of spring-mass, disk-shaft, and other lumped systems that possess several degrees of freedom. Though it is a trial-and-error method, it is rather general in that it may be used to calculate all of the natural frequencies of a system, and in that it may also be applied to find specific frequencies of systems which have damping and externally-applied forces or moments. Let us first derive the Holzer method equations for systems that are unsupported at its two ends. The equations of motion for the three-degree-of-freedom system shown in Fig. 4–1 are

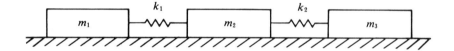

Fig. 4-1

$$m_1\ddot{x}_1 = -k_1(x_1 - x_2)$$
$$m_2\ddot{x}_2 = k_1(x_1 - x_2) - k_2(x_2 - x_3)$$
$$m_3\ddot{x}_3 = k_2(x_2 - x_3)$$

Addition of these three equations of motion gives

$$m_1\ddot{x}_1 + m_2\ddot{x}_2 + m_3\ddot{x}_3 = 0$$

which means that the sum of the inertia forces for this unsupported system must equal zero. If the system is vibrating at one of its principal modes, then the motion is harmonic and $\ddot{x}_i = -\omega^2 x_i$, where ω is the natural frequency that corresponds to this mode and x_i is the displacement amplitude of mass m_i. Thus, for an unsupported n-degree-of-freedom system,

$$\sum_{i=1}^{n} m_i\ddot{x}_i = 0$$

and substitution of $\ddot{x}_i = -\omega^2 x_i$ for a principal-mode vibration shows that

$$\omega^2 \sum_{i=1}^{n} m_i x_i = 0 \qquad (4\text{-}4)$$

if ω is a natural frequency of the system. Solving the first equation of motion for displacement x_2 gives

$$x_2 = x_1 + \frac{m_1 \ddot{x}_1}{k_1} = x_1 - \frac{m_1 \omega^2 x_1}{k_1} \qquad (4\text{-}5)$$

Solving the second equation of motion for displacement x_3 gives

$$x_3 = x_2 + \frac{m_2 \ddot{x}_2}{k_2} - \frac{k_1(x_1 - x_2)}{k_2}$$

Substitution of $\ddot{x}_2 = -\omega^2 x_2$ and $k_1(x_1 - x_2) = -m_1 \ddot{x}_1 = m_1 \omega^2 x_1$ gives

$$x_3 = x_2 - \frac{\omega^2(m_1 x_1 + m_2 x_2)}{k_2} \qquad (4\text{-}6)$$

For an n-degree-of-freedom, spring-mass system, Eq. (4-6) can be rewritten in the following more general form:

$$x_i = x_{i-1} - \frac{\omega^2}{k_{i-1}} \sum_{j=1}^{i-1} m_j x_j \qquad (4\text{-}7)$$

where $2 \leq j \leq n$. The Holzer method for an undamped and unsupported spring-mass system is based on Eqs. (4-4) and (4-7). Using the analogy table in Sec. 2.4, the Holzer method equations for an undamped and unsupported disk-shaft system possessing n degrees of freedom are

$$\omega^2 \sum_{i=1}^{n} I_i \theta_i = 0 \qquad (4\text{-}8)$$

$$\theta_i = \theta_{i-1} - \frac{\omega^2}{k_{t_{i-1}}} \sum_{j=1}^{i-1} I_j \theta_j \qquad (4\text{-}9)$$

where ω is a natural frequency and θ_i is the angular-displacement amplitude of disk i.

To apply Eqs. (4-4) and (4-7) for a spring-mass system, we must first assume natural frequency ω. We let $x_1 = 1$, and then calculate x_2 to x_n using Eq. (4-7). If the assumed value for ω turns out to be a natural frequency, then the calculated values of x_1 to x_n will satisfy Eq. (4-4), which states that the sum of the inertia forces equals zero. If Eq. (4-4) is not satisfied, we must assume another value of ω and repeat this procedure. Note that this method does not furnish the ω-values to use for the next trial, as an iterative method does. We know that we are approaching a natural frequency when the sum of the inertia forces approaches zero. A plot of $\omega^2 \sum_{i=1}^{n} m_i x_i$ versus the assumed ω-values, after several trial calculations are performed, should help in selecting new trial values. That is, if this plotted curve matches the exact curve, then the natural frequencies occur where this curve inter-sects the horizontal axis, where $\omega^2 \sum_{i=1}^{n} m_i x_i = 0$.

If we have a forced vibration of an unsupported system, the sum of the inertia forces must equal the externally-applied force. Thus, if the spring-mass system is unsupported and has a harmonic external force, then the Holzer method may be applied by using Eq. (4–7) together with the following equation:

$$\omega^2 \sum_{i=1}^{n} m_i x_i = F_E \tag{4–10}$$

where F_E is the amplitude of the externally-applied force and ω is the frequency of the applied harmonic force. In this case, we would not use the Holzer method to calculate the natural frequencies, but to obtain a plot of external-force-amplitude F_E versus the applied fre-quency ω. In this case, the Holzer method gives the displacement configuration and the amplitude of the applied external force for different applied frequencies. Since resonant amplitudes vary with the amount of damping in the system, the Holzer method is useful for the analysis of linearly-damped systems undergoing a forced vibration. Such an analysis will require the use of complex numbers because the damping force, for a linear damper between masses i and $i + 1$, is proportional to the relative velocity $(\dot{x}_{i+1} - \dot{x}_i)$, and because velocity $\dot{x}_i$ equals $i\omega x_i$ during a harmonic motion.

The remainder of this section will deal with the calculation of natural frequencies for other types of spring-mass systems that have no external forces. If the spring-mass system contains n masses and is rigidly supported at one end, we let $x_1 = 1$ and apply Eq. (4–7) to calculate displacements x_2 to x_{n+1}. We start these calculations at the unsupported mass, which we shall denote as m_1. For this n-degree-of-freedom system, the term x_n is the displacement of the last mass, and the term x_{n+1} is the displacement of the rigid support. Equation (4–4) does not hold for this case, since the sum of the inertia forces does not equal zero. We know that we have assumed a natural-frequency value for ω if the calculated value of x_{n+1}, the displacement of the rigid support, equals zero. If $x_{n+1} \neq 0$, we assume another value of ω and repeat the procedure. A plot of x_{n+1} versus the assumed ω-values, after a few trials are performed, will help us choose the next ω-values to assume. If the spring-mass system is fixed at both ends, the procedure is the same except that displacements x_2 to x_{n+1} are calculated from

$$x_i = x_{i-1} + \frac{1}{k_i} [k_1 x_1 - \omega^2 \sum_{j=1}^{i-1} m_j x_j] \tag{4–11}$$

Let us now briefly summarize the Holzer method for calculating the natural frequencies of an undamped spring-mass system. For each trial, we assume a value for natural frequency ω and let the amplitude of mass m_1 equal unity. The inertia forces and the amplitudes of the other masses are calculated, where Eq. (4–7) is used if one end is unsupported and Eq. (4–11) is used if both ends are rigidly supported. If both ends are unsupported, Eq. (4–4) must be satisfied. If one end is rigidly supported, displacement x_{n+1} of this support must equal zero. A plot of $\omega^2 \sum_{i=1}^{n} m_i x_i$ or x_{n+1} versus the assumed ω-values will help us to estimate the natural frequencies of the system. Section 7.19 furnishes a computer program that applies the Holzer method to systems where one or both ends are unsupported. This program can be easily modified to handle systems where both ends are rigidly supported.

Illustrative Example 4.4.   Use the Holzer method to verify the natural frequencies computed in Ill. Ex. 3.1 of Sec. 3.2.

Solution.   Since Fig. 3-1 in Sec. 3.2 shows that this spring-mass system is fixed at both ends, Eq. (4–11) must be used to compute x_2 and x_3, where x_3 is the displacement of the lower rigid support. The following table shows the Holzer method calculations for $\omega = 5.00$ rad/sec, where x_i is calculated by Eq. (4–11).

Location	m_i	k_i	x_i	$\omega^2 m_i x_i$	$\sum \omega^2 m_i x_i$
1	0.100	2.50	1.00	2.50	2.50
2	0.100	3.50	1.00	2.50	5.00
3	∞	2.50	0		

Since $x_3 = 0$, one natural frequency is 5.00 rad/sec, and the mode shape is given by $x_1 = x_2 = 1.00$, which were also obtained in Ill. Ex. 3.1. The following equations show the Holzer-method calculations for $\omega = 9.76$ rad/sec:

$$x_2 = 1.00 + \frac{1}{3.50}[(2.50)(1.00) - (9.76)^2(0.100)(1.00)] = -1.00$$

$$x_3 = -1.00 + \frac{1}{2.50}\{2.50 - (9.76)^2[(0.100)(1.00) + (0.100)(-1.00)]\} = 0$$

where Eq. (4–11) was applied twice. Since $x_3 = 0$, the other natural frequency is 9.76 rad/sec, and the mode shape is given by $x_1 = 1.00$ and $x_2 = -1.00$, which were also obtained in Ill. Ex. 3.1.

Illustrative Example 4.5.   For the spring-mass system in Fig. 4-1, suppose that $m_1 = 0.100$ lb-sec^2/in., $m_2 = 0.200$ lb-sec^2/in., $m_3 = 0.400$ lb-sec^2/in., $k_1 = 1.00$ lb/in., and $k_2 = 2.00$ lb/in. Use the Holzer method to determine whether 1.25 rad/sec and 5.00 rad/sec are natural frequencies for this system.

Solution. Since this spring-mass system is unsupported at both ends, both Eqs. (4-4) and (4-7) must be satisfied if the assumed ω-value is a natural frequency. The following table shows the Holzer-method calculations for $\omega = 1.25$ rad/sec, where x_i is calculated by Eq. (4-7).

Location	m_i	x_i	$\omega^2 m_i x_i$	$\sum \omega^2 m_i x_i$	k_i
1	0.100	1.000	0.156	0.156	1.00
2	0.200	0.844	0.264	0.420	2.00
3	0.400	0.634	0.396	0.816	

Since $\sum \omega^2 m_i x_i \neq 0$, Eq. (4-4) is not satisfied and 1.25 rad/sec is not a natural frequency. The first of the following tables shows the Holzer-method calculations for $\omega = 5.00$ rad/sec.

Location	m_i	x_i	$\omega^2 m_i x_i$	$\sum \omega^2 m_i x_i$	k_i
1	0.100	1.000	2.500	2.500	1.00
2	0.200	-1.500	-7.500	-5.000	2.00
3	0.400	1.000	10.000	5.000	

Location	m_i	x_i	$\omega^2 m_i x_i$	$\sum \omega^2 m_i x_i$	k_i
1	0.100	1.000	1.600	1.600	1.00
2	0.200	-0.600	-1.920	-0.320	2.00
3	0.400	-0.440	-2.810	-3.130	

Since $\sum \omega^2 m_i x_i \neq 0$, 5.00 rad/sec is not a natural frequency. Next we shall try $\omega = 4.00$ rad/sec, where the Holzer-method calculations are given by the last table. Since this system has three degrees of freedom, there are three natural frequencies. We must keep assuming ω values until we find three values which satisfy Eq. (4-4), which states that the sum of the inertia forces equals zero. Since $\sum \omega^2 m_i x_i > 0$ for both $\omega = 1.25$ rad/sec and $\omega = 5.00$ rad/sec, and since $\sum \omega^2 m_i x_i < 0$ for $\omega = 4.00$ rad/sec, we know that one natural frequency lies between 1.25 and 4.00 rad/sec and that another natural frequency lies between 4.00 rad/sec and 5.00 rad/sec.

Illustrative Example 4.6. For the spring-mass system in Ill. Ex. 4.2, suppose that $W_1 = 5.00$ lb, $W_2 = 21.0$ lb, $k_1 = 4.00$ lb/in., and $k = 7.00$ lb/in. Use the Holzer method to check the Rayleigh-method approximation of the lowest natural frequency that was computed in this example.

Solution. Substitution of the given numerical values into the equations in Ill. Ex. 4.2 gives

$$x_1 = \frac{W_1 + W_2}{k_1} = \frac{26.0}{4.00} = 6.50 \text{ in.}$$

$$x_2 = x_1 + \frac{W_2}{k} = 6.50 + \frac{21.0}{7.00} = 9.50 \text{ in.}$$

$$\omega = \sqrt{\frac{(386)[(5.00)(6.50) + (21.0)(9.50)]}{(5.00)(6.50)^2 + (21.0)(9.50)^2}} = 6.51 \text{ rad/sec}$$

Since this system is unsupported at the bottom end and rigidly supported at the top end (as seen in Fig. 3-1 when spring k_2 is removed), the Holzer-method calculations must start at the unsupported bottom end. The Holzer-method calculations for $\omega = 6.51$ rad/sec are given in the following table, where x_i is calculated by Eq. (4-7) and where $m_1 = 5.00/386 = 0.0129$ and $m_2 = 21.0/386 = 0.0545$. Note that the Holzer table terms start from the free end, instead of the fixed end, and that x_0 is the displacement of the rigid surface. Since displacement x_0 is very close to zero, 6.51 rad/sec is close to a natural-frequency value, and the mode-shape ratio x_2/x_1 is close to 1.000/0.670 or 1.49.

Location	m_i	x_i	$\omega^2 m_i x_i$	$\Sigma \, \omega^2 m_i x_i$	k_i
2	0.0545	1.000	2.310	2.310	7.00
1	0.0129	0.670	0.365	2.675	4.00
0	∞	0.001			

Illustrative Example 4.7. Write the Holzer–method equations for determining the natural frequencies of the disk-shaft system in Fig. 3-4 in Sec. 3.2.

Solution. Since this disk-shaft system is unsupported at one end and rigidly supported at the other end, we apply Eq. (4-9) to calculate the θ_i-values, where we must start at the free end. Thus, we start by assuming a value for frequency ω, set θ_3 (the angular displacement of disk I_3) equal to unity, and then perform the following calculations where θ_0 is the angular displacement of the rigid support. Note that these equations are subscripted differently than Eq. (4-9) in order to agree with Fig. 3-4.

$$\theta_2 = \theta_3 - \frac{\omega^2}{k_{t3}} (I_3 \theta_3)$$

$$\theta_1 = \theta_2 - \frac{\omega^2}{k_{t2}} (I_3 \theta_3 + I_2 \theta_2)$$

$$\theta_0 = \theta_1 - \frac{\omega^2}{k_{t1}} (I_3 \theta_3 + I_2 \theta_2 + I_1 \theta_1)$$

We will have assumed a natural-frequency value for ω when the calculated value of θ_0, the displacement of the rigid support, equals zero. If we made up a table for performing these Holzer-method calculations, then the table titles would be location, I_i, θ_i, $\omega^2 I_i \theta_i$, $\Sigma \omega^2 I_i \theta_i$, and k_{ti}, instead of the Holzer table titles given in Ill. Ex. 4.6.

4.4   The Myklestad or Prohl Method

The Holzer method is a tabular method that determines the natural frequencies (including those of higher modes) of systems that consist of concentrated masses connected by massless springs (or of disks connected by massless elastic shafts). Sec. 1.2 of this text showed how multi-story buildings and other actual systems may be simplified to spring-mass systems for specific vibration analyses. The method in this section, which was simultaneously developed by N. O. Myklestad and M. A. Prohl, is a tabular method that determines the natural frequencies (including those of higher modes) of systems that consist of concentrated masses connected by massless beam segments. Though they are massless, these beam segments possess the elastic properties of a real beam. The *Myklestad* method is similar to the Holzer method in that it is a trial-and-error procedure where a value for frequency ω is assumed and tabular calculations are carried out along this beam system to determine whether the four boundary conditions for the beam system are met. The tabular calculations for the Myklestad method (which compute the shear force, bending moment, slope, and deflection along this beam system) are much more involved than the Holzer-method calculations, which makes it desirable to have a digital computer program for applying this more involved method.

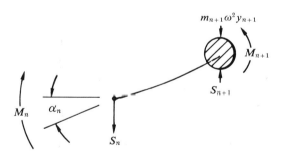

Fig. 4-2

Suppose that the beam system is vibrating at a principal mode, which means that it is vibrating with harmonic motion at one of its natural frequencies. A segment of the beam system between stations n and $n + 1$ is shown in Fig. 4–2, where slope α_n, bending moments M_n and M_{n+1}, shear forces S_n and S_{n+1}, and inertia force $m_{n+1}\omega^2 y_{n+1}$ are shown. The massless beam segment is of length l_n, and this figure does not show the vertical deflections y_n and y_{n+1}, which are positive upward and which are measured from the equilibrium position. Since the slope and deflection at station n are $\alpha_n \sin \omega t$ and $y_n \sin \omega t$ during a principal-mode vibration, Fig. 4–2 shows an extreme deflection condition. Using Fig. 4–2 as a free-body diagram, a force and moment balance gives

$$S_{n+1} = S_n + m_{n+1}\omega^2 y_{n+1} \qquad (4\text{–}12)$$

$$M_{n+1} = M_n - S_n l_n \qquad (4\text{–}13)$$

Let u_{Fn} denote the vertical deflection at station $n + 1$, measured from a tangent at station n, due to a unit force at station n; and let u_{Mn} denote the amount of this vertical deflection that is caused by a unit moment at station n. Similarly, v_{Fn} and v_{Mn} are the slopes at station $n + 1$, measured from a tangent at station n, due to a unit force and unit moment at station n. Thus, from the geometry of the system shown in Fig. 4–2, we have

$$\alpha_{n+1} = \alpha_n - S_n v_{Fn} + M_n v_{Mn}$$

$$y_{n+1} = y_n - \alpha_n l_n + S_n u_{Fn} - M_n u_{Mn}$$

From beam theory it can be shown that $v_{Mn} = l_n/(EI)_n$, $v_{Fn} = u_{Mn} = l_n^2/2(EI)_n$, and $u_{Fn} = l_n^3/6(EI)_n$ for a uniform beam segment. Thus, the previous two equations can be rewritten as follows if all of the beam segments in the system are uniform:

$$\alpha_{n+1} = \alpha_n - \frac{S_n l_n^2}{2(EI)_n} + \frac{M_n l_n}{(EI)_n} \qquad (4\text{–}14)$$

$$y_{n+1} = y_n - \alpha_n l_n + \frac{S_n l_n^3}{6(EI)_n} - \frac{M_n l_n^2}{2(EI)_n} \qquad (4\text{–}15)$$

The first part of the procedure for this method is to assume a value for natural frequency ω. We choose station 1 to be an unsupported or a simply-supported end, but not a fixed (or rigidly-supported) end. Thus, we start our calculations at the free or unsupported end for a cantilever beam. We let the slope α_1 at station 1 be the unknown in our calculations. That is, if station 1 is a simply-supported end, deflection $y_1 = 0$ and bending moment $M_1 = 0$, and we set shear force $S_1 = 1$. If station 1 is a free (or unsupported) end, we know from our strength of materials course that $M_1 = S_1 = 0$, and we let the unknown deflection $y_1 = 1$, as a reference value. It should be noted that $S_1 = m_1\omega^2$ (since $y_1 = 1$) at a free end if there is a mass m_1 at station 1.

The next part of the procedure is to calculate the values of S_{n+1}, M_{n+1}, α_{n+1}, and y_{n+1} for the other stations in the beam system using Eqs. (4–12) to (4–15). Let the symbol N denote the last station in this system. Since slope α_1 was carried as an unknown in these calculations, we will end up with equations of the form

$$S_N = d_1\alpha_1 + e_1 \qquad \text{(4-16)}$$
$$M_N = d_2\alpha_1 + e_2 \qquad \text{(4-17)}$$
$$\alpha_N = d_3\alpha_1 + e_3 \qquad \text{(4-18)}$$
$$y_N = d_4\alpha_1 + e_4 \qquad \text{(4-19)}$$

where the d_i-and e_i-terms are numerically calculated values. We can now compute slope α_1 from the boundary conditions at station N. If station N is rigidly supported, we know from our strength of materials course that $\alpha_N = y_N = 0$. Thus, we can calculate α_1 from Eq. (4-18), which gives $\alpha_1 = -e_3/d_3$ for $\alpha_N = 0$. If the assumed value of ω is a natural frequency of the system, then the computed value of y_N will equal zero when $\alpha_1 = -e_3/d_3$. If this does not happen, we assume another value for frequency ω and repeat the procedure. A plot of the computed values of y_N versus assumed frequency ω, after a few trial calculations are performed, can be used to help select new trial values to obtain $y_N = 0$.

If station N is unsupported, the boundary conditions are $S_N = M_N = 0$. Use of Eq. (4-16) gives $\alpha_1 = -e_1/d_1$ for $S_N = 0$. Thus, the computed value for M_N will equal zero when $\alpha_1 = -e_1/d_1$ if the assumed value for ω is a natural frequency of the system. If station N is simply supported, the boundary conditions are $M_N = y_N = 0$. Since Eq. (4-17) gives $\alpha_1 = -e_2/d_2$ for $M_N = 0$, then $y_N = 0$ when $\alpha_1 = -e_2/d_2$ and when ω is a natural frequency of the system. When we have converged upon a natural-frequency value, we will also obtain the deflection, slope, shear force, and bending moment values along the beam system during that particular principal-mode vibration.

The Myklestad method has also been extended to cover the analysis of rotating elastic components, such as propellers, turbine blades, etc., where engineering judgment must be used to redefine the rotating component to be a system of point masses connected by uniform beam segments. For this analysis, the vibration is assumed to be perpendicular to the plane of rotation, and the values of S, M, α, and y are computed along the beam system, as before. The effect of rotation is included in the equations for these four terms. For example, the inertia force component, $-m_{n+1}\Omega^2 r_{n+1}\alpha_{n+1}$, is added to the equation for S_{n+1}, where Ω is the rotational angular velocity. The Myklestad method has also been extended to cover the analysis of airplane wings and other coupled flexure-torsion vibration problems. For this type of problem, the values of S, M, α, y, θ (the angle of twist), and T (the twisting torque) are computed at successive stations along the length of the beam system. The details of these two extended Myklestad analyses can be found in other more complete references for this method.

Illustrative Example 4.8.   Consider the 4-ft long cantilever beam with two masses (weighing 38.6 lb each) that is shown in Fig. 4-3, where $EI = 2.00 \times 10^5$ lb-in². Use the Myklestad method to determine whether its fundamental natural frequency is close to that for a uniform cantilever beam.

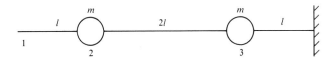

Fig. 4-3

Solution.   From Fig. 4-3, we see that $l = 4/4 = 1.00$ ft $= 12$ in. and that $m = 38.6/386 = 0.100$. In Ill. Ex. 4.3, it is given that the fundamental natural frequency of a uniform cantilever beam is given by

$$\omega = 3.52\sqrt{\frac{g(EI)}{wL^4}}$$

Since $w = 2(38.6)/4 = 19.3$ lb/ft, substitution of numerical values (using ft for the length dimension) gives

$$\omega = 3.52\sqrt{\frac{(32.2)(2.00 \times 10^5)}{(19.3)(4.00)^4(144)}} = 10.6 \text{ rad/sec}$$

for the equivalent uniform cantilever beam. Starting our calculations at station 1, the free end, we have

$$S_1 = M_1 = 0$$

and we let $y_1 = 1.00$ and carry α_1 as an unknown. Use of Eqs. (4-12) to (4-15) give

$$y_2 = 1.00 - \alpha_1(12) + 0 - 0 = 1.00 - 12.0\alpha_1$$
$$\alpha_2 = \alpha_1 - 0 + 0 = \alpha_1$$
$$M_2 = M_1 - S_1 l = 0 - 0 = 0$$
$$S_2 = S_1 + m\omega^2 y_2 = 0 + (0.100)(10.6)^2(1.00 - 12.0\alpha_1)$$
$$= 11.1 - 13.3\alpha_1$$

for the values at station 2, where the first mass is located, and these equations give

$$y_3 = (1.00 - 12.0\alpha_1) - \alpha_1(24.0) + \frac{(11.1 - 133\alpha_1)(24)^3}{6(2.00 \times 10^5)} - 0$$
$$= 1.128 - 37.54\alpha_1$$
$$\alpha_3 = \alpha_1 - \frac{(11.1 - 133\alpha_1)(24)^2}{2(2.00 \times 10^5)} + 0 = 0.0160 + 0.192\alpha_1$$
$$M_3 = M_2 - S_2(2l) = 0 - (11.1 - 133\alpha_1)(24)$$
$$= -266 + 3200\alpha_1$$
$$S_3 = (11.1 - 133\alpha_1) + (0.100)(10.6)^2(1.128 - 37.54\alpha_1)$$
$$= 23.7 - 551\alpha_1$$

for station 3, where the second mass is located. At station 4, where the rigid wall is located, use of Eqs. (4-14) and (4-15) give

$$\alpha_4 = (0.0160 + 0.192\alpha_1) - \frac{(23.7 - 551\alpha_1)(12.0)^2}{2(2.00 \times 10^5)}$$

$$+ \frac{(-266 + 3200\alpha_1)(12.0)}{2.00 \times 10^5} = -0.0085 + 0.583\alpha_1$$

$$y_4 = (1.128 - 37.54\alpha_1) - (12.0)(0.0160 + 0.192\alpha_1)$$

$$+ \frac{(23.7 - 551\alpha_1)(12.0)^3}{6(2.00 \times 10^5)} - \frac{(-266 + 3200\alpha_1)(12.0)^2}{2(2.00 \times 10^5)}$$

$$= 1.368 - 48.98\alpha_1$$

Thus, use of $\alpha_4 = 0$ gives $\alpha_1 = 0.0085/0.583 = 0.0146$ rad. Since use of $y_4 = 0$ gives $\alpha_1 = 1.368/48.98 = 0.0280$ rad, we should assume another natural-frequency value to continue this problem.

4.5　Dunkerly and Stodola Methods

The *Dunkerly equation* is useful for approximating the fundamental natural frequency of systems whose second natural-frequency values are much larger than the lowest natural-frequency value. For an undamped, spring-mass system that has n degrees of freedom, the coefficient of $(1/\omega^2)^{n-1}$ in the characteristic equation is as follows, where elastic stiffness k_{ie} is defined to be the force required to cause a unit displacement of mass m_i:

$$\frac{m_1}{k_{1e}} + \frac{m_2}{k_{2e}} + \cdots + \frac{m_n}{k_{ne}} = \frac{1}{\omega_{11}^2} + \frac{1}{\omega_{22}^2} + \cdots + \frac{1}{\omega_{nn}^2}$$

Since the term ω_{ii} equals $\sqrt{k_{ie}/m_i}$, ω_{ii} is the system natural frequency when mass m_i is the only mass in the system. If ω_1 to ω_n are the n roots of the characteristic equation, then the characteristic equation can be written as

$$\left(\frac{1}{\omega^2} - \frac{1}{\omega_1^2}\right)\left(\frac{1}{\omega^2} - \frac{1}{\omega_2^2}\right) \cdots \left(\frac{1}{\omega^2} - \frac{1}{\omega_n^2}\right) = 0$$

If the previous equation is multiplied out, the coefficient of $(1/\omega^2)^{n-1}$ is

$$\frac{1}{\omega_1^2} + \frac{1}{\omega_2^2} + \frac{1}{\omega_3^2} + \cdots + \frac{1}{\omega_n^2} \approx \frac{1}{\omega_1^2}$$

where the approximation holds because we assumed that ω_2, ω_3, etc. are much larger than ω_1. Thus, equating the two coefficients of $(1/\omega^2)^{n-1}$ for the two different forms of the characteristic equation gives the following approximation, which is known as *Dunkerly's equation*:

$$\frac{1}{\omega_1^2} \approx \frac{1}{\omega_{11}^2} + \frac{1}{\omega_{22}^2} + \cdots + \frac{1}{\omega_{nn}^2} \qquad \text{(4–20)}$$

Suppose that we had a uniform beam that had two applied loads P_1 and P_2. We can let ω_{11} be the natural frequency of the beam (when it is unloaded) due only to its weight of w lb per unit length; we can let ω_{22} be the beam natural frequency due only to load P_1 when the beam is considered to be weightless; and we can let ω_{33} be the natural frequency of the beam when the only load is P_2. We can now apply Eq. (4–20), the Dunkerly equation, to approximate ω_1, the fundamental natural frequency of the beam system when all of the loads are applied. The Dunkerly equation is also useful for vibration testing when an eccentric shaker is attached to the tested structure, since the measured frequency ω_1 will include the effects of both ω_{11}, the frequency of the structure, and ω_{22}, the frequency of the eccentric exciter. The use of Eq. (4–20) for vibration testing is exemplified in Ill. Ex. 4.9.

The *Stodola method*, which is exemplified in Ill. Ex. 4.10, is limited in application to undamped systems, and, like the Rayleigh method, is generally used only for calculating the lowest natural frequency of a system. It is a physical method in that the system deflections are assumed, and because this method computes the inertia and spring forces and a new set of deflections from these assumed deflections. Since the new set of deflections is used for the assumed values for the next trial, this is an iterative method which converges to a solution when the computed deflections become equal to those of the assumed deflections. A much more widely used iteration method for calculating natural frequencies is the *matrix-iteration method*, which is discussed in Sec. 4.9.

Illustrative Example 4.9. An eccentric shaker that weighs 2.00 lb is attached to a structure at a point where the influence coefficient is 0.00500 in./lb. If one of the measured natural frequencies is 60 rad/sec, determine the corresponding natural frequency of the structure.

Solution. In Eq. (4–20) let ω_1 be the measured frequency and let ω_{22} denote the natural frequency of the eccentric shaker when on the structure and when there are no other masses. Since ω_{22} equals $\sqrt{1/m_2\alpha_{22}}$ (where m_2 is the mass of the shaker and α_{22} is the influence coefficient at the point of application), use of Dunkerly's equation gives

$$\frac{1}{\omega_1^2} = \frac{1}{\omega_{11}^2} + m_2\alpha_{22}$$

where ω_{11} is the natural frequency of the structure when the shaker is not attached. Substitution of the given numerical values in the previous equation gives

$$\frac{1}{(60)^2} = \frac{1}{\omega_1^2} + \left(\frac{2.00}{386}\right)(0.00500)$$

Solving this equation, we find that ω_1, the corresponding natural frequency of the structure, equals 63.0 rad/sec.

Illustrative Example 4.10. Repeat Ill. Ex. 4.6 using the Stodola method.

Solution. For our first trial, we shall use the static deflections, which are defined in Ill. Ex. 4.2, as our first assumptions for deflections x_1 and x_2. In Ill. Ex. 4.6, we see that these static deflections are $x_1 = 6.50$ in. and $x_2 = 9.50$ in. Thus, $x_2/x_1 = 9.50/6.50 = 1.47$, and the inertia forces are

$$-m_1\ddot{x}_1 = m_1\omega^2 x_1 = \frac{5.00}{386}(\omega^2)(6.50) = 0.0841\omega^2$$

$$-m_2\ddot{x}_2 = m_2\omega^2 x_2 = \frac{21.0}{386}(\omega^2)(9.50) = 0.517\omega^2$$

The spring force equals the sum of the inertia forces below that particular spring. Thus, the force in the first spring is $0.601\omega^2$, and the force in the second spring is $0.517\omega^2$. The deflections for the two springs are

$$0.601\frac{\omega^2}{k_1} = \frac{0.601\omega^2}{4.00} = 0.150\omega^2$$

$$0.517\frac{\omega^2}{k_2} = \frac{0.517\omega^2}{7.00} = 0.074\omega^2$$

Thus, $x_1 = 0.150\omega^2$ and $x_2 = (0.150 + 0.074)\omega^2 = 0.224\omega^2$, and $x_2/x_1 = 0.224/0.150 = 1.48$. Since the calculated x_2/x_1-ratio almost equals the ratio for the assumed values, we can stop the iteration here. Thus, $x_2 = 1.50\,x_1$ for this first mode, and the natural frequency can be computed from

$$x_1 = 6.50 = 0.150\omega^2$$

which gives $\omega = \sqrt{6.50/0.150} = 6.59$ rad/sec, and which is close to the Rayleigh method value given in Ill. Ex. 4.6.

4.6 Review of Matrix Arithmetic

Since the rest of this chapter uses matrix notation, this section is included for review purposes. A *matrix* is merely a rectangular grouping of numbers, which are arranged in rows and columns to form an array. Each number is called an *element* of the matrix. Matrices do not have a specific value, as do determinants, but they are very useful as a shorthand notation for carrying out certain types of calculations. The following matrix:

$$\begin{bmatrix} -5 & 3 & 7 \\ 2 & 0 & -9 \end{bmatrix}$$

consists of two rows, three columns, and six elements. The first column consists of -5 and 2, while the second row consists of 2, 0, and -9. Note that the previous array is enclosed in brackets to denote that it is a matrix. The *size* of a matrix is specified by its number of rows and columns. Thus, the above matrix is of size two-by-three. In general, a matrix of m rows and n columns (i.e., size m by n) can be represented as follows:

$$[B] = \begin{bmatrix} B_{11} & B_{12} & \cdots & B_{1n} \\ B_{21} & B_{22} & \cdots & B_{2n} \\ \cdot\cdot & \cdot\cdot & \cdots\cdot\cdot \\ B_{m1} & B_{m2} & \cdots & B_{mn} \end{bmatrix}$$

In the above matrix, each element is identified by two subscripts, where the first subscript denotes the row and the second denotes the column in which this element belongs. Thus, element B_{46} belongs to the fourth row and the sixth column. Note that the *row number* is counted from the top and that the *column number* is counted from the left.

A *square matrix* is a matrix that has the same number of rows and columns (i.e., $m = n$). It should be emphasized that a square matrix is not the same thing as a determinant. For example, the following determinant, which is represented by a pair of vertical lines, has a numerical value of 10, while square matrices do not have a numerical value.

$$|A| = \begin{vmatrix} 6 & -4 \\ -2 & 3 \end{vmatrix} = (6)(3) - (-4)(-2) = 10$$

The *principal diagonal* of a square matrix $[B]$ consists of elements $B_{11}, B_{22}, B_{33}, \ldots, B_{mm}$. A *diagonal matrix* is a square matrix where all of the elements are zero, except those on the principal diagonal. We shall represent a diagonal matrix by the notation $\{B\}$. A *unit matrix*, which is denoted by $[I]$, is a diagonal matrix where all of the elements on the principal diagonal equal unity. Thus, the following matrix is a three-by-three unit matrix:

$$\begin{bmatrix} 1 & 0 & 0 \\ 0 & 1 & 0 \\ 0 & 0 & 1 \end{bmatrix}$$

The above matrix is also a square matrix and a diagonal matrix. A *column matrix* has only one column and is denoted by $\{B\}$, while a *row matrix* has only one row and is denoted by $\lfloor B \rfloor$.

Two matrices are equal only if the corresponding elements of both matrices are equal. Thus, $[A] = [B]$ only if $A_{ij} = B_{ij}$ for all values of *i* and *j*. This also means that the two matrices must have the same number of rows and columns. Next we shall discuss the basic matrix arithmetic operations: addition, subtraction, multiplication, and inversion.

Two matrices can be added or subtracted only if they are of the same size. To add two matrices, add the corresponding elements. Thus, $[A] + [B] = [C]$ implies that $C_{ij} = A_{ij} + B_{ij}$ for all *i* and *j* combinations. For example,

$$\begin{bmatrix} 4 & 2 \\ -1 & 5 \end{bmatrix} + \begin{bmatrix} 1 & 7 \\ 3 & 2 \end{bmatrix} = \begin{bmatrix} 4+1 & 2+7 \\ -1+3 & 5+2 \end{bmatrix} = \begin{bmatrix} 5 & 9 \\ 2 & 7 \end{bmatrix}$$

It can be seen that $[A] + [B] = [B] + [A]$. To subtract two matrices, subtract the corresponding elements. Thus, $[A] - [B] = [C]$ implies that $C_{ij} = A_{ij} - B_{ij}$ for all values of *i* and *j*. As an example,

$$\begin{bmatrix} 4 & 2 \\ -1 & 5 \end{bmatrix} - \begin{bmatrix} 1 & 7 \\ 3 & 2 \end{bmatrix} = \begin{bmatrix} 4-1 & 2-7 \\ -1-3 & 5-2 \end{bmatrix} = \begin{bmatrix} 3 & -5 \\ -4 & 3 \end{bmatrix}$$

There are two types of multiplication. If a matrix $[A]$ is multiplied by a constant *k* (i.e., a scalar term), then each element in the original matrix is multiplied by that scalar. Thus, $k[A] = [B]$ implies that $B_{ij} = kA_{ij}$ for all values of *i* and *j*. For example,

$$4\begin{bmatrix} 4 & 2 \\ -1 & 5 \end{bmatrix} = \begin{bmatrix} 16 & 8 \\ -4 & 20 \end{bmatrix}$$

The second type of multiplication involves the multiplication of two matrices, and this is not always possible. If we can obtain the product $[A][B]$, then the number of columns in $[A]$ must equal the number of rows of $[B]$. Usually $[A][B]$ does not equal $[B][A]$, so the order of the two matrices is important. Now let us suppose that it is possible to obtain $[C] = [A][B]$. The element C_{ij} (which is in the *i*th row and *j*th column of the product matrix $[C]$) is obtained by taking the sum of the products of the elements of row *i* of $[A]$ and the corresponding elements of column *j* of $[B]$. If matrix $[A]$ is of size *m* by *n* and matrix $[B]$ is of size *n* by *l*, this may be expressed in mathematical form by

$$C_{ij} = \sum_{k=1}^{n} A_{ik} B_{ik} \tag{4-21}$$

where *k* is a dummy index. Matrix $[C]$ will be of size *m* by *l*. That is, the number of rows in product matrix $[C]$ equals the number of rows in matrix $[A]$, and the number of columns of $[C]$ equals the number of columns in $[B]$. As a numerical example, let us consider

$$\begin{bmatrix} 1 & 3 & 4 \\ 2 & 0 & 5 \end{bmatrix} \begin{bmatrix} 4 & 0 \\ 0 & 2 \\ 1 & 5 \end{bmatrix} = \begin{bmatrix} 8 & 26 \\ 13 & 25 \end{bmatrix}$$

Let us represent the above matrix multiplication symbolically by $[A][B] = [C]$. Multiplication is possible because the number of columns of $[A]$ equals the number of rows of $[B]$ (i.e., both equal three). To obtain element C_{11}, we use the first row of $[A]$ and the first column of $[B]$, and perform the following calculation:

$$C_{11} = (1)(4) + (3)(0) + (4)(1) = 8$$

The element C_{12} is obtained using row 1 of $[A]$ and column 2 of $[B]$, and performing the following calculation:

$$C_{12} = (1)(0) + (3)(2) + (4)(5) = 26$$

Elements C_{21} and C_{22} are obtained using row 2 of $[A]$ and performing the following calculations:

$$C_{21} = (2)(4) + (0)(0) + (5)(1) = 13$$
$$C_{22} = (2)(0) + (0)(2) + (5)(5) = 25$$

As special cases, it can be shown that any column matrix pre-multiplied by any matrix (e.g., $[A]\{B\}$) results in a column matrix, and also that

$$[A][I] = [A] \tag{4-22}$$
$$[I][A] = [A] \tag{4-23}$$

where $[A]$ represents any square matrix, and $[I]$ is a unit matrix. The previous equations are analogous to the scalar equations: $1 \cdot k = k$ and $k \cdot 1 = k$. Thus, we see an analogy between the unit matrix and the number unity.

Matrix division is never possible, and the matrix operation that is most similar in concept is *matrix inversion*. Matrix inversion is possible only with a square matrix. Another requirement is that the determinant of the square matrix (e.g., $|A|$) must now equal zero. This might be considered to be analogous to division by zero. The inverse of a matrix $[A]$ is written as $[A]^{-1}$, and the definition of this *matrix inverse* is given by the following equation:

$$[A][A]^{-1} = [I] \tag{4-24}$$

That is, the multiplication of a square matrix with its inverse results in the unit matrix $[I]$. It can also be shown that $[A]^{-1}[A] = [I]$. Thus, a matrix inverse can be considered to be analogous to the reciprocal of a scalar, since $k \cdot k^{-1} = k^{-1}$. $k = 1$ if k^{-1} is the reciprocal of scalar k.

Methods for finding the inverse of a matrix, which are classified into the two broad categories of *elimination* and *iteration,* are briefly discussed in Sec. 4.7.

If $[A][X] = [B]$, then $[X] = [A]^{-1}[B]$, which is somewhat analogous to matrix division. To prove this result, if $[A][X] = [B]$, then premultiplication of both sides by $[A]^{-1}$ gives

$$[A]^{-1}[A][X] = [A]^{-1}[B]$$

Since $[A]^{-1}[A][X] = [I][X]$ and since $[I][X] = [X]$, we obtain the result that $[X] = [A]^{-1}[B]$. This result is used in Sec. 4.7 to show the usefulness of a matrix inverse for solving simultaneous, linear algebraic equations.

4.7   Methods for Matrix Inversion and Solution of Simultaneous Equations

We shall discuss the solution of simultaneous, linear algebraic equations along with matrix inversion in this section because the Gauss and Gauss-Jordan elimination methods can be used to solve both types of problems. A set of simultaneous, linear algebraic equations can be expressed in the matrix form of $[A]\{x\} = \{d\}$. As a simplified example, the following three linear algebraic equations:

$$a_{11}x_1 + a_{12}x_2 + a_{13}x_3 = d_1 \qquad \textbf{(4–25a)}$$
$$a_{21}x_1 + a_{22}x_2 + a_{23}x_3 = d_2 \qquad \textbf{(4–25b)}$$
$$a_{31}x_1 + a_{32}x_2 + a_{33}x_3 = d_3 \qquad \textbf{(4–25c)}$$

can be expressed by the following matrix equation, which can be verified by matrix multiplication:

$$\begin{bmatrix} a_{11} & a_{12} & a_{13} \\ a_{21} & a_{22} & a_{23} \\ a_{31} & a_{32} & a_{33} \end{bmatrix} \begin{Bmatrix} x_1 \\ x_2 \\ x_3 \end{Bmatrix} = \begin{Bmatrix} d_1 \\ d_2 \\ d_3 \end{Bmatrix}$$

Cramer's rule is a rather inefficient method for solving simultaneous, linear algebraic equations when the number of equations is large because of the large determinants that must be evaluated. The *Gaussian elimination method* uses algebraic methods to eliminate the unknowns. For a simplified example, let us consider the three linear algebraic equations given by Eqs. (4–25a, b, c). The first step of this method is to eliminate the term x_1 from all of the equations but the first equation. We can eliminate x_1 in the ith equation by multiplying the first equation by (a_{i1}/a_{11}) and subtracting the result from the ith equation. Thus, the new elements a'_{ij} and d'_i in the ith equation are given by

$$a'_{ij} = a_{ij} - \left(\frac{a_{i1}}{a_{11}}\right)a_{1j} \qquad (4\text{--}26)$$

$$d'_i = d_i - \left(\frac{a_{i1}}{a_{11}}\right)d_1 \qquad (4\text{--}27)$$

and the new equations are as follows:

$$a_{11}x_1 + a_{12}x_2 + a_{13}x_3 = d_1$$
$$a'_{22}x_2 + a'_{23}x_3 = d'_2$$
$$a'_{32}x_2 + a'_{33}x_3 = d'_3$$

We can now eliminate x_2 algebraically from the third of the new set of equations by multiplying the second equation by (a'_{32}/a'_{22}) and subtracting this result from the third equation. Thus, the new elements in the third equation are given by

$$a''_{33} = a'_{33} - \left(\frac{a'_{32}}{a'_{22}}\right)a'_{23}$$

$$d''_3 = d'_3 - \left(\frac{a'_{32}}{a'_{22}}\right)d'_2$$

and the new equations will be of the following triangular form:

$$a_{11}x_1 + a_{12}x_2 + a_{13}x_3 = d_1$$
$$a'_{22}x_2 + a'_{23}x_3 = d'_2$$
$$a''_{33}x_3 = d''_3$$

Now we can solve the third of the previous set of equations for x_3, then solve the second equation for x_2, and then solve the first equation for x_1. Since they were involved in divisions, we do not want coefficients a_{11} or a'_{22} to equal zero. We can always rearrange the set of equations so that this does not happen because switching the order of a pair of equations will not affect the solutions for x_1, x_2, and x_3.

The *Gauss-Jordan method* is a procedure that goes a few steps further by eliminating all variables but x_1 in the first equation, eliminating all variables but x_2 in the second equation, etc., so that we end up with the solutions for the x_i-unknowns. To permit the Gaussian or the Gauss-Jordan methods to be performed by a digital computer, we can form a matrix composed of the a_{ij}-and d_i-elements as follows. It is an augmented matrix for $[A]$ because we added an extra column composed of the d_i-elements to matrix $\lfloor A \rfloor$.

$$\begin{bmatrix} a_{11} & a_{12} & a_{13} & d_1 \\ a_{21} & a_{22} & a_{23} & d_2 \\ a_{31} & a_{32} & a_{33} & d_3 \end{bmatrix}$$

Since the equations that compute the new elements are of a set form, they can be programmed to compute the matrix

$$\begin{bmatrix} a_{11} & a_{12} & a_{13} & d_1 \\ 0 & a'_{22} & a'_{23} & d'_2 \\ 0 & a'_{32} & a'_{33} & d'_3 \end{bmatrix}$$

and finally the matrix

$$\begin{bmatrix} a_{11} & a_{12} & a_{13} & d_1 \\ 0 & a'_{22} & a'_{23} & d'_2 \\ 0 & 0 & a''_{33} & d''_3 \end{bmatrix}$$

from which x_1, x_2, and x_3 may be obtained. If necessary, matrix rows may be rearranged just as the algebraic equations can also be rearranged. The Gauss-Jordan method, which can also be programmed by use of generalized equations as shown in Ill. Ex. 4.12, ends up with a matrix of the following form, where the solutions are $x_1 = d'''_1$, $x_2 = d'''_2$, and $x_3 = d'''_3$.

$$\begin{bmatrix} 1 & 0 & 0 & d'''_1 \\ 0 & 1 & 0 & d'''_2 \\ 0 & 0 & 1 & d'''_3 \end{bmatrix}$$

The Gaussian elimination method and the Gauss-Jordan method are exemplified in Ill. Exs. 4.11 and 4.12. There are many other numerical methods for solving simultaneous, linear algebraic equations (including iterative techniques), and their descriptions can be found in numerical analysis texts.

Now let us discuss matrix inversion. The reader has learned a matrix inversion method in a previous algebra course where the first step in the procedure was to replace each element of the matrix with its co-factor. This method is a good hand-calculation method when we want to invert a square matrix of size 2 or 3, but it is a very inefficient method for large matrices because each calculation of a co-factor requires the evaluation of a determinant. The determinant of the whole matrix must also be evaluated. Thus, if this method is used to invert a matrix of size n, then $(n^2 + 1)$ determinants must be evaluated. There are many other methods that calculate an inverse by direct means, and which are more efficient in comparison. Some of these better known methods are the *Gauss, Gauss-Jordan, Crout, Choleski*, and *Doolittle* matrix inversion methods. The *Gauss-Jordan matrix inversion method* is an elimination method that uses the same mathematical operations as those for the Gauss-Jordan method for solving simultaneous, linear algebraic equations. The difference is that each starts with a different augmented matrix. A matrix is augmented when additional columns are added to it. A similar statement can be made for the Gauss matrix inversion method with regard to the Gauss elimination method for solving simultaneous algebraic equations. If we wanted to invert a matrix $[A]$ of size 3 by the Gauss-Jordan method, the initial augmented matrix would be as follows, where the nine a_{ij}-terms are the nine elements of matrix $[A]$:

$$\begin{bmatrix} a_{11} & a_{12} & a_{13} & 1 & 0 & 0 \\ a_{21} & a_{22} & a_{23} & 0 & 1 & 0 \\ a_{31} & a_{32} & a_{33} & 0 & 0 & 1 \end{bmatrix}$$

This method performs specific operations to calculate successive augmented matrices. The details for this method are exemplified in Ill. Ex. 4.13. When the method is completed, the last augmented matrix will be of the form

$$\begin{bmatrix} 1 & 0 & 0 & e_{11} & e_{12} & e_{13} \\ 0 & 1 & 0 & e_{21} & e_{22} & e_{23} \\ 0 & 0 & 1 & e_{31} & e_{32} & e_{33} \end{bmatrix}$$

and the inverse matrix is

$$\begin{bmatrix} e_{11} & e_{12} & e_{13} \\ e_{21} & e_{22} & e_{23} \\ e_{31} & e_{32} & e_{33} \end{bmatrix}$$

That is, the Gauss-Jordan method, in effect, ends at a condition where the first three columns of the *initial* augmented matrix are multiplied by their inverse and where the last three columns of the initial matrix are also multiplied by this inverse. A comparison of the initial and the final augmented matrices should show this effect. The equations used to calculate the successive augmented matrices are the same as those used to calculate the successive matrices when the Gauss-Jordan method is used to solve a set of linear algebraic equations. These equations are given and are verified in Ill. Ex. 4.12, which solves a set of linear algebraic equations. These equations are applied in Ill. Ex. 4.13 to invert a matrix. To summarize the calculation procedure, this method obtains the inverse of a square matrix of size n in n steps or reductions. For the kth step, the elements in the kth row, for the next augmented matrix, are calculated from the equation

$$a'_{kj} = \frac{a_{kj}}{a_{kk}} \qquad (4\text{–}28)$$

and the other elements for the next augmented matrix are calculated from the equation

$$a'_{ij} = a_{ij} - a_{ik}a'_{kj} \qquad (4\text{–}29)$$

where a_{ij}, a_{kj}, a_{ik}, and a_{kk} are elements of the previous matrix, and where a'_{ij} and a'_{kj} are computed elements for the next augmented matrix. As can be verified from Ill. Ex. 4.13, when a matrix of size 3 is inverted by the Gauss-Jordan method, the first step obtains an augmented matrix of the following form, when Eqs. (4–28) and (4–29) are applied:

$$\begin{bmatrix} 1 & b_{12} & b_{13} & b_{14} & 0 & 0 \\ 0 & b_{22} & b_{23} & b_{24} & 1 & 0 \\ 0 & b_{32} & b_{33} & b_{34} & 0 & 1 \end{bmatrix}$$

and the second step obtains an augmented matrix of the following form when these two equations are applied:

$$\begin{bmatrix} 1 & 0 & d_{13} & d_{14} & d_{15} & 0 \\ 0 & 1 & d_{23} & d_{24} & d_{25} & 0 \\ 0 & 0 & d_{33} & d_{34} & d_{35} & 1 \end{bmatrix}$$

The inverse matrix is obtained in the third step by these two equations, as is shown in Ill. Ex. 4.13. The numerous arithmetic operations that are required when a large matrix is inverted by a direct method can cause significant roundoff errors in the results. These errors can be corrected by following a direct-method solution with an iterative type of solution. Thus, the better matrix inverse digital-computer subprograms combine a direct solution method with an iterative method. One iterative method applies the following equation to successively compute a new inverse $[A^{-1}]_{i+1}$ from a previously computed inverse $[A^{-1}]_i$, until these two matrices are equal and where $[I]$ is a unit matrix.

$$[A^{-1}]_{i+1} = [A^{-1}]_i [2[I] - [A][A^{-1}]_i] \qquad (4\text{-}30)$$

Illustrative Example 4.11. Solve the following three algebraic equations for x_1, x_2, and x_3 using the Gaussian elimination method.

$$x_1 + 0.20x_2 + 0.50x_3 = 2.00$$
$$0.20x_1 + x_2 + 0.30x_3 = 1.00$$
$$0.50x_1 + 0.30x_2 + x_3 = 3.00$$

Solution. The initial augmented matrix for this problem is as follows:

$$\begin{bmatrix} 1.00 & 0.20 & 0.50 & 2.00 \\ 0.20 & 1.00 & 0.30 & 1.00 \\ 0.50 & 0.30 & 1.00 & 3.00 \end{bmatrix}$$

Using Eqs. (4-26) and (4-27), we obtain

$$a'_{21} = 0.20 - \frac{0.20}{1.00}(1.00) = 0$$

$$a'_{22} = 1.00 - \frac{0.20}{1.00}(0.20) = 0.96$$

$$a'_{23} = 0.30 - \frac{0.20}{1.00}(0.50) = 0.20$$

$$d'_2 = 1.00 - \frac{0.20}{1.00}(2.00) = 0.60$$

$$a'_{31} = 0.50 - \frac{0.50}{1.00}(1.00) = 0$$

$$a'_{32} = 0.30 - \frac{0.50}{1.00}(0.20) = 0.20$$

$$a'_{33} = 1.00 - \frac{0.50}{1.00}(0.50) = 0.75$$

$$d'_3 = 3.00 - \frac{0.50}{1.00}(2.00) = 2.00$$

Thus, the second augmented matrix is

$$\begin{bmatrix} 1.00 & 0.20 & 0.50 & 2.00 \\ 0 & 0.96 & 0.20 & 0.60 \\ 0 & 0.20 & 0.75 & 2.00 \end{bmatrix}$$

Since calculation shows that $a''_{33} = 0$, $a''_{33} = 0.708$, and $d''_{33} = 1.875$, the third augmented matrix is:

$$\begin{bmatrix} 1.000 & 0.200 & 0.500 & 2.000 \\ 0 & 0.960 & 0.200 & 0.600 \\ 0 & 0 & 0.708 & 1.875 \end{bmatrix}$$

Thus, $x_3 = 1.875/0.708 = 2.65$. The other two solutions are

$$x_2 = [0.600 - 0.200(2.65)]/0.960 = 0.0720$$
$$x_1 = [2.000 - 0.500(2.65) - 0.200(0.072)] = 0.661$$

Illustrative Example 4.12. Solve the problem in Ill. Ex. 4.11 using the Gauss-Jordan elimination method.

Solution. Since both procedures start out the same, the first two augmented matrices will be the same as those obtained by Gaussian elimination. In this method, we also divide the first row by a_{11} so that a'_{11} will always equal unity. To calculate the third augmented matrix, we use the following equations, where it should be noted that we first divide the second row by a'_{22} so that a''_{22} will equal unity. For convenience, we have now represented elements d_1, d_2, and d_3 by a_{14}, a_{24}, and a_{34}, respectively.

$$a''_{2j} = \frac{a'_{2j}}{a'_{22}}$$
$$a''_{1j} = a'_{1j} - a_{12}a''_{2j}$$
$$a''_{3j} = a'_{3j} - a_{32}a''_{2j}$$

Use of the previous equations gives us the following third augmented matrix:

$$\begin{bmatrix} 1.000 & 0 & 0.458 & 1.875 \\ 0 & 1.000 & 0.208 & 0.625 \\ 0 & 0 & 0.708 & 1.875 \end{bmatrix}$$

To calculate the fourth augmented matrix, we use the following equations, where it should be noted that we first divide the third row by a_{33}'' so that a_{33}''' will equal unity:

$$a_{3j}''' = \frac{a_{3j}''}{a_{33}''}$$
$$a_{1j}''' = a_{1j}'' - a_{13}'' a_{3j}'''$$
$$a_{2j}''' = a_{2j}'' - a_{23}'' a_{3j}'''$$

Use of the previous equations gives us the following fourth augmented matrix:

$$\begin{bmatrix} 1.000 & 0 & 0 & 0.066 \\ 0 & 1.000 & 0 & 0.073 \\ 0 & 0 & 1.000 & 2.648 \end{bmatrix}$$

Thus, the solutions are $x_1 = 0.666$, $x_2 = 0.073$, and $x_3 = 2.648$. From this example, it can be seen that the operations for calculating the $(k + 1)$th augmented matrix are to first divide the kth row by element a_{kk} so that the next value of a_{kk} will equal unity. That is, $a_{kj}' = a_{kj}/a_{kk}$. The other elements can then be calculated from the following equation, where a_{ij}' now denotes the elements of the $(k + 1)$th matrix, and a_{ij} now denotes the elements of the kth augmented matrix:

$$a_{ij}' = a_{ij} - a_{ik} a_{kj}' = a_{ij} - a_{ik} \frac{a_{kj}}{a_{kk}}$$

Illustrative Example 4.13. Invert the following matrix using the Gauss-Jordan method.

$$\begin{bmatrix} 4 & 1 & 2 \\ 6 & 8 & 7 \\ 2 & 5 & 3 \end{bmatrix}$$

Solution. The initial augmented matrix is

$$\begin{bmatrix} 4.0 & 1.0 & 2.0 & 1.0 & 0 & 0 \\ 6.0 & 8.0 & 7.0 & 0 & 1.0 & 0 \\ 2.0 & 5.0 & 3.0 & 0 & 0 & 1.0 \end{bmatrix}$$

As stated previously, for the kth step (i.e., for the $(k + 1)$th augmented matrix), the elements of the kth row are calculated by the equation

$$d'_{kj} = \frac{a_{kj}}{a_{kk}}$$

and the other elements are calculated by the equation

$$a'_{ij} = a_{ij} - a_{ik}d'_{kj}$$

Thus, for the first step, the equations are $a'_{ij} = a_{1j}/a_{11}$ for the first row and $a'_{ij} = a_{ij} - a_{11}a'_{ij}$ for the other rows. Since $a_{11} = 4$, we have $a'_{ij} = a_{1j}/4$. Applying these equations, the second augmented matrix is

$$\begin{bmatrix} 1.00 & 0.25 & 0.50 & 0.25 & 0 & 0 \\ 0 & 6.50 & 4.00 & -1.50 & 1.00 & 0 \\ 0 & 4.50 & 2.00 & -0.50 & 0 & 1.00 \end{bmatrix}$$

For the second step, the equations are $a'_{2j} = a_{2j}/a_{22}$ for the second row and $a'_{ij} = a_{ij} - a_{i2}a'_{2j}$ for the other two rows. Applying these equations, the third augmented matrix is

$$\begin{bmatrix} 1.000 & 0 & 0.346 & 0.306 & -0.038 & 0 \\ 0 & 1.000 & 0.615 & -0.231 & 0.154 & 0 \\ 0 & 0 & -0.770 & 0.540 & -0.692 & 1.000 \end{bmatrix}$$

For the third step, the equations are $a'_{3j} = a_{3j}/a_{33}$ for the third row and $a'_{ij} = a_{ij} - a_{i3}a'_{3j}$ for the other two rows. Applying these equations, the fourth augmented matrix is

$$\begin{bmatrix} 1.000 & 0 & 0 & 0.550 & -0.350 & 0.450 \\ 0 & 1.000 & 0 & 0.200 & -0.400 & 0.800 \\ 0 & 0 & 1.000 & -0.700 & 0.900 & -1.300 \end{bmatrix}$$

Thus, the inverse of the given matrix was found to be the following matrix:

$$\begin{bmatrix} 0.550 & -0.350 & 0.450 \\ 0.200 & -0.400 & 0.800 \\ -0.700 & 0.900 & -1.300 \end{bmatrix}$$

The previous matrix is the inverse of the given matrix because matrix multiplication shows that

$$
\begin{bmatrix}
4.00 & 1.00 & 2.00 \\
6.00 & 8.00 & 7.00 \\
2.00 & 5.00 & 3.00
\end{bmatrix}
\begin{bmatrix}
0.550 & -0.350 & 0.450 \\
0.200 & -0.400 & 0.800 \\
-0.700 & 0.900 & -1.300
\end{bmatrix}
$$

$$
=
\begin{bmatrix}
1.000 & 0 & 0 \\
0 & 1.000 & 0 \\
0 & 0 & 1.000
\end{bmatrix}
$$

4.8   Methods for Calculating Eigenvalues

Equations (3–1a, b) in Sec. 3.2 can be expressed in the following matrix form:

$$
\begin{bmatrix} m_1 & 0 \\ 0 & m_2 \end{bmatrix}
\begin{Bmatrix} \ddot{x}_1 \\ \ddot{x}_2 \end{Bmatrix}
+
\begin{bmatrix} k_1 + k & -k \\ -k & k_2 + k \end{bmatrix}
\begin{Bmatrix} x_1 \\ x_2 \end{Bmatrix}
=
\begin{Bmatrix} 0 \\ 0 \end{Bmatrix}
\qquad \textbf{(4–31)}
$$

where these equations are the equations of motion for the spring-mass system in Fig. 3–1. The equations of motion for any undamped, free vibration can be written in the matrix form

$$
[m]\{\ddot{x}\} + [k]\{x\} = 0 \qquad \textbf{(4–32)}
$$

where $[m]$ is called the *inertia matrix*, $[k]$ is called the *stiffness matrix*, and $\{x\}$ is the *displacement matrix*. In Eq. (4–31) we see that

$$
[m] = \begin{bmatrix} m_1 & 0 \\ 0 & m_2 \end{bmatrix}
$$

$$
[k] = \begin{bmatrix} k_1 + k & -k \\ -k & k_2 + k \end{bmatrix}
$$

It may be noted that the inertia matrix $[m]$ is a diagonal matrix only when there is no dynamic coupling in the equations of motion for the system. The inverse of a diagonal matrix is another diagonal matrix where the inverse diagonal elements are the reciprocals of the original diagonal elements. This can be verified by matrix multiplication as follows:

$$\begin{bmatrix} d_{11} & 0 & 0 \\ 0 & d_{22} & 0 \\ 0 & 0 & d_{33} \end{bmatrix} \begin{bmatrix} \dfrac{1}{d_{11}} & 0 & 0 \\ 0 & \dfrac{1}{d_{22}} & 0 \\ 0 & 0 & \dfrac{1}{d_{33}} \end{bmatrix} = \begin{bmatrix} 1 & 0 & 0 \\ 0 & 1 & 0 \\ 0 & 0 & 1 \end{bmatrix}$$

New let us pre-multiply Eq. (4–32) by $[m]^{-1}$, the inverse of the inertia matrix, to obtain

$$[m]^{-1}[m]\{\ddot{x}\} + [m]^{-1}[k]\{x\} = 0$$

Since $[m]^{-1}[m] = [I]$, and if we define matrix $[H]$ to equal $[m]^{-1}[k]$, the previous equation becomes

$$\{\ddot{x}\} + [H]\{x\} = 0 \qquad\qquad (4\text{–}33)$$

where $[H]$ is called the *dynamic matrix*. For the spring-mass system in Fig. 3–1,

$$[H] = [m]^{-1}[k] = \begin{bmatrix} \dfrac{1}{m_1} & 0 \\ 0 & \dfrac{1}{m_2} \end{bmatrix} \begin{bmatrix} k_1 + k & -k \\ -k & k_2 + k \end{bmatrix}$$

$$- \begin{bmatrix} \dfrac{k_1 + k}{m_1} & \dfrac{-k}{m_1} \\ \dfrac{-k}{m_2} & \dfrac{k_2 + k}{m_2} \end{bmatrix} \qquad\qquad (4\text{–}34)$$

Now let us discuss and define eignevalues. The *eigenvalues* (also called *characteristic values*) of a matrix are the roots of a special determinant equation. That is, the eigenvalues of a square matrix $[A]$ are the roots of the equation

$$|[A] - \lambda[I]| = 0 \qquad\qquad (4\text{–}35)$$

If matrix $[A]$ is of size 3 by 3, then the above equation may be written as follows:

$$\begin{vmatrix} A_{11} - \lambda & A_{12} & A_{13} \\ A_{21} & A_{22} - \lambda & A_{23} \\ A_{31} & A_{32} & A_{33} - \lambda \end{vmatrix} = 0$$

Since the above determinant, when evaluated, results in a polynomial in λ, the above equation is a polynomial equation which is called the *characteristic equation*. The roots of this characteristic equation (i.e., polynomial) are called the *characteristic numbers* or *eigenvalues* of matrix $[A]$. Let us illustrate that Eq. (4–33) describes a physical problem that requires the determination of eigenvalues in order to obtain the natural frequencies for the problem solution. At a principal mode, the system is vibrating in harmonic motion at a natural frequency ω, so that $\ddot{x}_i = -\omega^2 x_i$ and $\{\ddot{x}\} = -\omega^2\{x\}$. Thus, for a principal-mode vibration, Eq. (4–33) becomes

$$-\omega^2\{x\} + [H]\{x\} = [[H] - \omega^2[I]]\{x\} = 0$$

Since $\{x\} \neq 0$, the previous equation has a solution only if the determinant of the matrix that is a coefficient of $\{x\}$ (which we shall call $[H - \omega^2 I]$) equals zero. Thus, for an n-degree-of-freedom system, we have

$$|[H - \omega^2 I]| = \begin{vmatrix} H_{11} - \omega^2 & H_{12} & \cdots & H_{1n} \\ H_{21} & H_{22} - \omega^2 & \cdots & H_{2n} \\ \vdots & \vdots & & \vdots \\ H_{n1} & H_{n2} & \cdots & H_{nn} - \omega^2 \end{vmatrix} = 0$$

where H_{ij} is an element of the dynamic matrix which equals $[[m]^{-1}[k]]$. For the spring-mass system in Fig. 3–1, substitution of Eq. (4–34) in Eq. (4–36) gives

$$\begin{vmatrix} \dfrac{k_1 + k}{m_1} - \omega^2 & -\dfrac{k}{m_1} \\ -\dfrac{k}{m_2} & \dfrac{k_2 + k}{m_2} \end{vmatrix} = 0$$

Comparison of the previous equation with Eq. (3–4) in Sec. 3.2 shows that the previous equation is equivalent to Eq. (3–4) if the first and second rows of the previous equation are multiplied by m_1 and m_2, respectively. Thus, Eq. (4–36) is the characteristic equation for the vibration of an undamped system that possesses n degrees of freedom from which we can determine its n natural frequencies. More important, the n eigenvalues of dynamic matrix $[H]$ are equal to the squares of the n natural frequencies, and matrix $[H]$ can be computed from the equations of motion for the system.

It is seen that only square matrices have eigenvalues and that an nth-order square matrix has n eigenvalues. This is because the characteristic equation is a polynomial of degree n in λ, as can be shown when the determinant in Eq. (4–35) is expanded. There is also an *eigenvector* corresponding to each discrete eigenvalue, where the eigenvector associated with eigenvalue λ_i of matrix $[A]$ is defined to be the column matrix $\{x\}$ that satisfies the following equation:

$$[[A] - \lambda_i[I]]\{x\} = 0 \qquad (4\text{–}37)$$

It can be seen that one procedure for determining the eigenvalues of a matrix consists first of determining the characteristic polynomial equation and then computing the roots of this polynomial, as shown in Ill. Ex. 4.14. Complex roots or eigenvalues usually have complex eigenvectors associated with them. For obtaining the characteristic polynomial of large matrices, which involves the expansion of a large determinant, numerical methods such as the *Leverrier-Faddeev* method may be used. The reader is referred to numerical analysis texts for details. Methods for obtaining the roots of polynomial equations were discussed in Sec. 3.3. There are also special methods for finding specific eigenvalues of a matrix (e.g., the smallest or the largest one), and there are some methods that obtain the eigenvalues from the matrix itself without first determining the characteristic equation. The first part of Ill. Ex. 4.14 exemplifies how the eigenvalues of a matrix may be determined by obtaining the roots of a polynomial equation and how the corresponding eigenvectors are obtained by solving a set of simultaneous, linear algebraic equations. The last part of this example shows how the largest eigenvalue and its eigenvector may be determined directly from the matrix itself by using a matrix-iteration technique.

Since there are digital computer subprograms available that calculate the eigenvalues of a matrix, it is seen that these subprograms can be very useful in professional practice to calculate the natural frequencies of vibration for complex structures. Figure 3–1 and Eqs. (3–1), (4–31), and (4–32) illustrate and define k_{ij}, an element of the stiffness matrix. Since it is easier to measure the influence coefficient α_{ij} than the stiffness k_{ij} of a complex structure and since it can be shown that

$[\alpha] = [k]^{-1}$, we can calculate the natural frequencies from the eigenvalues of matrix $[H]$, where $[H] = [m]^{-1}[\alpha]^{-1}$. Methods for obtaining the inverse of a matrix were discussed in Sec. 4.8, and these methods are also available as digital-computer subprograms. Let us show that $[\alpha] = [k]^{-1}$. From the definition of an influence coefficient, we have

$$\{x\} = [\alpha]\{F\}$$

where $[\alpha]$ is the matrix of the α_{ij}-influence-coefficients and $\{F\}$ is the matrix of the statically-applied forces. The static-force-balance equations when static forces are applied to this system can be written in matrix form as $\{F\} = [k]\{x\}$. Substitution of this equation for $\{F\}$ in the previous equation for $\{x\}$ gives

$$\{x\} = [\alpha][k]\{x\}$$

from which we see that $[\alpha][k] = [I]$ or $[\alpha] = [k]^{-1}$.

From Eqs. (3–39a, b, c) in Sec. 3.8, we see that the equations of motion for the free vibration of an undamped system, when at a principal mode, can be written in the following matrix form using influence coefficients:

$$\{x\} = \omega^2[\alpha]\{mx\}$$

Using the definition of an eigenvalue, and using Eq. (3–40) in Sec. 3.8, we see that the eigenvalues of the following matrix are equal to the squares of the reciprocals of the natural frequencies (i.e., $\lambda_i = 1/\omega_i^2$):

$$[\alpha][m] = \begin{bmatrix} \alpha_{11}m_1 & \alpha_{12}m_2 & \cdots & \alpha_{1n}m_n \\ \alpha_{21}m_1 & \alpha_{22}m_2 & \cdots & \alpha_{2n}m_n \\ \vdots & \vdots & & \vdots \\ \alpha_{n1}m_1 & \alpha_{n2}m_2 & \cdots & \alpha_{nn}m_n \end{bmatrix} \tag{4-38}$$

The right-hand side of the previous equation shows that inertia matrix $[m]$ must be a diagonal matrix (i.e., no dynamic coupling) if we wish to obtain the natural frequencies from the eigenvalues of matrix $[\alpha m] = [\alpha]\{m\}$.

If we wished to consider more general systems, the equations of motion for the forced vibration of a linearly-damped system having n degrees of freedom can be written in matrix form for any coordinate system as follows:

$$[m]\{\ddot{q}\} + [c]\{\dot{q}\} + [k]\{q\} = \{Q\} \tag{4-39}$$

where $[c]$ is the *linear damping matrix*, $\{q\}$ is the *coordinate matrix*, and $\{Q\}$ is the *generalized force matrix* that corresponds to the generalized coordinates $\{q\}$. If the system is lightly damped, has no external forces, and is stable, then the roots of its characteristic equation (i.e., eigenvalues) will be complex with negative real parts and will occur in complex conjugate pairs. The characteristic roots for such damped, free vibrations was previously discussed in more general fashion in Sec. 3.4.

Illustrative Example 4.14. Find the eigenvalues and the eigenvectors of the following matrix:

$$\begin{bmatrix} 1.0 & 0.2 & 0.5 \\ 0.2 & 1.0 & 0.3 \\ 0.5 & 0.3 & 1.0 \end{bmatrix}$$

Solution. The three eigenvalues for this matrix may be obtained by solving the following determinant equation, as shown by Eq. (4-35);

$$\begin{bmatrix} 1.0 - \lambda & 0.2 & 0.5 \\ 0.2 & 1.0 - \lambda & 0.3 \\ 0.5 & 0.3 & 1.0 - \lambda \end{bmatrix} = 0$$

Expansion of the determinant gives the following third degree polynomial equation:

$$f(\lambda) = \lambda^3 - 3\lambda^2 + 2.62\lambda + 0.68 = 0$$

We shall solve for one root using the Newton-Rhapson method, which is

$$\lambda_{n+1} = \lambda_n - \frac{f(\lambda)}{f'(\lambda)}$$

Assuming that $\lambda = 1.000$ gives $\lambda_1 = 1.000$ and

$$\lambda_2 = 1.000 - \frac{f(1.000)}{f'(1.000)} = 1.000 - \frac{(-0.060)}{(-0.380)} = 0.842$$

$$\lambda_3 = 0.842 - \frac{f(0.842)}{f'(0.842)} = 0.842 - \frac{(-0.050)}{(-0.305)} = 0.828$$

Another calculation gives $\lambda_4 = 0.828$, which means that one root of $f(\lambda)$ is 0.828. Division of $f(\lambda)$ by $(\lambda - 0.828)$ gives

$$g(\lambda) = \lambda^2 - 2.172\lambda + 0.8216$$

The roots of $g(\lambda)$ can be easily obtained by use of the quadratic formula. The results are $\lambda = 1.685$ and $\lambda = 0.488$. The three eigenvalues are, thus, equal to 0.488, 0.828, and 1.685.

An eigenvector is defined to be the set of x_i-values that satisfy the following matrix equation for a specific value of eigenvalue λ_i:

$$[[A] - \lambda_i[I]]\{x\} = 0$$

The previous equation can be rewritten in the following form:

$$[A]\{x\} = \lambda_i\{x\}$$

Insertion of the a_{ij}-elements for matrix $[A]$ and preforming a matrix multiplication gives

$$x_1 + 0.2x_2 + 0.5x_3 = \lambda_1 x_1$$
$$0.2x_1 + x_2 + 0.3x_3 = \lambda_1 x_2$$
$$0.5x_1 + 0.3x_2 + x_3 = \lambda_1 x_3$$

For $\lambda_i = 1.685$, these equations become

$$x_1 + 0.2x_2 + 0.5x_3 = 1.685x_1$$
$$0.2x_1 + x_2 + 0.3x_3 = 1.685x_2$$
$$0.5x_1 + 0.3x_2 + x_3 = 1.685x_3$$

Solution of these three algebraic equations, after setting $x_1 = 1.000$, gives $x_2 = 0.757$ and $x_3 = 1.064$. Thus, we have the eigenvector that is associated with an eigenvalue of 1.685. Repeating this procedure for $\lambda_i = 0.828$ gives $x_1 = 1.00$, $x_2 = -1.91$, and $x_3 = 0.420$. For $\lambda_i = 0.487$, we obtain $x_1 = 1.00$, $x_2 = 0.274$, and $x_3 = -1.14$.

Let us now illustrate how we may solve for the eigenvalues and eigenvectors by use of a relatively simple iteration precedure. A version of this iteration method is given in Sec. 4.9, which is very commonly used in vibration analysis to find the natural frequencies of spring-mass, disk-shaft, and complex elastic structures. Since an eigenvector can be defined to be the $\{x\}$-matrix that satisfies the following equation, where λ is an eigenvalue for matrix $[A]$:

$$\{r\} \equiv [A]\{x\} = \lambda\{x\}$$

we can use the following iteration equation, where $\{x\}_{n+1}$ is the next approximation for $\{x\}$ and is calculated from the previous approximation $\{x\}_n$:

$$\{x\}_{n+1} = [A]\{x\}_n$$

The iteration is completed when we come to a condition where

$$\{x\}_{n+1} = k\{x\}_n$$

When this condition occurs, k is an eigenvalue for matrix $[A]$ and $\{x\}_{n+1}$ is its eigenvector. To apply this iteration technique to this problem, let us first assume that $x_1 = x_2 = x_3 = 1$. Thus, the first iteration gives

$$\{x\}_1 = \begin{bmatrix} 1.0 & 0.2 & 0.5 \\ 0.2 & 1.0 & 0.3 \\ 0.5 & 0.3 & 1.0 \end{bmatrix} \begin{Bmatrix} 1.00 \\ 1.00 \\ 1.00 \end{Bmatrix} = \begin{Bmatrix} 1.70 \\ 1.50 \\ 1.80 \end{Bmatrix} = 1.70 \begin{Bmatrix} 1.00 \\ 0.88 \\ 1.06 \end{Bmatrix}$$

The values of $\{x\}_{n+1}$ are normalized so that one x_i-element equals unity before it is utilized in the next iteration. This is done to prevent large magnitudes.

The second iteration gives

$$\{x\}_2 = \begin{bmatrix} 1.0 & 0.2 & 0.5 \\ 0.2 & 1.0 & 0.3 \\ 0.5 & 0.3 & 1.0 \end{bmatrix} \begin{Bmatrix} 1.00 \\ 0.88 \\ 1.06 \end{Bmatrix} = \begin{Bmatrix} 1.72 \\ 1.40 \\ 1.82 \end{Bmatrix} = 1.72 \begin{Bmatrix} 1.00 \\ 0.817 \\ 1.06 \end{Bmatrix}$$

The ninth iteration gives

$$\{x\}_9 = [A] \begin{Bmatrix} 1.00 \\ 0.755 \\ 1.06 \end{Bmatrix} = \begin{Bmatrix} 1.68 \\ 1.27 \\ 1.79 \end{Bmatrix} = 1.68 \begin{Bmatrix} 1.00 \\ 0.760 \\ 1.06 \end{Bmatrix}$$

Since the tenth iteration gives

$$\{x\}_{10} = [A] \begin{Bmatrix} 1.00 \\ 0.760 \\ 1.06 \end{Bmatrix} = \begin{Bmatrix} 1.68 \\ 1.27 \\ 1.79 \end{Bmatrix} = 1.68 \begin{Bmatrix} 1.00 \\ 0.760 \\ 1.06 \end{Bmatrix}$$

one eigenvalue for matrix $[A]$ equals 1.68, and its corresponding eigenvector is given by $x_1 = 1.00$, $x_2 = 0.760$, and $x_3 = 1.06$. This method produces the largest eigenvalue, which for some problems is all that is desired. For other problems, however, we only desire to find the smallest eigenvalue. This can be done by devising an iteration procedure to solve the following matrix equation for $(\lambda - f)$:

$$[[A] - f[I]]\{x\} = (\lambda - f)\{x\}$$

where constant f equals the largest eigenvalue, or approximately so. Thus, $|\lambda - f|$ becomes a maximum when λ equals the smallest eigenvalue. Another iteration technique is the *Jacobi method*, whose details may be found in numerical analysis references.

4.9   The Matrix-Iteration Method

In Sec. 4.8 we saw that the natural frequencies of an undamped system undergoing a free vibration can be determined from the eigenvalues of dynamic matrix $[H]$, where

$$[H] = [m]^{-1}[k] = [m]^{-1}[\alpha]^{-1} \qquad \textbf{(4-40)}$$

where $[\alpha] = [k]^{-1}$, and where the eigenvalues equal the squares of the natural frequncies (i.e., $\lambda_i = \omega_i^2$). We saw that the natural frequencies ω_i can also be calculated from the eigenvalues λ_i of the following matrix, where $\lambda_i = 1/\omega_i^2$:

$$[\alpha m] = [\alpha][m] = [k]^{-1}[m] \qquad (4\text{-}41)$$

Also in Sec. 4.8, we saw that the equations of motion for the free vibration of an undamped system, when at a principal mode, can be written in the following matrix form using influence coefficients:

$$\{x\} = \omega^2[\alpha]\{mx\} = \omega^2[\alpha][m]\{x\} = \omega^2[\alpha m]\{x\} \qquad (4\text{-}42)$$

Because of the form of Eq. (4-42), we can apply an iteration method to compute the values of ω^2 that is similar to the iteration method given in the last part of Ill. Ex. 4.14. The iteration equation is $\{x\}_{n+1} = \omega^2[\alpha m]\{x\}_n$, and we will converge to a solution when $\{x\}_{n+1} = d\{x\}_n$, where d is a constant. For the first cycle of the iteration, we must assume the values for column matrix $\{x\}$, which is the configuration of the principal mode whose natural frequency we wish to find. That is, $\{x\}_1$ contains the assumed values for $x_1, x_2, \cdots, x_n$; and we compute matrix $\{x\}_2$ from $\{x\}_2 = \omega^2[\alpha m]\{x\}_1$. The next cycle of the iteration computes $\{x\}_3$ from the equation $\{x\}_3 = \omega^2[\alpha m]\{x\}_2$, and further values of $\{x\}_{n+1}$ are computed in a similar fashion until we reach a point where $\{x\}_{n+1}$ is proportional to $\{x\}_n$. When the iteration has converged to this point, we can compute the lowest natural frequency of the system from the given matrix equation. To prevent large numbers, it is best to normalize the trial value for $\{x\}_n$ so that it equals the previously calculated $\{x\}_{n+1}$-matrix divided by one of its x_i-elements so that one element in the trial $\{x\}_n$-matrix will equal unity. For the next higher modes and natural frequencies, the orthogonality principle is used to obtain a new set of matrix equations, where this iteration method is repeated to obtain these higher natural frequencies. The details for this iteration method (and the use of the orthogonality principle) are much more clearly shown in the illustrative examples that are furnished in this section.

It should be noted that the lowest natural frequency and mode shape are obtained when the iteration method is applied to the principal-mode equation of motion $\{x\} = \omega^2[\alpha m]\{x\}$. If the iteration method is applied to the principal-mode equation of motion $\{x\} = (1/\omega^2)[H]\{x\}$, which can be obtained from Eq. (4-33) by letting $\{\ddot{x}\} = -\omega^2\{x\}$, then the highest natural frequency and mode shape (i.e., the lowest value of $1/\omega^2$) will be obtained. In the beginning of this section, note that the eigenvalues of matrix $[H]$ equal the inverse of the eigenvalues for matrix $[\alpha m]$.

Illustrative Example 4.15. Calculate the natural frequencies for the spring-mass system in Ill. Ex. 4.6 using the matrix-iteration method.

Solution. Let us use the static displacements as our first approximation for the values of x_1 and x_2 for the lowest principal mode. Use of these values from Ill. Ex. 4.6 gives $x_2/x_1 = 9.50/6.50 = 1.47$. If we wish to normalize the value of x_1 to unity, then we have $x_1 = 1.00$ and $x_2 = 1.47$. Using the results of Ill. Ex. 3.18 in Sec. 3.8, the influence coefficients are

$$\alpha_{11} = \alpha_{12} = \alpha_{21} = \frac{1}{k_1} = \frac{1}{4.00} = 0.250$$

$$\alpha_{22} = \frac{1}{k_1} + \frac{1}{k} = \frac{1}{4.00} + \frac{1}{7.00} = 0.393$$

Since $m_1 = 5.00/386 = 0.0129$ and $m_2 = 21.0/386 = 0.0545$, we have

$$[\alpha m] = [\alpha][m] = \begin{bmatrix} 0.250 & 0.250 \\ 0.250 & 0.393 \end{bmatrix} \begin{bmatrix} 0.0129 & 0 \\ 0 & 0.0545 \end{bmatrix}$$

and the iteration equation is

$$\begin{Bmatrix} x_1 \\ x_2 \end{Bmatrix} = \omega^2 [\alpha m]\{x\} = \omega^2 \begin{bmatrix} 0.00322 & 0.0136 \\ 0.00322 & 0.0213 \end{bmatrix} \begin{Bmatrix} x_1 \\ x_2 \end{Bmatrix}$$

Use of $x_1 = 1.00$ and $x_2 = 1.47$ for first trial values gives

$$\begin{Bmatrix} x_1 \\ x_2 \end{Bmatrix} = \omega^2 \begin{bmatrix} 0.00322 & 0.0136 \\ 0.00322 & 0.0213 \end{bmatrix} \begin{Bmatrix} 1.00 \\ 1.47 \end{Bmatrix}$$

$$= \omega^2 \begin{Bmatrix} 0.0232 \\ 0.0346 \end{Bmatrix} = 0.0232\,\omega^2 \begin{Bmatrix} 1.00 \\ 1.48 \end{Bmatrix}$$

Since this first iteration gives

$$\begin{Bmatrix} 1.00 \\ 1.47 \end{Bmatrix} = 0.0232\omega^2 \begin{Bmatrix} 1.00 \\ 1.48 \end{Bmatrix}$$

we are close to convergence to a solution where the first or lowest mode shape is given by $x_1 = 1.00$ and $x_2 = 1.48$, and where the lowest natural frequency can be computed from $0.0232\omega^2 = 1$ to give $\omega = 6.57$ rad/sec. It should be noted that these results are close in value to the Stodola method results in Ill. Ex. 4.10.

To obtain the second natural frequency and mode shape, we can either use the orthogonality principle or we can iterate the equation $\{x\} = (1/\omega^2)[H]\{x\}$. The orthogonality equation for two degrees of freedom is given by Eq. (3-47) of Sec. 3.10. Since our first mode results show that $A_1 = 1.00$ and $B_1 = 1.48$, substitution in Eq. (3-47) gives

$$(0.0129)(1.00)A_2 + (0.0545)(1.48)B_2 = 0.0129A_2 + 0.0809B_2 = 0$$

whose solution is $A_2 = -6.27B_2$. The two equations $A_2 = -6.27B_2$ and $B_2 = B_2$ can be written in matrix form as

$$\begin{Bmatrix} A_2 \\ B_2 \end{Bmatrix} = \begin{bmatrix} 0 & -6.27 \\ 0 & 1 \end{bmatrix} \begin{Bmatrix} A_2 \\ B_2 \end{Bmatrix} \equiv [T_1] \begin{Bmatrix} A_2 \\ B_2 \end{Bmatrix}$$

To calculate the natural frequency and mode shape for the second mode, we iterate the matrix equation $\{x\} = \omega^2[\alpha m][T_1]\{x\}$, where matrix $[T_1]$, which suppresses the effect of the first mode, was shown to be obtained from the orthogonality principle. Substitution of numerical values gives the following iteration equation for the second mode:

$$\begin{Bmatrix} x_1 \\ x_2 \end{Bmatrix} = \omega^2 \begin{bmatrix} 0.00322 & 0.0136 \\ 0.00322 & 0.0213 \end{bmatrix} \begin{bmatrix} 0 & -6.27 \\ 0 & 1 \end{bmatrix} \begin{Bmatrix} x_1 \\ x_2 \end{Bmatrix}$$

$$= \omega^2 \begin{bmatrix} 0 & -0.0066 \\ 0 & 0.0011 \end{bmatrix} \begin{Bmatrix} x_1 \\ x_2 \end{Bmatrix}$$

Since $A_2 = -6.27B_2$ from the orthogonality equation, then $x_1 = -6.27x_2$ or $x_2/x_1 = -0.160$. Substitution of $x_1 = 1.000$ and $x_2 = -0.160$ in the previous equation gives

$$\begin{Bmatrix} x_1 \\ x_2 \end{Bmatrix} = \omega^2 \begin{bmatrix} 0 & -0.0066 \\ 0 & 0.0011 \end{bmatrix} \begin{Bmatrix} 1.000 \\ -0.160 \end{Bmatrix} = \omega^2 \begin{Bmatrix} 0.00106 \\ -0.000176 \end{Bmatrix}$$

$$= 0.00106\omega^2 \begin{Bmatrix} 1.000 \\ -0.164 \end{Bmatrix}$$

Since we are close to convergence, we obtain $x_1 = 1.000$ and $x_2 = -0.164$, where these x_1-and x_2-solutions define the shape of the second mode, and the second natural frequency can be determined from $0.00106\omega^2 = 1$ to give $\omega = 30.8$ rad/sec.

As stated before, we could also obtain the second mode results by iterating the matrix equation $\{x\} = (1/\omega^2)[H]\{x\}$. Since the equations of motion for this system are

$$\ddot{x}_1 + (k_1 + k)x_1 - kx_2 = \ddot{x}_1 + 11.0x_1 - 7.00x_2 = 0$$
$$\ddot{x}_2 + kx_2 - kx_1 = \ddot{x}_2 - 7.00x_1 + 7.00x_2 = 0$$

then $k_{11} = 11.0$, $k_{12} = k_{21} = -7.00$, and $k_{22} = 7.00$. Thus, the dynamic matrix $[H]$ for this problem is

$$[H] \equiv [m]^{-1}[k] = \begin{bmatrix} 1/0.0129 & 0 \\ \\ 0 & 1/0.0545 \end{bmatrix} \begin{bmatrix} 11.0 & -7.00 \\ \\ -7.00 & 7.00 \end{bmatrix}$$

$$= \begin{bmatrix} 854 & -542 \\ \\ -129 & 129 \end{bmatrix}$$

and the iteration equation is

$$\begin{Bmatrix} x_1 \\ x_2 \end{Bmatrix} = \frac{1}{\omega^2} \begin{bmatrix} 854 & -542 \\ -129 & 129 \end{bmatrix} \begin{Bmatrix} x_1 \\ x_2 \end{Bmatrix}$$

Use of the previously calculated second mode results, $x_1 = 1.00$ and $x_2 = -0.164$, gives

$$\begin{Bmatrix} x_1 \\ x_2 \end{Bmatrix} = \frac{1}{\omega^2} \begin{bmatrix} 854 & -542 \\ -129 & 129 \end{bmatrix} \begin{Bmatrix} 1.000 \\ -0.164 \end{Bmatrix} = \frac{1}{\omega^2} \begin{Bmatrix} 943 \\ -150 \end{Bmatrix}$$

$$= \frac{943}{\omega^2} \begin{Bmatrix} 1.000 \\ -0.160 \end{Bmatrix}$$

Since this first iteration has almost converged to a solution, the configuration of the second mode shape is specified by $x_1 = 1.000$ and $x_2 = -0.160$, and the second natural frequency can be computed from $943/\omega^2 = 1$ to obtain $\omega = 30.8$ rad/sec.

Illustrative Example 4.16.   For the three-degree-of-freedom spring-mass system in Ill. Ex. 3.18 of Sec. 3.8, suppose that masses m_1, m_2, and m_3 were all equal and that both springs had the same spring constant value. Use the matrix-iteration method to determine the three natural frequencies and mode shapes for this spring-mass system.

Solution.   Let m denote the value of the three equal masses and let k denote the value of the two equal spring constants. From Eqs. (3-43), (3-44), and (3-45) in Ill. Ex. 3.18, the influence coefficients for this spring-mass system are

$$\alpha_{11} = \alpha_{12} = \alpha_{13} = \alpha_{21} = \alpha_{31} = \frac{1}{k}$$

$$\alpha_{22} = \alpha_{23} = \alpha_{32} = \frac{2}{k}$$

$$\alpha_{33} = \frac{3}{k}$$

Thus, the $[\alpha m]$-matrix is

$$[\alpha m] \equiv [\alpha][m] = \begin{bmatrix} \dfrac{1}{k} & \dfrac{1}{k} & \dfrac{1}{k} \\[2mm] \dfrac{1}{k} & \dfrac{2}{k} & \dfrac{2}{k} \\[2mm] \dfrac{1}{k} & \dfrac{2}{k} & \dfrac{3}{k} \end{bmatrix} \begin{bmatrix} m & 0 & 0 \\ 0 & m & 0 \\ 0 & 0 & m \end{bmatrix}$$

$$= (m/k) \begin{bmatrix} 1 & 1 & 1 \\ 1 & 2 & 2 \\ 1 & 2 & 3 \end{bmatrix}$$

and the iteration equation is

$$\begin{Bmatrix} x_1 \\ x_2 \\ x_3 \end{Bmatrix} = \omega^2[\alpha m]\{x\} = (\omega^2 m/k) \begin{bmatrix} 1 & 1 & 1 \\ 1 & 2 & 2 \\ 1 & 2 & 3 \end{bmatrix} \begin{Bmatrix} x_1 \\ x_2 \\ x_3 \end{Bmatrix}$$

If we use $x_1 = 1.00$, $x_2 = 1.50$, and $x_3 = 2.00$ as the trial values for the first cycle of this iteration, we obtain

$$\begin{Bmatrix} x_1 \\ x_2 \\ x_3 \end{Bmatrix} = (\omega^2 m/k) \begin{bmatrix} 1 & 1 & 1 \\ 1 & 2 & 2 \\ 1 & 2 & 3 \end{bmatrix} \begin{Bmatrix} 1.00 \\ 1.50 \\ 2.00 \end{Bmatrix} = (\omega^2 m/k) \begin{Bmatrix} 4.50 \\ 8.00 \\ 10.00 \end{Bmatrix}$$

$$= (4.50\ m\omega^2/k) \begin{Bmatrix} 1.00 \\ 1.78 \\ 2.22 \end{Bmatrix}$$

For the second cycle of this iteration, we use the computed values $x_1 = 1.00$, $x_2 = 1.78$, and $x_3 = 2.22$ to obtain

$$\begin{Bmatrix} x_1 \\ x_2 \\ x_3 \end{Bmatrix} = (\omega^2 m/k) \begin{bmatrix} 1 & 1 & 1 \\ 1 & 2 & 2 \\ 1 & 2 & 3 \end{bmatrix} \begin{Bmatrix} 1.00 \\ 1.78 \\ 2.22 \end{Bmatrix}$$

$$= (\omega^2 m/k) \left\{ \begin{array}{c} 5.00 \\ 9.00 \\ 11.22 \end{array} \right\} = (5.00\,\omega^2 m/k) \left\{ \begin{array}{c} 1.00 \\ 1.80 \\ 2.24 \end{array} \right\}$$

Since we are close to convergence to a solution, the first or lowest mode shape is given by $x_1 = 1.00$, $x_2 = 1.80$, and $x_3 = 2.24$, and the lowest natural frequency can be computed from $5.00\,\omega^2 m/k = 1$ to give $\omega = \sqrt{k/(5m)}$.

To obtain the second natural frequency and mode shape, we shall first use Eq. (3-48a) in Sec. 3.10 to apply the orthogonality principle as follows:

$$m_1 A_1 A_2 + m_2 B_1 B_2 + m_3 C_1 C_2 = m(1.00)A_2 + m(1.80)B_2 + m(2.24)C_2$$

where the results of our first iteration gave $A_1 = 1.00$, $B_1 = 1.80$, and $C_1 = 2.24$ for our three amplitudes for the first mode. Solving the previous equation for A_2, we obtain

$$A_2 = -1.80B_2 - 2.24C_2$$

The previous equation, along with $B_2 = B_2$ and $C_2 = C_2$, can be written in matrix form as

$$\left\{ \begin{array}{c} A_2 \\ B_2 \\ C_2 \end{array} \right\} = \left[\begin{array}{ccc} 0 & -1.80 & -2.24 \\ 0 & 1 & 0 \\ 0 & 0 & 1 \end{array} \right] \left\{ \begin{array}{c} A_2 \\ B_2 \\ C_2 \end{array} \right\} \equiv [T_1] \left\{ \begin{array}{c} A_2 \\ B_2 \\ C_2 \end{array} \right\}$$

To calculate the natural frequency and mode shape for the second mode, we iterate the matrix equation $\{x\} = \omega^2[\alpha m][T_1]\{x\}$, where matrix $[T_1]$, which is obtained from the orthogonality principle, is given above. Substitution of numerical values gives the following iteration equation for the second mode:

$$\left\{ \begin{array}{c} x_1 \\ x_2 \\ x_3 \end{array} \right\} = (\omega^2 m/k) \left[\begin{array}{ccc} 1 & 1 & 1 \\ 1 & 2 & 2 \\ 1 & 2 & 3 \end{array} \right] \left[\begin{array}{ccc} 0 & -1.80 & -2.24 \\ 0 & 1 & 0 \\ 0 & 0 & 1 \end{array} \right] \left\{ \begin{array}{c} x_1 \\ x_2 \\ x_3 \end{array} \right\}$$

$$= (\omega^2 m/k) \left[\begin{array}{ccc} 0 & -0.80 & -1.24 \\ 0 & 0.20 & -0.24 \\ 0 & 0.20 & 0.76 \end{array} \right] \left\{ \begin{array}{c} x_1 \\ x_2 \\ x_3 \end{array} \right\}$$

If we use $x_1 = 1.00$, $x_2 = 0.500$, and $x_3 = -1.00$ as trial values for the first cycle of this iteration, we obtain

$$\begin{Bmatrix} x_1 \\ x_2 \\ x_3 \end{Bmatrix} = (\omega^2 m/k) \begin{bmatrix} 0 & -0.80 & -1.24 \\ 0 & 0.20 & -0.24 \\ 0 & 0.20 & 0.76 \end{bmatrix} \begin{Bmatrix} 1.00 \\ 0.500 \\ -1.00 \end{Bmatrix}$$

$$= (\omega^2 m/k) \begin{Bmatrix} 0.84 \\ 0.34 \\ -0.66 \end{Bmatrix} = (0.84\omega^2 m/k) \begin{Bmatrix} 1.00 \\ 0.404 \\ -0.785 \end{Bmatrix}$$

For the second cycle of this iteration, we use the computed values $x_1 = 1.00$, $x_2 = 0.404$, and $x_3 = -0.785$ to obtain

$$\begin{Bmatrix} x_1 \\ x_2 \\ x_3 \end{Bmatrix} = (\omega^2 m/k) \begin{bmatrix} 0 & -0.80 & -1.24 \\ 0 & 0.20 & -0.24 \\ 0 & 0.20 & 0.76 \end{bmatrix} \begin{Bmatrix} 1.00 \\ 0.404 \\ -0.785 \end{Bmatrix}$$

$$= (\omega^2 m/k) \begin{Bmatrix} 0.627 \\ 0.264 \\ -0.500 \end{Bmatrix} = (0.627\omega^2 m/k) \begin{Bmatrix} 1.00 \\ 0.424 \\ -0.800 \end{Bmatrix}$$

For the third cycle of this iteration, we use the computed values $x_1 = 1.00$, $x_2 = 0.424$, and $x_3 = -0.800$ to obtain

$$\begin{Bmatrix} x_1 \\ x_2 \\ x_3 \end{Bmatrix} = (\omega^2 m/k) \begin{bmatrix} 0 & -0.80 & -1.24 \\ 0 & 0.20 & -0.24 \\ 0 & 0.20 & 0.76 \end{bmatrix} \begin{Bmatrix} 1.00 \\ 0.424 \\ -0.800 \end{Bmatrix}$$

$$= (\omega^2 m/k) \begin{Bmatrix} 0.653 \\ 0.276 \\ -0.523 \end{Bmatrix} = (0.653\omega^2 m/k) \begin{Bmatrix} 1.00 \\ 0.423 \\ -0.803 \end{Bmatrix}$$

Since we are close to convergence to a solution, the second mode shape is given by $x_1 = 1.00$, $x_2 = 0.423$, and $x_3 = -0.803$, and the second natural frequency can be computed from $0.653\omega^2 m/k = 1$ to give $\omega = \sqrt{1.53k/m}$.

To obtain the highest natural frequency and mode shape, we will use the orthogonality principle rather than iterate the matrix equation $\{x\} = (1/\omega^2)[H]\{x\}$. Substitution of the computed first and second mode amplitudes into Eqs. (3-48b, c) of Sec. 3.10 gives

$$m(1.00)A_3 + m(1.80)B_3 + m(2.24)C_3 = 0$$
$$m(1.00)A_3 + m(0.423)B_3 + m(-0.803)C_3 = 0$$

where the results of our last iteration gave $A_2 = 1.00$, $B_2 = 0.423$, and $C_2 = -0.803$ for our three amplitudes for the second mode. An algebraic solution of the two orthogonality equations gives $A_3 = 1.80C_3$ and $B_3 = -2.26C_3$. These two equations, along with $C_3 = C_3$, can be written in matrix form as

$$\begin{Bmatrix} A_3 \\ B_3 \\ C_3 \end{Bmatrix} = \begin{bmatrix} 0 & 0 & 1.80 \\ 0 & 0 & -2.26 \\ 0 & 0 & 1.00 \end{bmatrix} \begin{Bmatrix} A_3 \\ B_3 \\ C_3 \end{Bmatrix} \equiv [T_2] \begin{Bmatrix} A_3 \\ B_3 \\ C_3 \end{Bmatrix}$$

To calculate the natural frequency and mode shape for the third mode, we iterate the following matrix equation:

$$\{x\} = \omega^2[\alpha m][T_1][T_2]\{x\}$$

where matrix $[T_2]$, which is obtained from the orthogonality principle, is given above. Substitution of numerical values gives the following iteration equation for the third mode, where the resulting matrix equation is free of any first-two-mode effects:

$$\begin{Bmatrix} x_1 \\ x_2 \\ x_3 \end{Bmatrix} = (\omega^2 m/k) \begin{bmatrix} 0 & -0.80 & -1.24 \\ 0 & 0.20 & -0.24 \\ 0 & 0.20 & 0.76 \end{bmatrix} \begin{bmatrix} 0 & 0 & 1.80 \\ 0 & 0 & -2.26 \\ 0 & 0 & 1.00 \end{bmatrix} \begin{Bmatrix} x_1 \\ x_2 \\ x_3 \end{Bmatrix}$$

$$= (\omega^2 m/k) \begin{bmatrix} 0 & 0 & 0.557 \\ 0 & 0 & -0.692 \\ 0 & 0 & 0.308 \end{bmatrix} \begin{Bmatrix} x_1 \\ x_2 \\ x_3 \end{Bmatrix}$$

From our last orthogonality results, we know that

$$x_3/x_1 = C_3/A_3 = 1/1.80 = 0.555$$
$$x_2/x_1 = B_3/A_3 = -2.26C_3/1.80C_3 = -1.25$$

Thus, we use $x_1 = 1.00$, $x_2 = -1.25$, and $x_3 = 0.555$ in the iteration equation to obtain

$$\begin{Bmatrix} x_1 \\ x_2 \\ x_3 \end{Bmatrix} = (\omega^2 m/k) \begin{bmatrix} 0 & 0 & 0.557 \\ 0 & 0 & -0.692 \\ 0 & 0 & 0.308 \end{bmatrix} \begin{Bmatrix} 1.00 \\ -1.25 \\ 0.555 \end{Bmatrix}$$

$$= (\omega^2 m/k) \begin{Bmatrix} 0.309 \\ -0.385 \\ 0.171 \end{Bmatrix} = (0.309\omega^2 m/k) \begin{Bmatrix} 1.00 \\ -1.24 \\ 0.555 \end{Bmatrix}$$

Since the previous iteration equation shows a convergence, our third and highest mode shape is given by $x_1 = 1.00$, $x_2 = -1.24$, and $x_3 = 0.555$, and the highest natural frequency can be computed from $0.309\ \omega^2 m/k = 1$ to obtain $\omega = \sqrt{3.24k/m}$.

Illustrative Example 4.17. Use the matrix-iteration method to verify the Rayleigh-method approximation for the lowest natural frequency in Ill. Ex. 4.1.

Solution. From Ill. Ex. 4.1, our influence coefficients are

$$\alpha_{11} = y_{11}/300 = 0.0239$$
$$\alpha_{12} = y_{12}/200 = 0.0233$$
$$\alpha_{21} = y_{21}/300 = 0.0233$$
$$\alpha_{22} = y_{22}/200 = 0.0266$$

Since $m_1 = 300/g$ and $m_2 = 200/g$, the $[\alpha m]$-matrix is

$$[\alpha m] \equiv [\alpha][m] = \frac{1}{386} \begin{bmatrix} 0.0239 & 0.0233 \\ 0.0233 & 0.0266 \end{bmatrix} \begin{bmatrix} 300 & 0 \\ 0 & 200 \end{bmatrix}$$

$$= \frac{1}{386} \begin{bmatrix} 7.18 & 4.65 \\ 6.98 & 5.31 \end{bmatrix}$$

Since Ill. Ex. 4.1 showed that the two static deflections had the ratio $y_2/y_1 = 12.29/11.83 = 1.04$, we shall use $y_1 = 1.00$ and $y_2 = 1.04$ for our trial values for the first mode. Since the iteration equation for this mode is $\{y\} = \omega^2[\alpha m]\{y\}$, the first cycle of this iteration gives

$$\begin{Bmatrix} y_1 \\ y_2 \end{Bmatrix} = \frac{\omega^2}{386} \begin{bmatrix} 7.18 & 4.65 \\ 6.98 & 5.31 \end{bmatrix} \begin{Bmatrix} 1.00 \\ 1.04 \end{Bmatrix}$$

$$= \frac{\omega^2}{386} \begin{Bmatrix} 12.03 \\ 12.50 \end{Bmatrix} = \frac{12.03\omega^2}{386} \begin{Bmatrix} 1.00 \\ 1.04 \end{Bmatrix}$$

which shows that this iteration has converged. Thus, $12.03\omega^2/386 = 1$ from which we obtain $\omega^2 = 386/12.03$ or $\omega = 5.70$ rad/sec, which agrees with the result in Ill. Ex. 4.1.

PROBLEMS

4.1. Suppose that the loaded, massless beam in Ill. Ex. 4.1 also supports a 500-lb load that is located at the center of the beam.. Approximate the lowest natural frequency using the Rayleigh method.

4.2. Derive Eq. (4–3) in Ill. Ex. 4.2 which is the Rayleigh method for a spring-mass system.

4.3. Verify the result in Ill. Ex. 4.3 using the exact deflection $y = (y_m \sin \pi x/L) \sin \omega t$.

4.4. Verify the fundamental frequency for a uniform cantilever beam that is given in Ill. Ex. 4.3 using the Rayleigh method and the static-deflection curve.

4.5. Divide the uniform simply-supported beam in Ill. Ex. 4.3 into four equal massless lengths. Assume that there is a concentrated load of magnitude $wL/4$ at the center of each of these massless equal lengths, and calculate the lowest natural frequency by the Rayleigh method. Compare this result with that in Ill. Ex. 4.3.

4.6. Repeat Prob. 4.5 for a uniform cantilever beam.

4.7. Repeat Probs. 4.5 and 4.6 using two equal lengths.

4.8. Suppose that the loaded beam in Ill. Ex. 4.1 is a cantilever beam instead of being simply supported at both ends. Approximate the lowest natural frequency by the Rayleigh method.

4.9. Approximate the lowest natural frequency for the three-degree-of-freedom, spring-mass system in Ill. Ex. 3.18 using the Rayleigh method.

4.10. Complete Ill. Ex. 4.5 to determine the three natural frequencies for this system.

4.11. Suppose that the three masses in Fig. 4–1 were all equal and that both springs had equal spring-constant values. Calculate the three natural frequencies using the Holzer method.

4.12. Calculate the second natural frequency in Ill. Ex. 4.6 using the Holzer method.

4.13 Suppose that the three disks in Fig. 3–4 in Sec. 3.2 had equal moments of inertia and that the three shafts had equal torsional spring-constant values. Calculate the three natural frequencies using the Holzer method.

4.14. Verify the computed natural frequencies and mode shapes in Ill. Ex. 4.16 using the Holzer method.

4.15. Complete Ill. Ex. 4.8 to determine the lowest natural frequency by the Myklestad method.

4.16. Repeat Ill. Ex. 4.8 for the case in which this loaded·beam is simply supported at both ends. Compare this result with that in Ill. Ex. 4.3 for a uniformly loaded beam that is simply supported at both ends.

4.17. In Ill. Ex. 4.8, suppose that this beam is subdivided into four equal lengths and that there is a 19.3-lb mass located at the centers of these four subdivided lengths. Calculate the lowest natural frequency by the Myklestad method, and compare this result with that for Ill. Ex. 4.8 and that for a uniformly loaded cantilever beam.

4.18. Calculate the second natural frequency of the beam system in Ill. Ex. 4.8 using the Myklestad method.

4.19. Suppose in Fig. 4–3 that EI equals 5.00×10^5 lb-in.^{2}, the left-hand mass weighs 20 lb, and the right-hand mass weighs 60 lb. Calculate the lowest natural frequency by the Myklestad method.

4.20. Calculate the lowest natural frequency of the simply-supported beam system in Ill. Ex. 4.1 by the Myklestad method when $EI = 2.00 \times 10^5$ lb–in.^{2}

4.21. Suppose in Ill. Ex. 4.9 that the influence coefficient is unknown. Determine this influence coefficient and the lowest natural frequency of the structure if the addition of 1.20 lb to the shaker lowers the measured frequency to 40 rad/sec.

4.22. Suppose that a uniformly loaded cantilever beam, where this loading is 8 lb/ft and where the beam length is 12 ft, supports a 20-lb load at its center. Use the Dunkerly method to approximate the lowest natural frequency of this beam system.

4.23. Repeat Prob. 4.22 if this loaded beam is simply supported at both ends.

4.24. Repeat Prob. 4.22 if there is also a 30-lb load at the free end.

4.25. Verify the lowest natural frequency and mode shape that was computed in Ill. Ex. 3.2 in Sec. 3.2 using the Stodola method.

4.26. Verify the lowest natural frequency and mode shape that was computed in Ill. Ex. 4.16 using the Stodola method.

4.27. Calculate $[A] + [B]$, $[A] - [B]$, $[A][B]$, $[B][A]$, $|A|$, and $|B|$ when matrices $[A]$ and $[B]$ are as follows:

$$[A] = \begin{bmatrix} 4 & 2 & 7 \\ 3 & 0 & -6 \\ 1 & -5 & 2 \end{bmatrix}, \quad [B] = \begin{bmatrix} -2 & 3 & -5 \\ -4 & 7 & 0 \\ 0 & 8 & 4 \end{bmatrix}$$

4.28. Repeat Prob. 4.27 when matrix $[B]$ is as follows. State which arithmetic operations are impossible and furnish the reason.

$$[B] = \begin{bmatrix} 6 & -4 \\ -2 & 7 \\ 3 & 5 \end{bmatrix}$$

4.29. Calculate the inverse of matrices $[A]$ and $[B]$ in Prob. 4.27 using the Gauss-Jordan method.

4.30. Calculate the eigenvalues of matrix $[A]$ in Prob. 4.27 using the Newton-Raphson method and the quadratic formula.

4.31. Repeat Prob. 4.30 using the iteration method that is given in Ill. Ex. 4.14.

4.32. Use the matrix-iteration method to verify the natural frequencies and mode shapes that were computed in Ill. Ex. 3.1 in Sec. 3.2.

4.33. Compute the three natural frequencies of the spring-mass system in Ill. Ex. 4.5 using the matrix-iteration method.

4.34. Repeat Prob. 4.13 using the matrix-iteration method.

4.35. Calculate the highest natural frequency in Ill. Ex. 4.16 by iterating the matrix equation $\{x\} = (1/\omega^2)[H]\{x\}$.

4.36. For the spring-mass system in Ill. Ex. 3.18, suppose that the k_1-, k_2-, and k_3-values are 6.00, 2.00, and 2.00 lb/in., respectively, and that m_1g-,m_2g,-and m_3g-weight values are 20 lb, 10 lb, and 5 lb, respectively. Calculate the three natural frequencies and mode shapes using the matrix-iteration method.

4.37. Repeat Prob. 4.36 if $m_1g = 10$ lb, $m_2g = 20$ lb, $m_3g = 30$ lb, $k_1 = 6.00$ lb/in., $k_2 = 4.00$ lb/in., and $k_3 = 2.00$ lb/in.

4.38. Use the Holzer method to verify your matrix-iteration calculations for Probs. 4.36 and 4.37.

4.39. For the three-degree-of-freedom, spring-mass system in Fig. 3–3, suppose that $m_1 = m_2 = m_3$, and that $k_1 = k_2 = k_3 = k_4$. Calculate the three natural frequencies and mode shapes using the matrix-iteration method. Verify your results using the Holzer method.

4.40. Repeat Prob. 4.39 if $k_1 = 6.00$ lb/in., $k_2 = k_3 = k_4 = 2.00$ lb/in., $m_1g = 20$ lb, $m_2g = 10$ lb, and $m_3g = 5$ lb.

4.41. Repeat Prob. 4.39 if $m_1g = 10$ lb, $m_2g = 20$ lb, $m_3g = 30$ lb, $k_1 = 6.00$ lb/in., $k_2 = 4.00$ lb/in., $k_3 = 2.00$ lb/in., and $k_4 = 1.00$ lb/in.

4.42. In Fig. 3–3, suppose that the lower end of spring k_3 is connected to a mass m_4, instead of to a rigid surface, and that mass m_4 is unsupported at its lower end. If all four masses have an equal weight mg and if all four spring constants equal k, calculate the four natural frequencies and mode shapes using the matrix-iteration method. Verify your results using the Holzer method.

4.43. Repeat Prob. 4.42 if $k_1 = 6.00$ lb/in., $k_2 = k_3 = k_4 = 2.00$ lb/in., $m_1g = 20$ lb, $m_2g = 10$ lb, and $m_3g = m_4g = 5$ lb.

4.44. Repeat Prob. 4.42 if k_1, k_2, k_3, and k_4 equal 6.00, 4.00, 2.00, and 1.00 lb/in., respectively, and m_1g, m_2g, m_3g, and m_4g equal 10, 20, 30, and 40 lb, respectively.

4.43. Solve Prob. 4.5 using the matrix-iteration method.

4.46. Solve Prob. 4.6 using the matrix-iteration method.

5   Vibration of Systems with Distributed Mass and Elasticity

5.1   Introduction

Up to this point, this text has been analyzing lumped dynamic systems composed of concentrated masses connected by massless springs (or beam segments) or of disks connected by massless, elastic shafts. This chapter deals with the vibration of systems that have continuously distributed mass and elasticity. The treatment is introductory in that we will consider only undamped systems where the body material is homogeneous, isotropic, and elastic (that is, it obeys Hooke's Law).

A distributed system can be considered to be composed of an infinite number of infinitesimal mass-particles. Since an infinite number of coordinates are needed to locate the infinite number of particles, a distributed system has an infinite number of degrees of freedom and natural frequencies. A condition of resonance occurs when a distributed system is excited at a natural frequency, and each natural frequency has a unique mode shape. For a principal-vibration mode, each particle of the system undergoes harmonic motion at the corresponding natural frequency. As for a lumped-mass system, a free vibration of a distributed system can be resolved to be a linear combination of the principal modes. Certain sets of initial and boundary conditions can cause a principal-mode vibration to occur, but impact (i.e., sudden) loads and other transient forces usually excite many or all of the natural frequencies, so that the resultant motion is a rather complicated one that is a superposition of the excited modes.

We shall see that the vibration of distributed systems are governed by partial differential equations, rather than by ordinary differential equations. This chapter will consider classical (i.e., separation of variables), Laplace transform, and numerical methods of solution. For review purposes, it may be noted that the total first derivative df/dx for the one-variable function $f(x)$ is defined by

$$\frac{df}{dx} \equiv f'(x) = \lim_{\Delta x \to 0} \frac{f(x + \Delta x) - f(x)}{\Delta x} \tag{5-1}$$

providing the given limit exists. If we are given a function of two variables $f(x, y)$, we can have two first partial derivatives. One specifies the rate of change of the value of the function $f(x, y)$ when x is varied and y is held constant, while the other specifies the rate of change when y is varied and x is held constant. They are defined by

$$\frac{\partial f}{\partial x} = \lim_{\Delta x \to 0} \frac{f(x + \Delta x, y) - f(x, y)}{\Delta x} \tag{5-2}$$

$$\frac{\partial f}{\partial y} = \lim_{\Delta y \to 0} \frac{f(x, y + \Delta y) - f(x, y)}{\Delta y} \tag{5-3}$$

To simplify writing, we can use subscript notation when writing these partial derivatives. The notation is $f_x = \partial f/\partial x$ and $f_y = \partial f/\partial y$. The subscripted terms f_{xx} and f_{yy} are used to denote the second partial derivatives, $\partial^2 f/\partial x^2$ and $\partial^2 f/\partial y^2$, respectively.

5.2 Transverse Vibrations of a String and Membrane

Let us first consider the transverse vibrations of an elastic string. Several assumptions are made throughout the derivation for this problem, but it can be seen that they are all physically reasonable. Let us consider the oscillations that take place in a perfectly flexible string of length L that is tightly stretched by a constant tension force of T

pounds. Let the vibrating motion take place in only one plane, the xy-plane, and let the x-axis be directed along the undisturbed string position. Assume pure transverse vibrations; that is, the horizontal displacement at any point on the string is negligible when compared to the vertical displacement, which we shall denote as $\bar{y}$.

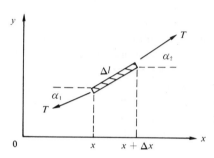

Fig. 5-1

Consider a very small length Δl of the string, as shown in Fig. 5–1. Assume that the deflections $\bar{y}$ are so small, when compared to the initial string length L, that the change in length of the string has no effect on tension T. Thus, the forces acting at the ends of the element are each of magnitude T. Since Δl approximately equals Δx, the mass of the element equals $w\Delta x/g$, where w is the weight per unit length of the string at point x. Assume that w is small enough so that the gravitational force $w\Delta x$ on the element is negligible when compared to tension T. If frictional effects are neglected, the only significant forces acting on the element are the tension forces. Since the string is perfectly flexible, the transmitted tension force will be tangential to the string. Upon application of Newton's second law in the y-direction, we obtain

$$\frac{w\Delta x}{g}\frac{\partial^2 \bar{y}}{\partial t^2} = T \sin \alpha_2 - T \sin \alpha_1$$

Since the deflections are very small, α_2 and α_1 are small enough so that we may use the approximations $\sin \alpha_2 = \tan \alpha_2$ and $\sin \alpha_1 = \tan \alpha_1$. Since $\tan \alpha_2$ equals $(\partial \bar{y}/\partial x)$ at location $(x + \Delta x)$ and since $\tan \alpha_1$ equals $(\partial \bar{y}/\partial x)$ at location x, we obtain upon substitution

$$\frac{w}{g}\frac{\partial^2 \bar{y}}{\partial t^2} = T\left[\frac{(\partial \bar{y}/\partial x)_{x+\Delta x} - (\partial \bar{y}/\partial \bar{x})_x}{\Delta x}\right]$$

Taking the limit as Δx approaches zero, we obtain the following equation, where Eq. (5–2) was applied:

$$\frac{\partial^2 \bar{y}}{\partial t^2} = \frac{Tg}{w} \frac{\partial^2 \bar{y}}{\partial x^2}$$

If unit weight w is constant, then $Tg/w = a^2$ is constant and we have the following one-dimensional, time-varying equation (i.e., displacement $\bar{y}$ varies with both location x and time t):

$$\frac{\partial^2 \bar{y}}{\partial x^2} = \frac{1}{a^2} \frac{\partial^2 \bar{y}}{\partial t^2} \qquad (5\text{-}4)$$

Now let us consider the transverse vibrations of a flexible drumhead or membrane. Assume that the membrane is pulled evenly around its edge with tension T, so that because the membrane is perfectly flexible, the tension T per unit length is everywhere constant. As we did in the vibrating string problem, assume that the weight of the membrane is a negligible force, the vertical deflections $\bar{z}$ are small when compared to the size of the membrane, and the magnitudes of the lateral displacements are negligible when compared to the vertical displacements $\bar{z}$. The x, y-plane is located at the undisturbed membrane position. Taking an infinitesimal-sized element of dimensions Δx and Δy, the total forces on the sides of this element are shown in Fig. 5-2, which is a free-body diagram in the x, y-plane and is a top view of this element.

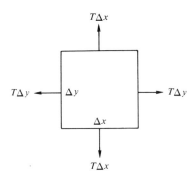

Fig. 5-2

A free-body diagram in the x, z-plane would be a side view of the element and would look similar to that shown in Fig. 5-1 for a vibrating string, except that the forces at the ends of the element are $T\Delta y$ instead of T. Thus, by reasoning analogous to the vibrating string problem, the net external vertical force in the x, z-plane is

$$T\Delta y \left[\left(\frac{\partial \bar{z}}{\partial x}\right)_{x+\Delta x} - \left(\frac{\partial \bar{z}}{\partial x}\right)_{x} \right]$$

Since a free-body diagram in the x, z-plane would show forces $T\Delta x$ at the ends and would otherwise be similar to Fig. 5–1, the net external vertical force in the y, z-plane is

$$T\Delta x\left[\left(\frac{\partial \bar{z}}{\partial y}\right)_{y+\Delta y} - \left(\frac{\partial \bar{z}}{\partial y}\right)_y\right]$$

Denoting w as the weight per unit area and applying Newton's second law in the vertical z-direction, we obtain

$$\frac{w}{g}\,\Delta x\Delta y\left(\frac{\partial^2 \bar{z}}{\partial t^2}\right) = T\Delta y\left[\left(\frac{\partial \bar{z}}{\partial x}\right)_{x+\Delta x} - \left(\frac{\partial \bar{z}}{\partial x}\right)_x\right]$$
$$+ T\Delta x\left[\left(\frac{\partial \bar{z}}{\partial y}\right)_{y+\Delta y} - \left(\frac{\partial \bar{z}}{\partial y}\right)_y\right]$$

Dividing the above equation by the term $(\Delta x\Delta y)$, taking limits as Δx and Δy approach zero, and applying Eqs. (5–2) and (5–3), we obtain

$$T\left[\frac{\partial^2 \bar{z}}{\partial x^2} + \frac{\partial^2 \bar{z}}{\partial y^2}\right] = \frac{w}{g}\frac{\partial^2 \bar{z}}{\partial t^2}$$

Letting $a^2 = Tg/w$, we obtain the following two-dimensional, time-varying equation (i.e., displacement $\bar{z}$ varies with location (x, y) and time t):

$$\frac{\partial^2 \bar{z}}{\partial x^2} + \frac{\partial^2 \bar{z}}{\partial y^2} = \frac{1}{a^2}\frac{\partial^2 \bar{z}}{\partial t^2} \qquad (5\text{-}5)$$

5.3   Torsional Vibrations of a Rod

Consider a uniform elastic rod of circular cross-section vibrating torsionally about its longitudinal central axis. Let its cross-sectional area A, its specific weight (i.e., the weight per unit volume) w, its shear modulus of elasticity G, and its polar moment of inertia about the longitudinal central axis J all be of constant value along the length of the shaft. Let the x-axis be directed along the central axis of the shaft. Neglect the effects of friction and assume that all cross-sections remain plane during rotational vibration.

Consider a segment of the shaft Δx in length. Let $\theta(x, t)$ denote the angle of twist for the shaft at location x and time t. If the two ends of this segment are twisted at different angles, then we know that there is a resisting torque M_e due to shaft elasticity which is given by

$$M_e = k_t[\theta(x + \Delta x, t) - \theta(x, t)]$$

where k_t equals $GJ/\Delta x$. Taking the limit as Δx approaches zero, and applying Eq.(5–2), we have

$$M_e = \lim_{\Delta x \to 0} GJ\left[\frac{\theta(x + \Delta x, t) - \theta(x, t)}{\Delta x}\right] = GJ\frac{\partial\theta}{\partial x}$$

The moment of inertia for this segment is equal to the mass of the segment, which is $wA\Delta x/g$, times the square of the radius of gyration k. Since k^2 equals J/A, this moment of inertia equals $Jw\Delta x/g$. Applying the equation $\sum M = I(\partial^2\theta/\partial t^2)$ to this segment, we have

$$\frac{Jw\Delta x}{g}\frac{\partial^2\theta}{\partial t^2} = M(x + \Delta x, t) - M(x, t)$$

where $M(x, t)$ denotes the total moment at location x and time t. Taking the limit as Δx approaches zero, after the above equation is divided by Δx, and applying Eq. (5-2), we obtain

$$\frac{Jw}{g}\frac{\partial^2\theta}{\partial t^2} = \lim_{\Delta x \to 0}\frac{M(x + \Delta x, t) - M(x, t)}{\Delta x} = \frac{\partial M}{\partial x}$$

Since the only torque we are considering is that due to shaft elasticity, M equals $GJ\,\partial\theta/\partial x$ and $\partial M/\partial x$ equals $GJ\,\partial^2\theta/\partial x^2$. Thus, we have

$$\frac{w}{g}\frac{\partial^2\theta}{\partial t^2} = G\frac{\partial^2\theta}{\partial x^2}$$

Letting $a^2 = Gg/w$, we obtain the following equation for the vibration of this one-dimensional system:

$$\frac{\partial^2\theta}{\partial x^2} = \frac{1}{a^2}\frac{\partial^2\theta}{\partial t^2} \tag{5-6}$$

5.4 Longitudinal Vibrations of a Bar

Consider a bar that is composed of a homogeneous elastic material which satisfies Hooke's law. The bar is vibrating longitudinally along its length and the amplitude of the vibration is small when compared to the bar dimensions. It is assumed that plane sections remain plane during the vibration, and that the bar material has the same properties in all directions (i.e., is isotropic).

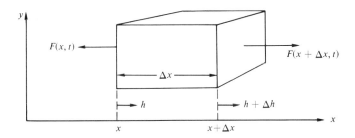

Fig. 5-3

Orient the x-axis along the length of the bar. Let us consider a very small length Δx of the bar, as shown in Fig. 5–3. At some later time, the face at x will be displaced h units and the face at $(x + \Delta x)$ will be displaced $(h + \Delta h)$ units. Thus, the new length of this element is

$$[(x + \Delta x) + (h + \Delta h)] - [x + h] = \Delta x + \Delta h$$

Since the change in length is Δh, the strain ϵ equals $\Delta h / \Delta x$. Letting Δx approach zero, we have

$$\epsilon = \frac{\partial h}{\partial x}$$

where a partial derivative resulted because displacement h varies with both time and location.

Let A be the cross-sectional area of the bar and let w be its specific weight (i.e., weight per unit volume). Thus, the mass of this element is $wA\Delta x/g$. Applying Newton's second law, we obtain

$$\frac{w}{g} A\Delta x \frac{\partial^2 h}{\partial t^2} = F(x + \Delta x, t) - F(x, t)$$

where the F-terms denote the forces acting at the ends of the element. Dividing the previous equation by Δx and letting Δx approach zero, we obtain

$$\frac{w}{g} A \frac{\partial^2 h}{\partial t^2} = \frac{\partial F}{\partial x}$$

Using Hooke's law, the modulus of elasticity E equals $F/A\epsilon$, where F/A is the stress for this problem. Solving for force F, we have

$$F = EA\epsilon = EA \frac{\partial h}{\partial x}$$

Thus, $\partial F/\partial x$ equals $EA\, \partial^2 h/\partial x^2$, so that

$$\frac{w}{g} \frac{\partial^2 h}{\partial t^2} = E \frac{\partial^2 h}{\partial x^2}$$

Letting $a^2 = Eg/w$, we obtain the following equation of motion for the vibration of this one-dimensional system:

$$\frac{\partial^2 h}{\partial x^2} = \frac{1}{a^2} \frac{\partial^2 h}{\partial t^2} \tag{5-7}$$

5.5   Solution by Separation of Variables

The *wave equation* is the name given to the following partial differential equation:

$$\frac{\partial^2 \phi}{\partial x^2} + \frac{\partial^2 \phi}{\partial y^2} + \frac{\partial^2 \phi}{\partial z^2} = \frac{1}{a^2} \frac{\partial^2 \phi}{\partial t^2} \tag{5-8}$$

Sections 5.2 to 5.4 showed that the vibration of a string, rod, and bar can be represented by the one-dimensional wave equation $\partial^2 \phi / \partial x^2 = (1/a^2)(\partial^2 \phi / \partial t^2)$, where ϕ represents a displacement, and Sec. 5.2 showed that a membrane vibration can be represented by the two-dimensional wave equation. A lossless electric transmission line, plane sound waves in a horn, sound waves in space, an electromagnetic field in empty space, and tidal waves in a long channel are other physical problems that are mathematically represented by the wave equation.

Suppose that we had two different physical systems that are governed by the one-dimensional wave equation $\phi_{xx} = \phi_{tt}/a^2$. The solutions for these two systems would be mathematically analogous if their boundary and initial conditions were mathematically analogous. If the system is a vibrating system, ϕ represents a displacement. Since displacement ϕ varies only with location x and time t, let us use the functional notation $\phi(x, t)$. Since the highest derivative with respect to location x in the one-dimensional wave equation is $\partial^2 \phi / \partial x^2$, we need two boundary conditions. If we had a system of length L that is firmly mounted at both ends so that these ends cannot move, the two boundary conditions would be $\phi(0, t) = 0$ and $\phi(L, t) = 0$. Since the highest derivative with respect to time t is $\partial^2 \phi / \partial t^2$, we need two initial conditions (i.e., two conditions at time zero) that could be the values of initial displacement $\phi(x, 0)$ and initial velocity $\partial \phi(x, 0)/\partial t$. For a similar reason, the vibrating membrane problem needs four boundary conditions (two for location x and two for location y) and two initial conditions.

The general solution of the one-dimensional wave equation $\partial^2 \phi / \partial x^2 = (\partial^2 \phi / \partial t^2)/a^2$ can be expressed by

$$\phi = f_1(x + at) + f_2(x - at) \tag{5-9}$$

where f_1 and f_2 are arbitrary functions which must satisfy the initial and boundary conditions for a specific problem. We know that Eq. (5-9) is a solution, because differentiation shows that $\partial^2 f_1 / \partial t^2 = a^2(\partial^2 f_1 / \partial x^2) = a^2 f_1''$ and $\partial^2 f_2 / \partial t^2 = a^2(\partial^2 f_2 / \partial x^2)$ regardless of what functions f_1 and f_2 are. It can be shown that function $f_1(x + at)$ represents a wave moving at velocity a in the negative x-direction, and that $f_2(x - at)$ represents a wave moving at velocity a in the positive x-direction. Sections 5.2 to 5.4 furnish the equations for a, the wave-propagation velocity, for four different vibration problems.

Let us now consider the separation-of-variables method, which is a more general solution technique. This is a rather powerful method that may be applied to the solution of certain types of linear partial differential equations. If we wish to solve a partial differential equation that has only two independent variables, x and t, then this method assumes that the solution is a product of two expressions, where one is a function of x alone and the other is a function of t alone. If this assumed solution form is valid, then the problem is simplified to that of solving two ordinary differential equations. Analogous statements can be made about this method when more than two independent variables are involved. It is easiest to discuss the application of this method by means of an example.

To illustrate the use of this method to solve a one-dimensional wave equation, let us analyze the small vibrations of a tightly stretched elastic string of length L that is fixed at both ends. Thus, from Eq. (5–4) the partial differential equation to solve is

$$\frac{\partial^2 \bar{y}}{\partial x^2} = \frac{1}{a^2} \frac{\partial^2 \bar{y}}{\partial t^2}$$

where $a^2 = Tg/w$. The boundary conditions are

$$\bar{y}(0, t) = 0$$
$$\bar{y}(L, t) = 0$$

If there is no initial velocity and if the string is given an initial displacement equal to $h \sin \pi x/L$, the initial conditions are

$$\bar{y}(x, 0) = h \sin \frac{\pi x}{L}$$

$$\frac{\partial \bar{y}}{\partial t}(x, 0) = 0$$

Since the independent variables are x and t, the separation of-variables method seeks a solution to this problem that is a product of two functions; one of x alone, and another of t alone. Thus, denoting these two functions as $X(x)$ and $T(t)$, we have

$$\bar{y}(x, t) = X(x)T(t)$$

Discarding the use of functional notation to make the equations more readable, we have the following equations for the first and second partial derivatives of $\bar{y}$:

$$\frac{\partial \bar{y}}{\partial x} = T\frac{dX}{dx}, \qquad \frac{\partial^2 \bar{y}}{\partial x^2} = T\frac{d^2 X}{dx^2}$$

$$\frac{\partial \bar{y}}{\partial t} = X\frac{dT}{dt}, \qquad \frac{\partial^2 \bar{y}}{\partial t^2} = X\frac{d^2 T}{dt^2}$$

After substituting the above two second partial derivatives in the vibrating string wave equation, we obtain

$$T \frac{d^2 X}{dx^2} = \frac{1}{a^2} X \frac{d^2 T}{dt^2}$$

which can be rewritten as

$$\frac{1}{X} \frac{d^2 X}{dx^2} = \frac{1}{a^2 T} \frac{d^2 T}{dt^2}$$

Note that the left-hand side of the above equation is a function of x alone, while the right-hand side is a function of t alone. Since this equation must hold for all values of x and for all values of t, it must be equal to a constant, which we shall call b. To verify the above statement, in the previous equation let x vary while t is fixed. Since the right-hand term is constant, the left-hand term is also constant. Thus, we have

$$\frac{1}{X} \frac{d^2 X}{dx^2} = b$$

$$\frac{1}{a^2 T} \frac{d^2 T}{dt^2} = b$$

which reduces the solution of the given partial differential equation to that of solving the following two ordinary differential equations,

$$\frac{d^2 X}{dx^2} - bX = 0$$

$$\frac{d^2 T}{dt^2} - a^2 b T = 0$$

If $b = 0$, we have by integration of the resulting differential equations

$$X = Ax + B$$
$$T = Ct + D$$

If b is positive, we have the solutions

$$X = Ae^{\sqrt{b}\, x} + Be^{-\sqrt{b}\, x}$$
$$T = Ce^{a\sqrt{b}\, t} + De^{-a\sqrt{b}\, t}$$

If b is negative, we have the solutions

$$X = A \cos \sqrt{-b}\, x + B \sin \sqrt{-b}\, x$$
$$T = C \cos a\sqrt{-b}\, t + D \sin a\sqrt{-b}\, t$$

Since we are solving a vibrating-string problem which implies periodic motion, only the last solution is physically valid. To emphasize that b is negative, let $b = -\lambda^2$. Thus, we have

$$\bar{y} = XT = (A \cos \lambda x + B \sin \lambda x)(C \cos a\lambda t + D \sin a\lambda t)$$

where the first factor specifies the mode shape and the second factor shows that the vibration frequency is $a\lambda$ radians per unit time.

The boundary condition $\bar{y}(0, t) = 0$ implies that $X(0)T(t) = 0$ for all values of t. Thus, $X(0) = A \cos 0 + B \sin 0 = 0$, so that $A = 0$. The boundary condition $\bar{y}(L, t) = 0$ implies that $x(L)T(t) = 0$, so that $X(L) = B \sin \lambda L = 0$. Thus, $\lambda L = n\pi$; that is, $\lambda = n\pi/L$, where $n = 1, 2, 3, 4, \ldots$. Note that the first part of the separation-of-variables procedure can be used to calculate the natural frequencies of a system. For this problem the natural frequencies equal $a\lambda$ radians per unit time. Summarizing our results, we have

$$\bar{y} = \left(B \sin \frac{n\pi}{L} x\right)\left(C \cos \frac{an\pi}{L} t + D \sin \frac{an\pi}{L} t\right) \qquad (5\text{--}10)$$

The initial condition $\partial \bar{y}(x, 0)/\partial t = 0$ implies that $X(x)dT(0)/dt = 0$, so that

$$\frac{dT}{dt}(0) = -Ca\lambda \sin 0 + Da\lambda \cos 0 = 0$$

Thus, $D = 0$, and we have

$$\bar{y} = BC \sin \lambda x \cos a\lambda t = G \sin \frac{n\pi}{L} x \cos \frac{an\pi}{L} t$$

Since $n = 1, 2, 3, \cdots$, and since for linear homogeneous partial differential equations the total solution is the sum (actually the linear combination) of all possible solutions, the total solution for $\bar{y}$ is the following infinite series:

$$\bar{y} = \sum_{n=1}^{\infty} G_n \sin \frac{n\pi x}{L} \cos \frac{an\pi t}{L} \qquad (5\text{--}11)$$

Note that each term of this series represents a principal-mode vibration whose natural frequency is $na/2L$ cycles/sec. At $t = 0$, Eq. (5–11) becomes

$$\bar{y}(x, 0) = \sum_{n=1}^{\infty} G_n \sin \frac{n\pi x}{L} = G_1 \sin \frac{\pi x}{L} + G_2 \sin \frac{2\pi x}{L} + \cdots$$

Thus, the initial condition $\bar{y}(x, 0) = h \sin \pi x/L$ implies that $G_1 = h$, and that all other G_n-values equal zero. Upon substitution of the G_n-values in Eq. (5–11), the final solution is the following first-mode vibration:

$$\bar{y} = h \sin \frac{\pi x}{L} \cos \frac{\pi a t}{L} \qquad (5\text{--}12)$$

The previous problem was chosen to have initial-condition values that make the solution of the problem especially simple. Now let us consider a more general case that consists of more vibration modes. If the given elastic string had the following arbitrary initial conditions, instead of those previously specified,

$$\bar{y}(x, 0) = f(x)$$

$$\frac{\partial \bar{y}}{\partial t}(x, 0) = 0$$

we would obtain Eq. (5–11) as we did before. The initial condition $\bar{y}(x, 0) = f(x)$ implies that

$$f(x) = \bar{y}(x, 0) = \sum_{n=1}^{\infty} G_n \sin \frac{n\pi x}{L} \qquad (5\text{–}13)$$

which we recognize as a Fourier sine series. Thus, the soluion for $\bar{y}$ is Eq. (5–11), where from Fourier series theory the G_n-coefficients are given by

$$G_n = \frac{2}{L} \int_0^L f(x) \sin \frac{n\pi x}{L} \, dx \qquad (5\text{–}14)$$

Now let us consider a case where there is an initial velocity, but no initial displacement. If the given elastic string had the following arbitrary initial conditions:

$$\bar{y}(x, 0) = 0$$

$$\frac{\partial \bar{y}}{\partial t}(x, 0) = g(x)$$

we would obtain Eq. (5–10), as we did before. The initial condition $\bar{y}(x, 0) = 0$ implies that $T(0) = C \cos 0 + D \sin 0 = 0$. Thus, $C = 0$ and we obtain the following infinite series for $\bar{y}$

$$\bar{y} = \sum_{n=1}^{\infty} H_n \sin \frac{n\pi x}{L} \sin \frac{an\pi t}{L} \qquad (5\text{–}15)$$

The initial condition $\partial \bar{y}(x, 0)/\partial t = g(x)$ implies that

$$g(x) = \sum_{n=1}^{\infty} \frac{an\pi}{L} H_n \sin \frac{n\pi x}{L}$$

which is a Fourier sine series, Thus, the solution for $\bar{y}$ is Eq. (5–15), where from Fourier series theory we have

$$H_n = \frac{2}{\pi an} \int_0^L g(x) \sin \frac{n\pi x}{L} \, dx \qquad (5\text{–}16)$$

The most general case is when there is both an initial displacement and an initial velocity. If the given string had the arbitrary initial conditions

$$\bar{y}(x, 0) = f(x)$$

$$\frac{\partial \bar{y}}{\partial t}(x, 0) = g(x)$$

it can be shown that the solution is

$$\bar{y} = \sum_{n=1}^{\infty} \left(G_n \cos\frac{an\pi t}{L} + H_n \sin\frac{an\pi t}{L} \right) \sin\frac{n\pi x}{L} \qquad (5\text{–}17)$$

where the coefficients G_n and H_n are given by the equations

$$G_n = \frac{2}{L} \int_0^L f(x) \sin\frac{n\pi x}{L} \, dx$$

$$H_n = \frac{2}{\pi a n} \int_0^L g(x) \sin\frac{n\pi x}{L} \, dx$$

It should be noted that this solution is the sum of the solutions of the two previous examples. That is, Eq. (5–17) is the sum of Eqs. (5–11) and (5–15), where Eqs. (5–14) and (5–16) specify G_n and H_n.

The procedure for solving the one-dimensional wave equation by separation of variables for other boundary conditions is similar. For problems involving the torsional vibration of an elastic rod, the boundary condition at a free or unsupported end is $\partial\theta/\partial x = 0$ since Sec. 5.3 shows that $M = GJ(\partial\theta/\partial x)$. For problems involving the longitudinal vibration of an elastic bar, the boundary condition at a free end is $\partial h/\partial x = 0$, since there is zero force at a free end, and since Sec. 5.4 shows that $F = EA(\partial h/\partial x)$. For solving a vibrating membrane problem, the solution procedure by separation of variables is the same, but it is harder to apply and the results are more involved. For this three-variable problem which solves for displacement $\bar{z}(x, y, t)$, consider a rectangular vibrating membrane of dimensions a and b. If this membrane has an initial displacement $f(x, y)$ but no initial velocity, the equation of motion and the initial and boundary conditions are as follows if the membrane is clamped at its four boundaries:

$$\frac{\partial^2 \bar{z}}{\partial x^2} + \frac{\partial^2 \bar{z}}{\partial y^2} = \frac{1}{\alpha^2}\frac{\partial^2 \bar{z}}{\partial t^2}$$

$$\bar{z}(0, y, t) = 0$$

$$\bar{z}(a, y, t) = 0$$

$$\bar{z}(x, 0, t) = 0$$

$$\bar{z}(x, b, t) = 0$$

$$\frac{\partial \bar{z}}{\partial t}(x, y, 0) = 0$$

$$\bar{z}(x, y, 0) = f(x, y)$$

Assuming that $\bar{z}(x, y, t) = X(x)\, Y(y)\, T(t)$, and applying the method of separation of variables, we obtain the following double-infinite-series solution:

$$\bar{z}(x, y, t) = \sum_{m=1}^{\infty} \sum_{n=1}^{\infty} a_{mn} \sin\frac{m\pi x}{a} \sin\frac{n\pi y}{b} \cos k_{mn} t \qquad (5\text{–}18)$$

where the coefficients are given by

$$k_{mn} = \alpha\pi \sqrt{\left(\frac{m}{a}\right)^2 + \left(\frac{n}{b}\right)^2} \tag{5-19}$$

$$a_{mn} = \frac{4}{ab} \int_0^a \int_0^b f(x, y) \sin\frac{m\pi x}{a} \sin\frac{n\pi y}{b} \, dx \, dy \tag{5-20}$$

Illustrative Example 5. 1.   Consider an elastic string that is 6 ft long, weighs 0.600 lb, and is stretched until the tension force is 5 lb throughout. If the string is fixed at both ends, has no initial displacement, and has an initial velocity of 2.05 ft/sec throughout the length of the string, find the variation of the string displacement with time.

Solution.   The partial differential equation to solve is

$$\frac{\partial^2 \bar{y}}{\partial x^2} = \frac{1}{a^2}\frac{\partial^2 \bar{y}}{\partial t^2}$$

where the boundary and initial conditions are

$$\bar{y}(0, t) = 0$$
$$\bar{y}(6, t) = 0$$
$$\bar{y}(x, 0) = 0$$
$$\frac{\partial \bar{y}}{\partial t}(x, 0) = 2.05$$

The solution of this problem is given by Eqs. (5-15) and (5-16), where $g(x) = 2.05$. Since $w = W/L$, and since $a^2 = Tg/w$, we have

$$a^2 = \frac{TgL}{W} = \frac{(5.00\text{ lb})\,(32.2\text{ ft/sec}^2)\,(6.00\text{ ft})}{0.600\text{ lb}} = 1610\text{ ft}^2/\text{sec}^2$$

so that $a = 40.1$ ft/sec. Thus, substitution in Eq. (5-16) gives

$$\begin{aligned}
H_n &= \frac{(2)(2.05)}{(3.14)(40.1)(n)}\int_0^6 \sin\frac{n\pi x}{6}\,dx \\
&= \frac{(2)(2.05)(6)}{(3.14)^2(40.1)(n^2)}\int_0^6 \sin u\,du \\
&= -\frac{0.0622}{n^2}\left[\cos\frac{n\pi x}{6}\right]_0^6 \\
&= -\frac{0.0622}{n^2}\left[(-1)^n - 1\right]
\end{aligned}$$

Thus, $H_n = 0$ if n is even, and $H_n = 0.1244/n^2$ if n is odd. Substitution in Eq. (5-15) gives

$$\begin{aligned}
\bar{y} &= 0.1244 \sum_{n=1}^{\infty}\frac{[\sin(2n-1)\pi x/6][\sin 40.1(2n-1)\pi t/6]}{(2n-1)^2} \\
&= 0.1244\left[\sin\frac{\pi x}{6}\sin\frac{40.1\pi t}{6} + \frac{1}{9}\sin\frac{\pi x}{2}\sin\frac{40.1\pi t}{2} + \cdots\right]
\end{aligned}$$

Illustrative Example 5.2.   Solve the problem in Ill. Ex. 5.1 when the elastic string also has an initial sinusoidal displacement that is equal to 0.250 $\sin \pi x/6$ ft.

Solution.   If the initial velocity were equal to zero, the solution would be given by Eq. (5-12). Substitution of numerical values in this equation gives

$$\bar{y} = 0.250 \sin \frac{\pi x}{6} \cos \frac{40.1\pi t}{6}$$

The complete solution to this problem is the sum of the solution given above and the solution given in Ill. Ex. 5. 1. Thus, the solution of this problem is

$$\bar{y} = 0.250 \sin \frac{\pi x}{6} \cos \frac{40.1\pi t}{6}$$

$$+ 0.1244 \sum_{n=1}^{\infty} \frac{[\sin(2n-1)\pi x/6][\sin 40.1(2n-1)\pi t/6]}{(2n-1)^2}$$

Illustrative Example 5.3.   Suppose that an elastic cylindrical rod of length L is rigidly mounted at one end and free at the other end. Solve for the motion $\theta(x, t)$ of this shaft if the free end is initially twisted an angle θ_0 and released at time zero with zero velocity.

Solution.   From Sec. 5. 3, the partial differential equation of motion is $\partial^2 \theta / \partial x^2 = (\partial^2 \theta / \partial t^2)/a^2$, and the boundary and initial conditions are

$$\theta(0, t) = 0$$

$$\frac{\partial \theta}{\partial x}(L, t) = 0$$

$$\theta(x, 0) = \frac{\theta_0 x}{L}$$

$$\frac{\partial \theta}{\partial t}(x, 0) = 0$$

Assuming that $\theta(x, t) = X(x)T(t)$ and applying the separation-of-variables method, we obtain the following equation, as before:

$$\theta = XT = (A \cos \lambda x + B \sin \lambda x)(C \cos \lambda at + D \sin \lambda at)$$

The boundary condition $\theta(0, t) = 0$ gives $X(0) = A \cos 0 = 0$ or $A = 0$. The boundary condition $\partial \theta(L, t)/\partial x = 0$ gives $dX(L)/dx = \lambda B \cos \lambda L = 0$. Since $\cos \lambda L = 0$, $\lambda = (2n - 1)\pi/2L$, where n is an integer. The initial condition $\partial \theta(x, 0)/\partial t = 0$ gives $dT(0)/dt = \lambda a D \cos 0 = 0$, so that $D = 0$. Thus, a combination of these results, for all values of n gives the solution

$$\theta(x, t) = \sum_{n=1}^{\infty} G_n \sin\left(\frac{2n-1}{2L}\right)\pi x \cos\left(\frac{2n-1}{2L}\right)\pi at \qquad (5\text{-}21)$$

where coefficient $G_n = B_n C_n$ must be determined. The initial condition $\theta(x, 0) = \theta_0 x/L$ gives the following Fourier sine series:

$$\theta(x, 0) = \sum_{n=1}^{\infty} G_n \sin\left(\frac{2n-1}{2L}\right)\pi x = \frac{\theta_0 x}{L}$$

Use of Fourier series theory shows that

$$G_n = \frac{2}{L} \int_0^L \left(\frac{\theta_0 x}{L}\right) \sin\left(\frac{2n-1}{2L}\right) \pi x \, dx$$

$$= \frac{(-1)^{n+1}(8\theta_0)}{(2n-1)^2\pi^2} \tag{5-22}$$

Thus, the solution for motion $\theta(x, t)$ is

$$\theta(x, t) = \frac{8\theta_0}{\pi^2}\left[\sin\frac{\pi x}{2L}\cos\frac{\pi at}{2L} - \frac{1}{9}\sin\frac{3\pi x}{2L}\cos\frac{3\pi at}{2L}\right.$$

$$\left. + \frac{1}{25}\sin\frac{5\pi x}{2L}\cos\frac{5\pi at}{2L} - \cdots\right] \tag{5-23}$$

5.6   Solution by Laplace Transforms

Usually we cannot apply the separation-of-variables method to solve for the vibration motion of a distributed system if a boundary condition varies with time. The Laplace transform method of solution is useful for some time-varying boundary conditions (e. g., step functions and simple pulse functions). Before we can solve partial differential equations by Laplace transforms, we must first obtain the transforms of partial derivatives. Consider a function of two independent variables $f(x, t)$. Its Laplace transform $F(x, s)$ is defined by

$$F(x, s) \equiv L\left\{f(x, t)\right\} = \int_0^\infty e^{-st}f(x, t)dt \tag{5-24}$$

where x is treated as a constant in the integration. For the inverse transform, we shall use the notation

$$f(x, t) = L^{-1}\{F(x, s)\} \tag{5-25}$$

To obtain $L\{\partial f/\partial t\}$, where x is kept constant, we have from Eq. (5-24)

$$L\left\{\frac{\partial f}{\partial t}\right\} = \int_0^\infty e^{-st}\frac{\partial f}{\partial t}\, dt$$

An integration by parts gives

$$\int_0^\infty e^{-st}\frac{\partial f}{\partial t}\, dt = e^{-st}f(x, t)\Big|_0^\infty + s\int_0^\infty e^{-st}f(x, t)dt$$

Since $e^{-st}f(x, t)$ approaches zero as t approaches infinity, if function $f(x, t)$ is Laplace transformable, we obtain

$$L\left\{\frac{\partial f}{\partial t}\right\} = sL\{f(x, t)\} - f(x, 0+)$$

$$= sF(x, s) - f(x, 0+) \tag{5-26}$$

which is analogous to the result that we had obtained in Sec. 2.10 for $L\{df/dt\}$. Calculating the Laplace transforms of the higher partial derivatives of $f(x, t)$ with respect to t, in the same manner as we previously did in Sec. 2. 10 for the one-variable function $f(t)$, we obtain

$$L\left\{\frac{\partial^2 f}{\partial t^2}\right\} = s^2 F(x, s) - sf(x, 0+) - \frac{\partial f}{\partial t}(x, 0+) \tag{5-27}$$

$$L\left\{\frac{\partial^n f}{\partial t^n}\right\} = s^n F(x, s) - s^{n-1} f(x, 0+)$$

$$- s^{n-2} \frac{\partial f}{\partial t}(x, 0+) - \cdots - \frac{\partial^{n-1} f}{\partial t^{n-1}}(x, 0+) \tag{5-28}$$

Now let us calculate the partial derivatives of $f(x, t)$ with respect to x. Since the function $f(x, t)$ must fulfill proper continuity and differentiability requirements, we have

$$\frac{\partial F(x, s)}{\partial x} = \frac{\partial}{\partial x} \int_0^\infty e^{-st} f(x, t) dt$$

$$= \int_0^\infty \frac{\partial}{\partial x} [e^{-st} f(x, t)] dt$$

$$= \int_0^\infty e^{-st} \frac{\partial f(x, t)}{\partial x} dt$$

Thus, we have obtained the following result:

$$L\left\{\frac{\partial f}{\partial x}\right\} = \frac{\partial F(x, s)}{\partial x} \equiv \frac{\partial F}{\partial x} \tag{5-29}$$

In like manner, we also obtain

$$L\left\{\frac{\partial^2 f}{\partial x^2}\right\} = \frac{\partial^2 F}{\partial x^2} \tag{5-30}$$

$$L\left\{\frac{\partial^n f}{\partial x^n}\right\} = \frac{\partial^n F}{\partial x^n} \tag{5-31}$$

Suppose that we wanted to solve the following simple, partial differential equation using Laplace transforms:

$$\frac{\partial f}{\partial t} = 12xt + 6x^2 + 3\cos 2x$$

where it is given that $f(x, 0) = 4 \sin x$. Denoting the Laplace transform of $f(x, t)$ by $F(x, s)$ or by F, the Laplace transforms of each term in the given differential equation are

$$L\left\{\frac{\partial f}{\partial t}\right\} = sF - 4 \sin x$$

$$L\{12xt\} = 12xL\{t\} = \frac{12x}{s^2}$$

$$L\{6x^2\} = 6x^2 L\{1\} = \frac{6x^2}{s}$$

$$L\{3 \cos 2x\} = \frac{3 \cos 2x}{s}$$

Thus, the Laplace transform of the given differential equation is

$$sF - 4 \sin x = \frac{12x}{s^2} + \frac{6x^2}{s} + \frac{3 \cos 2x}{s}$$

Solving the above equation for F, we obtain

$$F(x, s) = \frac{12x}{s^3} + \frac{1}{s^2}(6x^2 + 3 \cos 2x) + \frac{4 \sin x}{s}$$

Since the inverse transforms of the following terms are:

$$L^{-1}\left\{\frac{12x}{s^3}\right\} = 12xL^{-1}\left\{\frac{1}{s^3}\right\} = 12x\left(\frac{t^2}{2}\right) = 6xt^2$$

$$L^{-1}\left\{\frac{6x^2 + 3 \cos 2x}{s^2}\right\} = (6x^2 + 3 \cos 2x)t$$

$$L^{-1}\left\{\frac{4 \sin x}{s}\right\} = 4 \sin x\, L\left\{\frac{1}{s}\right\} = 4 \sin x$$

the solution of the given partial differential equation is

$$f(x, t) = 6xt^2 + 6x^2 t + 3t \cos 2x + 4 \sin x$$

which can easily be verified by calculating $\partial f/\partial t$ and $f(x, 0)$.

The previous example solved a very simple type of partial differential equation, but it did illustrate the solution procedure. Let us now summarize the general solution procedure when we wish to solve more general types of linear, partial differential equations by Laplace transforms. The first step in the solution procedure is to take the Laplace transform of the given partial differential equation, where the initial conditions are inserted. We will then obtain an ordinary differential equation, where $F(x, s) \equiv L\{f(x, t)\}$ is the dependent variable, and x is the independent variable. We must solve this ordinary differential equation to obtain $F(x, s)$. It is preferable to solve this equation by classical means. We could solve it by a second Laplace transformation, but this could cause confusing symbolism and could result in an involved procedure. The arbitrary functions and constants in the solution of $F(x, s)$ are determined from the given boundary conditions for the problem. We then obtain the desired solution for the function $f(x, t)$ by taking the inverse transform of $F(x, s)$.

The two illustrative examples in this section will further exemplify this solution procedure. These examples solve relatively simple types of vibration problems. Usually the application of this method is more involved than the separation-of-variables technique, but again it should be mentioned that the Laplace transform technique can be used to solve more types of partial differential equation problems. Often one of the most difficult parts of the procedure, when solving such distributed-system problems, is obtaining the inverse transform. This difficulty can sometimes be minimized when a more complete table of Laplace transforms is available.

Illustrative Example 5. 4.   Consider the vibration of an elastic string of length L, fixed at both ends, having an initial displacement $h \sin \pi x/L$ and no initial velocity. Find the time variation of the displacement at any point on the string.

Solution.   From Eq. (5-4), the partial differential equation to solve is the following one-dimensional wave equation:

$$\frac{\partial^2 \bar{y}}{\partial x^2} = \frac{1}{a^2} \frac{\partial^2 \bar{y}}{\partial t^2}$$

and the boundary and initial conditions are

$$\bar{y}(0, t) = 0$$
$$\bar{y}(L, t) = 0$$
$$\bar{y}(x, 0) = h \sin \frac{\pi x}{L}$$
$$\frac{\partial \bar{y}}{\partial t}(x, 0) = 0$$

Denoting the Laplace transform of $\bar{y}(x, t)$ by $Y(x, s)$ or by Y, application of Eqs. (5-27) and (5-30) shows that the Laplace transform of the given differential equation is

$$\frac{\partial^2 Y}{\partial x^2} = \frac{1}{a^2} \left[s^2 Y - s(h \sin \frac{\pi x}{L}) - 0 \right]$$

Since we have no partial derivatives of s, we end up with the ordinary differential equation

$$\frac{d^2 Y}{dx^2} - \frac{s^2}{a^2} Y = -\frac{hs}{a^2} \sin \frac{\pi x}{L}$$

The solution of the homogeneous differential equation $d^2 Y/dx^2 - (s^2/a^2) Y = 0$ is

$$Y_H = \cosh \frac{s}{a} x + B \sinh \frac{s}{a} x$$

Using the method of undetermined coefficients, we assume that the particular solution Y_P is of the form

$$Y_P = C \cos \frac{\pi x}{L} + D \sin \frac{\pi x}{L}$$

Upon substitution of Y_P into the ordinary differential equation, we obtain

$$C = 0$$
$$D = \frac{hs}{s^2 + a^2 \pi^2 / L^2}$$

and the solution for $Y(x, s)$ becomes

$$Y = Y_H + Y_P = A \cosh \frac{s}{a} x + B \sinh \frac{s}{a} x + \frac{hs}{s^2 + a^2\pi^2/L^2} \sin \frac{\pi x}{L}$$

Substitution of $x = 0$ in the above equation gives $Y(0, s) = A$. Since $\bar{y}(0, t) = 0$, $Y(0, s) = L\{\bar{y}(0, t)\} = 0$. Thus $A = 0$. In a similar manner, the boundary condition $\bar{y}(L, t) = 0$ implies that $Y(L, s) = B \sinh sL/a = 0$. Since sinh $r = 0$ only when $r = 0$, this means that $B = 0$. Thus, we have

$$Y(x, s) = \frac{hs}{s^2 + a^2\pi^2/L^2} \sin \frac{\pi x}{L}$$

Taking the inverse transform of $Y(x, s)$, we obtain the following solution:

$$y(x, t) = h \sin\frac{\pi x}{L} \cos \frac{\pi a t}{L} \qquad (5\text{-}32)$$

The above solution agrees with Eq. (5-12), which is a solution of the same problem that was obtained by using separation of variables.

Illustrative Example 5.5. Suppose that a semi-infinite rod lies along the positive half of the x-axis. If the left end of the rod is twisted an angle $f(t)$, where $f(t)$ is a known function of time, determine the timewise variation of angular displacement $\theta(x, t)$ along the rod.

Solution. We must solve the partial differential equation $\partial^2\theta/\partial x^2 = (\partial^2\theta/\partial t^2)/a^2$, where the boundary and initial conditions are

$$\theta(0, t) = f(t)$$
$$\theta(\infty, t) \text{ is finite}$$
$$\theta(x, 0) = 0$$
$$\frac{\partial \theta}{\partial t}(x, 0) = 0$$

If we let $T = T(x, s)$ denote $L\{\theta(x, t)\}$, the Laplace transform of this partial differential equation is as follows, where the two initial conditions were applied:

$$\frac{\partial^2 T}{\partial x^2} = \frac{1}{a^2}[s^2T - s(0) - 0]$$

which can be rewritten as the ordinary differential equation

$$\frac{d^2 T}{dx^2} - \frac{s^2}{a^2} T = 0$$

whose solution is

$$T \equiv T(x, s) = Ae^{sx/a} + Be^{-sx/a}$$

Since one boundary condition states that $\theta(\infty, t)$ is finite, $T(\infty, s) \equiv L\{\theta(\infty, t)\}$ is also finite. Since the equation for $T(x, s)$ shows that $T(\infty, s) = Ae^{\infty}$, then coefficient A must equal zero. Since $T(x, s) = Be^{-sx/a}$, the other boundary condition states that

$$Y(0, s) \equiv L\{\theta(0, t)\} = L\{f(t)\} = B$$

Since $B = L\{f(t)\}$, we have

$$Y(x, s) \equiv L\{\theta(x, t)\} = L\{f(t)\}\, e^{-sx/a}$$

and use of the second shifting theorem in Sec. 2.11 gives the following inverse transform:

$$\theta(x, t) = f\left(t - \frac{x}{a}\right) u\left(t - \frac{x}{a}\right) \tag{5-33}$$

which shows that the angular displacement at location x and time t (if $t > x/a$) equals the applied displacement at the left end at time $(t - x/a)$.

5.7   Solution by Numerical Methods

Two of the advantages of numerical-solution methods over analytical methods (such as separation of variables and Laplace transforms) are (1) a boundary condition that has a rather involved variation with time will not add much complication to their application, and (2) they are easy to program for digital computer solution. Section 7.26 furnishes a digital computer program that numerically solves the one-dimensional wave equation for arbitrary initial and boundary conditions. The first step in the numerical solution of a linear partial differential equation is to write each partial derivative, in the given equation, in finite-difference form. That is, the partial differential equation is written as a finite-difference equation. To simplify matters, we shall first learn to write total derivatives in finite-difference form.

Let $\Delta f(x)$ be the change in the value of a given function of one variable, $f(x)$, when x is increased by an amount Δx. Thus,

$$\Delta f(x) = f(x + \Delta x) - f(x)$$

where the term $\Delta f(x)$ is called the *first forward difference* of the function, relative to the positive increment Δx. The *second forward difference*, $\Delta^2 f(x)$, is given by

$$\begin{aligned}
\Delta^2 f(x) = \Delta[\Delta f(x)] &= \Delta f(x + \Delta x) - \Delta f(x)\\
&= [f(x + 2\Delta x) - f(x + \Delta x)] - [f(x + \Delta x) - f(x)]\\
&= f(x + 2\Delta x) - 2f(x + \Delta x) + f(x)
\end{aligned}$$

Since the limit of $\Delta f(x)/\Delta x$ as Δx approaches zero is defined to be the derivative df/dx, it is seen that if the interval Δx is very small, then we may approximate the derivative df/dx by

$$\frac{f(x + \Delta x) - f(x)}{\Delta x}$$

Since, for a derivative to exist at a point, the slopes must be equal on both sides of that point, we may also approximate df/dx by

$$\frac{f(x) - f(x - \Delta x)}{\Delta x}$$

Thus, we had shifted backwards one Δx-interval to approximate df/dx. The second derivative d^2f/dx^2 is defined to be $d(df/dx)/dx$. Thus, if the interval Δx is very small, we may approximate d^2f/dx^2 by $\Delta[\Delta f(x)/\Delta x]/\Delta x$, which is $\Delta^2 f(x)/(\Delta x)^2$. If we shift this result back one Δx-interval, we obtain the following approximation for d^2f/dx^2:

$$\frac{f(x + \Delta x) - 2f(x) + f(x - \Delta x)}{(\Delta x)^2}$$

In a similar manner, we may also express the partial derivatives of functions of several variables in finite-difference form. The results are analogous to those for total derivatives, since for a partial derivative all variables are held constant except one. Thus, for the two-variable function $\phi(x, t)$ we may approximate $\partial \phi/\partial t$ by

$$\frac{\phi(x, t + \Delta t) - \phi(x, t)}{\Delta t}$$

which will represent the partial derivative in the limit as Δt approaches zero. In a similar manner, we may approximate $\partial^2 \phi/\partial t^2$ by

$$\frac{\phi(x, t + \Delta t) - 2\phi(x, t) + \phi(x, t - \Delta t)}{(\Delta t)^2}$$

and $\partial^2 \phi/\partial x^2$ may be approximated by

$$\frac{\phi(x + \Delta x, t) - 2\phi(x, t) + \phi(x - \Delta x, t)}{(\Delta x)^2}$$

The last two approximations enable us to write the one-dimensional wave equation in finite-difference form. Now suppose that we had the three-variable function $\phi(x, y, t)$ which satisfies the two-dimensional wave equation. To write this partial differential equation in finite-difference form, we must first write the difference approximations for $\partial^2 \phi/\partial x^2$, $\partial^2 \phi/\partial y^2$, and $\partial^2 \phi/\partial t^2$. The difference approximation for $\partial^2 \phi/\partial x^2$ is

$$\frac{\phi(x + \Delta x, y, t) - 2\phi(x, y, t) + \phi(x - \Delta x, y, t)}{(\Delta x)^2}$$

Writing the one-dimensional wave equation $\phi_{xx} = \phi_{tt}/a^2$ (where $\phi_{xx} \equiv \partial^2 \phi/\partial x^2$ and $\phi_{tt} \equiv \partial^2 \phi/\partial t^2$) in finite-difference form, we have

$$\frac{\phi(x + \Delta x, t) - 2\phi(x, t) + \phi(x - \Delta x, t)}{(\Delta x)^2}$$
$$= \frac{\phi(x, t + \Delta t) - 2\phi(x, t) + \phi(x, t - \Delta t)}{a^2(\Delta t)^2}$$

If we let Δx equal $a(\Delta t)$, we obtain the following approximation equation for $\phi(x, t + \Delta t)$, where the accuracy of this approximation improves as Δx is made smaller. Note that $\Delta t = \Delta x/a$.

$$\phi(x, t + \Delta t) = \phi(x + \Delta x, t) + \phi(x - \Delta x, t) - \phi(x, t - \Delta t) \quad (5\text{-}34)$$

In a similar manner, if we write the two-dimensional wave equation $\phi_{xx} + \phi_{yy} = \phi_{tt}/a^2$ in finite-difference form and let $\Delta x = \Delta y = a(\Delta t)$, we will obtain the following approximation equation for $\phi(x, y, t + \Delta t)$:

$$\begin{aligned}
\phi(x, y, t + \Delta t) = {} & \phi(x + \Delta x, y, t) + \phi(x - \Delta x, y, t) \\
& + \phi(x, y + \Delta y, t) + \phi(x, y - \Delta y, t) \\
& - 2\phi(x, y, t) - \phi(x, y, t - \Delta t) \quad (5\text{-}35)
\end{aligned}$$

If the initial condition at a boundary differs from the boundary condition at an instant later (e. g., a step input at the boundary), then it is recommended that the average of these two values be used as the initial boundary value, in order to obtain the best results when solving partial differential equations by numerical techniques. In more general terms, it may be stated that if at time t there is a sudden change in a boundary value at a specified point, then the average of the two discontinuous values should be chosen as the boundary value at time t for that point. At a free end, $\partial\phi/\partial x = 0$. This boundary condition can be handled numerically by adding a fictitious length of thickness Δx. If the end is free at $x = L$, we let $\phi(L + \Delta x, t) = \phi(L, t)$ in our numerical calculations, where $\phi(L + \Delta x, t)$ is the displacement of the fictitious end at time t, so that $\partial\phi/\partial x = 0$ along the fictitious length.

Illustrative Example 5.6.   Consider an elastic string 10 ft long and fixed at both ends. Its value of a^2 or Tg/w is 16 ft²/sec². The string is held in the position shown in Fig. 5-4 for one sec, and is then released. Find the time variation of the string displacement up to 5 sec after the string is released from this initial position.

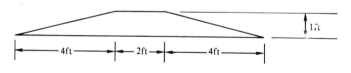

Fig. 5-4

Solution.   From Eq. (5-4) it is seen that this problem satisfies the one-dimensional wave equation. We shall calculate the displacement at five evenly-spaced intervals. Thus, $\Delta x = 2$ft. The choice of the Δx-interval specifies the size of the Δt-interval. Since $a = 4$ ft/sec, $\Delta t = \Delta x/a = 0.500$ sec.

Since the string is fixed at both ends, its displacement equals zero at $x = 0$ and at $x = 10$ ft for all values of time t. This result is shown in Table 5-1. The initial displacement, as given by Fig. 5-4, is also inserted in Table 5-1 for $t = 0$ and for $t = -0.500$ sec, since this displacement was maintained for one second. Since the initial and boundary conditions have now been inserted in Table 5-1, we can now apply Eq. (5-34) to calculate the other values of the string displacement $\bar{y}(x, t)$. For example

$$\bar{y}(2.00, 0.500) = \bar{y}(4.00, 0) + \bar{y}(0, 0) - \bar{y}(2.00, -0.500)$$
$$= 1.000 + 0 - 0.500 = 0.500$$

The results of these calculations are shown in Table 5-1 and from this table, it can be seen that an oscillation results. The accuracy of these results could be improved by using a smaller value for Δx, since numerical-solution techniques use approximation methods.

Table 5-1

x(ft) t(sec)	0	2.00	4.00	6.00	8.00	10.00
-0.500	0	0.500	1.000	1.000	0.500	0
0	0	0.500	1.000	1.000	0.500	0
0.500	0	0.500	0.500	0.500	0.500	0
1.000	0	0	0	0	0	0
1.500	0	-0.500	-0.500	-0.500	-0.500	0
2.000	0	-0.500	-1.000	-1.000	-0.500	0
2.500	0	-0.500	-1.000	-1.000	-0.500	0
3.000	0	-0.500	-0.500	-0.500	-0.500	0
3.500	0	0	0	0	0	0
4.000	0	0.500	0.500	0.500	0.500	0
4.500	0	0.500	1.000	1.000	0.500	0
5.000	0	0.500	1.000	1.000	0.500	0

5.8 Transverse Vibrations of a Beam

Let $w(x, t)$ denote a variable load per unit length acting along a beam. The mass of a beam segment Δx units long is approximately $A\rho(\Delta x)$, where cross-sectional area A and mass density ρ can vary with longitudinal distance x. Since the vertical acceleration of this segment is $\partial^2 y/\partial t^2$, Newton's second law gives, for any time t,

$$A\rho(\Delta x)\frac{\partial^2 y}{\partial t^2} = w(x, t)\Delta x + V(x, t) - V(x + \Delta x, t)$$

where $V(x, t)$ and $V(x + \Delta x, t)$ are the shear forces acting at the ends of this segment, which will vary with time. Dividing the above equation by Δx and taking the limit as Δx approaches zero, we obtain

$$A\rho \frac{\partial^2 y}{\partial t^2} = w(x, t) - \frac{\partial V}{\partial x}$$

where Eq. (5–2) was applied to obtain the term $\partial V/\partial x$. That is,

$$\frac{\partial V}{\partial x} = \lim_{\Delta x \to 0} \frac{V(x + \Delta x, t) - V(x, t)}{\Delta x}$$

In our strength of materials course, we obtained the following equation for a beam that is of constant material and cross-sectional area, where bending moment $M = \partial V/\partial x$. The equation is written in partial derivative form because here we have two variables, x and t.

$$M = \frac{\partial V}{\partial x} = EI \frac{\partial^4 y}{\partial x^4}$$

Thus, for these restrictions, we have upon substitution for $\partial V/\partial x$

$$EI \frac{\partial^4 y}{\partial x^4} = w(x, t) - A\rho \frac{\partial^2 y}{\partial t^2} \tag{5–36}$$

If the load $w(x, t)$ is negligible in comparison to the inertia force $A\rho(\partial^2 y/\partial t^2)$, we have

$$\frac{\partial^4 y}{\partial x^4} = -\frac{1}{a^2} \frac{\partial^2 y}{\partial t^2} \tag{5–37}$$

where a^2 equals $EI/A\rho$.

Equation (5–37) can be solved by separation of variables for different sets of boundary and initial conditions. Assuming that $y(x,t) = X(x)T(t)$, and applying this method, we obtain

$$T \frac{d^4 X}{dx^4} = -\frac{1}{a^2} X \frac{d^2 T}{dt^2}$$

which becomes upon rearrangement

$$\frac{1}{X} \frac{d^4 X}{dx^4} = -\frac{1}{a^2 T} \frac{d^2 T}{dt^2} = \lambda^4$$

where λ^4 is a positive constant in order to obtain vibratory motion. Thus, we have the two ordinary differential equations

$$\frac{d^4 X}{dx^4} - \lambda^4 X = 0$$

$$\frac{d^2 T}{dt^2} + a^2 \lambda^4 T = 0$$

whose solutions are

$$X(x) = A \cos \lambda x + B \sin \lambda x + C \cosh \lambda x + D \sinh \lambda x \quad \text{(5-38)}$$
$$T(t) = E \cos a\lambda^2 t + F \sin a\lambda^2 t \quad \text{(5-39)}$$

Since $y(x, t) = X(x) T(t)$, Eq. (5–38) specifies the shape of the principal-vibration modes. Since Eq. (5–39) shows that mode-shape frequency ω equals $a\lambda^2$ rad/sec, and since λ can be determined only by using the four boundary conditions for Eq. (5–38), the natural frequencies of vibration depend on the method of beam support.

Suppose that we have a beam of length L that is simply supported at both ends, has an initial displacement $f(x)$, and has no initial velocity. The simple supports at the ends of the beam give us four boundary conditions. The boundary condition $y(0, t) = 0$ implies that

$$X(0) = A + C = 0$$

The boundary condition $y(L, t) = 0$ implies that

$$X(L) = A \cos \lambda L + B \sin \lambda L + C \cosh \lambda L + D \sinh \lambda L = 0$$

The boundary condition $\partial^2 y(0, t)/\partial x^2 = 0$ implies that

$$\frac{d^2 X}{dx^2}(0) = -A\lambda^2 + C\lambda^2 = \lambda^2(C - A) = 0$$

The boundary condition $\partial^2 y(L, t)/\partial x^2 = 0$ implies that

$$\frac{d^2 X}{dx^2}(L) = -A\lambda^2 \cos\lambda L - B\lambda^2 \sin \lambda L + C\lambda^2 \cosh \lambda L + D\lambda^2 \sinh \lambda L = 0$$

Solving these four equations, which were obtained from the boundary conditions, we obtain

$$A = 0$$
$$B \sin \lambda L = 0$$
$$C = 0$$
$$D = \sin \lambda L = 0$$

The last of the previous four equations implies that $D = 0$, since the hyperbolic sine is not a periodic function. The second equation implies that $\lambda = n\pi/L$. The initial condition $\partial y(x, 0)/\partial t = 0$ implies that $dT(0)/dt = a\lambda^2 F = 0$. Thus, $F = 0$. Expressing our results as an infinite series, we obtain

$$y = \sum_{n=1}^{\infty} B_n \sin \frac{n\pi x}{L} \cos \frac{an^2\pi^2 t}{L^2} \quad \text{(5-40)}$$

where the B_n-coefficients are obtained from the initial condition

$$y(x, 0) = f(x) = \sum_{n=1}^{\infty} B_n \sin n\pi x/L.$$

Equation (5–40) shows that a simply-supported beam has an infinite number of natural frequencies and principal vibration modes. The natural frequencies equal $an^2\pi/2L^2$ cycles/sec, where n is an integer and $a = \sqrt{EI/A\rho}$. For a cantilever beam, which is fixed at one end and free at the other, it can be shown (by use of the four boundary conditions) that the values of λ can be determined from the equation

$$\cosh \lambda L \cos \lambda L = -1 \qquad (5\text{–}41)$$

For a beam that is fixed at one end and simply supported at the other, the four boundary conditions can be used to show that the values of λ can be determined from the equation

$$\tan \lambda L = \tanh \lambda L \qquad (5\text{–}42)$$

Since Eqs. (5–41) and (5–42) each have an infinite number of roots for λ and since natural frequency f_n equals $a\lambda^2/2\pi$ cycles/sec, the last two beams also have an infinite number of natural frequencies and principal vibration modes. The roots for λ for Eq. (5–41) or (5–42) can be obtained graphically or by use of a trial-and-error method (e. g., bisection or false-position method), as discussed in the last part of Sec. 7. 19. These roots are also tabulated in some vibration textbooks.

PROBLEMS

5.1. Calculate the wave-propagation velocity for the following systems:
  (a) An elastic string that weighs 0.0631 lb/ft, and has a 1.52-lb tension force.
  (b) An elastic membrane that weighs 0.108 lb/ft^2, and whose tension force is 0.678 lb/ft.
  (c) An elastic bar that weighs 71.5 lb/ft^3 and whose modulus of elasticity is 35.1×10^3 lb/ft^2.

5.2. Write the partial differential equation that describes each of the following systems. Clearly show your derivations.
  (a) A vibrating elastic string whose weight is not negligible in comparison to the tension force caused by the initial stretching of the string, and which has a vertical external force distribution of magnitude $P \sin \pi x/L$ lb/ft applied along its length.
  (b) The oscillation of a chain, whose weight per unit length is w, that is hung vertically from its supported top end.
  (c) A beam, of length L and of constant flexural rigidity, cross-sectional area, and density, that has a variable vertical load distribution of $F_0 e^{-at} \sin \pi x/L$ applied to it, where F_0 is in lb/ft.

5.3. A vibrating rectangular membrane, clamped at its boundaries, is e units long and d units wide. Locating one corner at $(x = 0, y = 0)$, its length along the positive x-axis, and its width along the positive y-axis, write the partial differential equation and its boundary conditions. If the initial velocity is zero and the initial displacement is either $xy(x - e)^2(y - d)^2$ or $xy(x - d)^2(y - e)^2$, write the initial conditions and state why you chose the displacement equation that you did. The membrane, which is under specific tension T, weighs w lb per unit area.

5.4. Write the partial differential equation, the boundary conditions, and the initial conditions for the following systems:

(a) A vibrating string of length L whose ends are fixed, is initially undisplaced, and whose initial velocity is $x(L - x)^2$. Express your comments if the initial velocity were $L(x - L)$.

(b) A uniform shaft of circular cross-section is rigidly mounted at one end and is unsupported at the other end. This shaft, which is 2.31 ft long, is initially untwisted, and has no initial velocity. The free end undergoes an oscillating angular displacement of $0.131 \sin 60t$ by means of an applied external moment.

(c) A long elastic bar of length L is rigidly mounted at one end and is free at the other end. The bar has no initial velocity, but its free end is stretched a distance d so that its initial displacement distribution is dx/L.

(d) An elastic bar of length L is rigidly supported at one end and is unsupported at the other end. There is no initial displacement or velocity, but an oscillating force is applied at the free end so that its displacement variation is given by $m \sin at$.

5.5. Comment on the validity of the following initial displacements for an elastic string of length L that is fixed at both ends: $A \tan (\pi x/L)$, $A[\cos (\pi x/L) - 1]$, $A \sin (\pi x/2L)$, and $A(x \sinh L - L \sinh x)$.

5.6. A uniform shaft of circular cross-section is vibrating torsionally about its longitudinal central axis. It is built-in at both ends, is L units long, and has a specific weight w and shear modulus G. If this shaft has zero initial velocity and is given an initial angular displacement $\theta_0(x)$, write the partial differential equation and initial and boundary conditions for this system. Also write the partial differential equation and initial and boundary conditions if the end is free at $x = L$. Comment on the use of $\theta_0(x) = Ax^2(L - x)^4$ for both means of support. Repeat for an initial angular displacement of Ax/L.

5.7. An elastic beam that has a modulus of elasticity E, length L, and constant specific weight w. If the beam is built-in at both ends, has zero initial velocity, and is given an initial longitudinal displacement $h(x)$, write the partial differential equation and initial and boundary conditions for this system. Write the partial differential equation and initial and boundary conditions if the given beam were a cantilever beam fixed at $x = 0$. Comment on the use of an initial displacement $h(x)$ equal to $x(L - x)$ for both means of beam support.

5.8. Consider a beam of length L whose load distribution may be neglected, and whose initial velocity is zero. If the initial transverse displacement $y_0(x)$ along the beam is equal to $x \sin \pi x/L$, write the partial differential equation and initial and boundary conditions for this system when the beam is fixed at both ends, and when the beam is simply supported at both ends. Specify which boundary conditions are violated by the initial displacement for both means of beam support.

5.9. In Sec. 5.5, we saw that the application of the two boundary conditions in the separation-of-variables procedure showed that sin $\lambda L = 0$ for a vibrating string (and also an elastic bar and rod) that is fixed at both ends. Since the natural frequencies equal $a\lambda$ radians per unit time, the frequency equation is given by sin $(\omega L/a) = 0$, where ω is a natural frequency. Use the boundary conditions (or means of support) to determine the frequency equation for the following systems:

(a) An elastic bar or rod that is unsupported at both ends.

(b) An elastic bar or rod rigidly supported at one end and unsupported at the other end.

(c) An elastic bar or rod in the vertical position, rigidly supported at the top, and having a concentrated mass of weight W and moment of inertia I at the other end.

5.10. Obtain the solution for longitudinal displacement $h(x,t)$ of a long elastic bar of length L for the following initial and boundary conditions. Use analogous results in Sec. 5.5 for your solutions.

(a) The bar is fixed at both ends, has no initial velocity, and has an initial displacement $d \sin \pi x/L$.

(b) The bar is fixed at both ends, has no initial velocity, and has an initial displacement $f(x)$.

(c) The bar is fixed at both ends, has no initial displacement, and has a constant initial velocity V along its length.

(d) The bar is fixed at one end and free at the other, and has no initial velocity, but the free end is pulled a distance m and released so that its initial displacement is mx/L.

5.11. Suppose that the center of the elastic string given in Ill. Ex. 5.1 is picked up a distance of 4 in. above the equilibrium position, so that the initial displacement is triangular, and is released without imparting any initial velocity. Find the variation of the string displacement with time.

5.12. Solve the problem in Ill. Ex. 5.1 when the elastic string also has an initial displacement that is equal to $(6x - x^2)$ ft.

5.13. Solve for the angular deflection $\theta(x,t)$ during the torsional vibration of a shaft of length L using the separation-of-variables method. It is given that the $x = 0$ end is fixed, the $x = L$ end is free, the initial angular displacement is $f(x)$, and the initial angular velocity is zero. At a free end, where there is no torque, $\partial\theta/\partial x = 0$.

5.14. Solve for the longitudinal displacement $h(x,t)$ during the longitudinal vibration of a uniform rod of length L using the separation-of-variables method. It is given that the $x = 0$ end is fixed, the $x = L$ end is free, the initial displacement is $f(x)$, and the initial velocity is zero. At a free end, where there is no external force, $\partial h/\partial x = 0$.

5.15. Suppose that an elastic bar of length L falls vertically from a free position and strikes a table with a velocity V, and then remains atop the table. Use the separation-of-variables method to find the longitudinal vibration variation of this bar.

5.16. A uniform elastic bar of length L is initially compressed a distance m by equal forces at its ends, and then these compressive forces are removed. Use the separation-of-variables method to find the longitudinal vibration variation of this bar.

5.17. Use the method given in Prob. 5.9 to obtain the frequency equation for the torsional vibration of an unsupported elastic rod which has a disk of moment of inertia I at each of its two free ends.

5.18. A uniform elastic rod of length L is rigidly supported at both ends. If the rod is twisted an angle γ at a distance c from the left end and suddenly released, use the separation-of-variables method to find the torsional vibration variation of this rod.

5.19. A uniform elastic rod of length L is rigidly supported at one end and unsupported at the other. There is no initial displacement or velocity, and there is a torque $T \sin \omega t$ applied to the free end. Find the torsional vibration variation of this rod.

5.20. For a given supersonic fluid flow above a sinusoidally-shaped wall, the following equations hold:

$$\frac{\partial^2 \phi}{\partial x^2} = \frac{1}{\lambda^2} \frac{\partial^2 \phi}{\partial y^2}$$

$$\frac{\partial \phi}{\partial y}(x, 0) = Uea \cos ax$$

where the wall shape is defined by the equation $y_{wall} = e \sin ax$. Use Eq. (5-9), the analytical solution $\phi(x, y) = f_1(x + \lambda_y) + f_2(x - \lambda y)$, to solve this problem for $\phi(x, y)$. It is given that $f_1(x + \lambda y) = 0$.

5.21. Write the Laplace transform of the partial differential equation and the boundary conditions for an elastic, vibrating membrane of rectangular shape that is clamped at its four edges.

5.22. A uniform elastic bar of length L is rigidly supported at the left end and is free at the other. A constant force F is suddenly applied at $t = 0$ and maintained at the free end. Find the Laplace transform of the longitudinal vibration displacement.

5.23. Repeat Prob. 5.22 if the applied force is an impulse function instead of a step function.

5.24. Suppose that a semi-infinite elastic bar lies along the positive half of the x-axis. If the left end is given a time-varying displacement $0.200 \sin 15t$, determine the timewise variation of the longitudinal displacement $h(x,t)$ along the bar.

5.25. Write the three-dimensional wave equation in finite-difference form.

5.26. Verify Eq. (5–35).

5.27. Repeat Ill. Ex. 5.6 using a Δx-interval of 1.00 ft.

5.28. Repeat Ill. Ex. 5.6 for the case in which the string is suddenly given its initial displacement and released.

5.29. Consider an elastic rod 10 ft long and initially at rest and undisplaced. Its value of a^2 or Gg/w is $100 \, ft^2/sec^2$. The right end is rigidly fastened, while the left end is given a periodic, rectangular-wave rotational displacement as shown in Fig. 2–16. This rectangular wave has an amplitude of 0.200 rad and a period of 2 sec. Using a Δx-interval of 2 ft, find the timewise variation of the shaft angular displacement for the first 3 sec using numerical methods.

5.30. Repeat Prob. 5.29, if there is no applied displacement at the left end, and if there is an initial angular displacement $0.0300x$ rad, where x is in ft, which is maintained for 2 sec before release.

5.31. Repeat Prob. 5.30 when the initial angular displacement is $0.0213x^{3/2} \sin (\pi x/10)$ rad, where x is in ft.

5.32. Use the separation-of-variables method to verify Eq. (5–41), from which the natural frequencies $a\lambda^2$ in rad/sec can be computed for a uniform cantilever beam.

5.33. Use the separation-of-variables method to verify Eq. (5–42). Write the frequency equation for this beam system.

5.34. Use the separation-of-variables method to obtain the frequency equation for a uniform beam that is rigidly supported at both ends.

5.35. Use the separation-of-variables method to obtain the frequency equation for a uniform beam that is unsupported at both ends.

5.36. Find the transverse vibrations in a uniform beam that is simply supported at both ends when the initial conditions are as follows:

(a)   A concentrated load P, located at $x = a$, that is suddenly removed.

(b)   A uniform load of w lb/ft that is suddenly removed.

5.37. Find the steady-state transverse-displacement variation in a uniform beam of length L that is simply supported at both ends, and which has a time-varying load of intensity $P \sin \omega t$ applied along its length.

6   Analog Solution
Methods

6.1   Review of Circuit Analysis

Electrical circuit elements are classified in two categories, active and passive. An *active circuit element* is a large source of energy. In this category we shall consider *voltage sources,* which maintain a voltage of a specified magnitude variation with time across its terminals, and *current sources,* which maintain a current of specified magnitude variation with time across its terminals. *Passive circuit elements* are capable of storing energy, and we shall consider the three most common types: resistors, capacitors, and inductors. The voltage drop, E_R, across a *linear resistor* is given by *Ohm's law,* which is

$$E_R = iR \tag{6-1}$$

where i is the current through the resistor in amperes and R is the resistance in ohms. Since resistors dissipate electrical energy by converting some of it into heat, there is a similarity between resistance and mechanical friction. *Capacitance* is a property based on the ability of the given circuit element (i.e., a capacitor) to store energy in an electric field. An uncharged capacitor becomes charged with electricity when it is connected to two points at different voltage levels. A *linear capacitor* is a circuit element whose voltage drop is given by the following formula:

$$E_c - E_c(0) = \frac{1}{C} \int_0^t i \, dt = \frac{q}{c} \tag{6-2}$$

where t is the time in seconds, i is the current in amperes, $E_c(0)$ is the initial capacitor voltage, q is the charge in coulombs, and C is a constant called capacitance that is measured in farads. By differentiation, the capacitor current i_c is found to be

$$i_c = C \frac{dE}{dt} \tag{6-3}$$

Inductance is a property based on the ability of the given circuit element (i.e., an inductor) to store energy in a magnetic field. An inductor offers opposition to a change in current. A *linear inductor* is a circuit element whose voltage drop E_L is given by the following equation:

$$E_L = L \frac{di}{dt} \tag{6-4}$$

where current i is in amperes, time t is in seconds, and L is a constant called inductance that is measured in henries. By integration, the inductor current i_L is found to be

$$i_L = \frac{1}{L} \int_0^t E \, dt + i_L(0) \tag{6-5}$$

where $i_L(0)$ is the initial value of the current (i.e., the value when time t equals zero).

So far we have talked about linear circuit elements. A resistor, capacitor, or inductor will be a *nonlinear circuit element* if its respective resistance, capacitance, or inductance varies with the current flow through it or with the voltage drop across its terminals. In this text we shall restrict ourselves to *lumped-element circuits,* where resistance, capacitance, and inductance in the curcuit are considered to exist only between the terminals of a passive circuit element. These types of circuits may be represented by ordinary differential equations. For the circuit diagrams of this text, Fig. 6-1 shows the symbolic representations of a voltage source of voltage E_g, a battery, a current source of current i_g amperes, a switch, a resistor of resistance R ohms, an inductor of inductance L henries, and a capacitor of capacitance C farads.

The procedure for writing the differential equations which describe

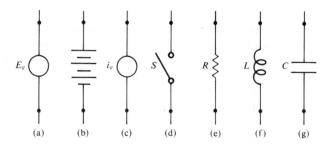

Fig. 6-1

the voltage or current variation with time for a circuit consisting of the lumped circuit elements mentioned in this section is based on Kirchoff's laws. *Kirchoff's voltage law* states that the algebraic sum of the voltage drops around a closed loop in an electric circuit is equal to zero. That is, the sum of the voltage drops is equal to the sum of the voltage rises in a given loop. *Kirchoff's current law* states that at any node (or junction) in an electric circuit the sum of the currents leaving is equal to the sum of the currents entering.

To be more specific, differential equations that describe the time-wise variation of the currents in an electrical circuit can be obtained by the *loop method*, a technique based on Kirchoff's voltage law. In order to use the method, first determine the number of independent loops for the given electrical circuit. Enough loops should be chosen so that together they contain the whole circuit. If all of the chosen loops are independent, then one important condition is that each of these loops covers a branch of the circuit that is not covered by any of the other loops. Second, assume clockwise directions for all of the currents in these independent loops (or let all of the loop currents be counterclockwise), where a current must flow through each circuit element. Also assume relative magnitudes for these currents. The third and final step is to write Kirchoff's voltage law for each loop, using the loop currents to determine the voltage across each circuit element. This procedure is illustrated in some of the example problems at the end of this section.

The differential equations that describe the timewise variation of the voltages in an electrical circuit can be obtained by the *node method*. This technique is based on Kirchoff's current law. The first step in this analysis is to determine the number of independent nodes, which we shall denote by the symbol N_n. If there are no transformers in the circuit so that the circuit consists of only one part, N_n equals the number of nodes in the circuit minus one. A *node* is defined to be a point where two or more circuit elements have a common connection. N_n is also equal to the number of circuit elements minus the number of independent loops in the circuit. One of the circuit nodes is chosen to be the reference node. The other nodes will then become the independent nodes, and their voltages are all taken with respect to the voltage of the reference node. Second, the relative magnitudes of the voltages at the circuit nodes are assumed, so that the directions of the currents can be obtained for this analysis. The reference node is often chosen to be at the lowest voltage. The third and final step is to write Kirchoff's current law for each independent node.

The loop method is easier to apply if there is a voltage source or transformer in the circuit, while the node method is easier to apply if there is a current source (for example, a transistor) in the circuit. If a circuit has a resistor of resistance R in parallel with a current source, whose magnitude is i_g, then they may be replaced by a voltage source, whose voltage is $i_g R$, connected in series with a resistor, whose resistance is R. This allows an application of the loop method to the circuit containing the current source. Conversely, if a voltage source, whose voltage is E_g, is connected in series with a resistor of resistance R, the node method may be applied to this circuit by replacing these two elements with a current source, whose current equals E_g/R, that is connected in parallel with a resistor of resistance R.

We shall consider only ideal transformers in this text. One of the illustrative examples in this section shows how circuits that have ideal transformers can be made equivalent (that is, analogous) to circuits that have fewer loops and no transformers. For an *ideal transformer*, mutual inductance is neglected, and, we have

$$\frac{E_1}{E_2} = \frac{i_2}{i_1} = \frac{N_1}{N_2} = a \qquad \qquad \textbf{(6-6)}$$

where subscript 1 denotes the primary coil, subscript 2 denotes the secondary coil, and N denotes the number of turns in the coil windings. The term a, which equals N_1/N_2, is called the *turns-ratio* of a transformer.

It is possible for a circuit analyzed by the loop method to be mathematically analogous to a non-equivalent circuit analyzed by the node method. Such circuits are said to be duals of each other, where one has voltage as the dependent variable and the other has current. In other words, *dual circuits* are two circuits that have the same mathematical form when the loop equations are written for one circuit and the node equations are written for the other circuit. These circuits are *exact dual circuits* if the coefficients of the analogous voltage and current quantities are numerically equal.

Illustrative Example 6.7 shows that we also have *dual circuit ele-*

ments; that is, a capacitor is the dual of an inductor, a resistor is the dual of a resistor, and an inductor is the dual of a capacitor. This example also shows that a voltage source and a current source are dual circuit elements. For exact dual elements, the magnitudes must be specified according to the manner prescribed in Ill. Ex. 6.7 (e.g., $C = L$, $L = C$, etc.)

The dual circuit of another circuit may be obtained by a set procedure. To find the dual of a given circuit if the circuit is planar (that is, if no wires cross each other), place a dot in each independent loop of the given circuit. Each dot will be a node of the dual. Place one dot outside the circuit. This will be the reference node of the dual. Draw and number the dots on another diagram. Draw lines on the original circuit between each dot such that each element (resistor, capacitor, etc.) is crossed once, and only once, by a line. At the same time, draw corresponding lines on the other diagram, inserting dual elements. As stated previously, the dual elements for a voltage source, capacitor, resistor, and inductor in one circuit are a current source, inductor, conductor (that is, resistor), and capacitor, respectively, in the other circuit. This second diagram is the dual circuit of the other circuit. A dual circuit obtained this way can be checked by writing the equations for both circuits. As stated before, a dual circuit is another type of analogy between electric circuit networks, where voltage and current are analogous quantities. This procedure is exemplified in Ill. Ex. 6.8.

Illustrative Example 6.1.   Consider the closed circuit in Fig. 6-2 composed of a voltage source whose voltage is E_g, a resistor whose resistance is R, a capacitor whose capacitance is C, and an inductor whose inductance is L that are all connected in series. The capacitor is uncharged when the circuit switch is closed. Write the differential equation that describes the timewise variation of current i in this circuit.

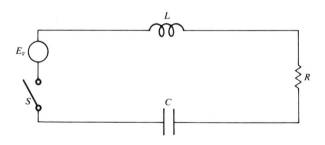

Fig. 6-2

Solution. Let time $t = 0$ at switch closure. Thus, since it was uncharged, the initial capacitor voltage $E_C(0)$ equals zero. Since this is a closed series circuit, it is a loop, and Kirchoff's voltage law may be applied. The voltage rise is that of the voltage source, and there are voltage drops across the resistor, capacitor, and inductor. Since this is a series circuit, the current i is constant throughout the circuit at a given instant of time. Thus,

$$E_g = E_R + E_C + E_L$$

Using Eqs. (6-1), (6-2), and (6-4), we have

$$L\frac{di}{dt} + Ri + \frac{1}{C}\int_0^t i\, dt = E_g$$

Illustrative Example 6.2. Consider a closed circuit composed of a current source, whose current is 10 sin 20 t amp, and a capacitor, whose capacitance is 0.00151 fd, connected in series. Write the differential equation for the timewise variation of voltage in this circuit.

Solution. Since we have a current source and only one other circuit element, it is easiest to use Kirchoff's current law. Letting the node be one of the two points between the current source and the capacitor, assume that current is entering this point from the current source. Then the current that is leaving this point goes through the capacitor. Thus, from Kirchoff's current law, i_g equals i_c. Using Eq. (6-3), we have

$$0.00151\frac{dE}{dt} = 10 \sin 20\, t$$

where E is the voltage drop across the capacitor.

Illustrative Example 6.3. Write the differential equations that describe the timewise variation of the currents in the electrical circuit illustrated in Fig. 6-3.

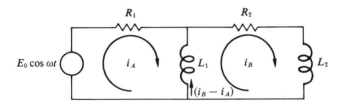

Fig. 6-3

Solution. Since we are asked to find the current variation, and since there is a voltage source in the circuit, it is best to use the loop method. There are two independent loops, and loop currents i_A and i_B have been assumed to be in a clockwise direction, as shown in Fig. 6-3. We shall assume that current i_B is larger than current i_A. There is a third loop in the circuit that goes along its outside boundaries. This is not an independent loop, since it is covered by the other two loops. If its loop equation were written along with the other two loop equations, it would be found that the resulting three equations would not all be independent. Currents i_A and i_B could be found, however, by solving any two of these three equations. Applying Kirchoff's voltage law to the two chosen loops, we have

$$R_1 i_A - L_1 \frac{d}{dt}(i_B - i_A) = E_0 \cos \omega t$$

$$L_1 \frac{d}{dt}(i_B - i_A) + R_2 i_B + L_2 \frac{di_B}{dt} = 0$$

where the minus sign in the first equation is due to the upward travel (which is counterclockwise in the first loop) of the net current $(i_B - i_A)$ in the circuit portion that is contained in both loops. This happened because we had assumed that i_B is larger than i_A. Rewriting these equations, we have

$$L_1 \frac{di_A}{dt} + R_1 i_A - L_1 \frac{di_B}{dt} = E_0 \cos \omega t$$

$$(L_1 + L_2)\frac{di_B}{dt} + R_2 i_B - L_1 \frac{di_A}{dt} = 0$$

Illustrative Example 6.4. Write the differential equations that describe the timewise variation of the currents in the two-loop circuit containing an ideal transformer, illustrated in Fig. 6-4. Also draw a one-loop analogous circuit. The capacitor is initially uncharged.

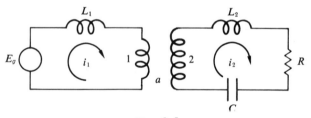

Fig. 6-4

Solution. Since we are asked to find the current variation and since there is a transformer in the circuit, it is best to use the loop method. The loop currents i_1 and i_2 are both assumed to be clockwise, as shown in Fig. 6-4. Applying Kirchoff's voltage law, we obtain

$$E_g = L_1 \frac{di_1}{dt} + E_1$$

$$E_2 = L_2 \frac{di_2}{dt} + R i_2 + \frac{1}{C} \int_0^t i_2 \, dt$$

where the primary voltage E_1 is treated as a voltage drop in the first loop, and the secondary voltage E_2 is treated as a voltage source in the second loop, since it is an induced voltage. From Eq. (6-6), we have $E_2 = E_1/a$ and $i_2 = ai_1$, where a is the transformer turns-ratio. Substituting these two equations into the second of the previous loop equations, we have the following two circuit equations:

$$E_g = L_1\frac{di_1}{dt} + E_1$$

$$\frac{E_1}{a} = L_2a\frac{di_1}{dt} + Rai_1 + \frac{a}{C}\int_0^t i_1\,dt$$

We can combine the two equations above into one as follows, by solving for E_1 in the first equation and substituting the result in the second equation:

$$E_g = (L_1 + a^2L_2)\frac{di_1}{dt} + Ra^2i_1 + \frac{a^2}{C}\int_0^t i_1\,dt$$

which is equivalent to the single-loop circuit illustrated in Fig. 6-5. This circuit, which is analogous to the given transformer circuit, can be used to solve for current i_1, using only one differential equation instead of two. Current i_2 can then be solved by using the algebraic equation $i_2 = ai_1$. From the results of this example, it can be seen that an analogous one-loop circuit can be obtained for a two-loop transformer circuit by multiplying the secondary loop resistances and inductances by a^2 and dividing the secondary loop capacitances by a^2, where a is the turns-ratio of the transformer. This is a type of analogy between electric circuit networks.

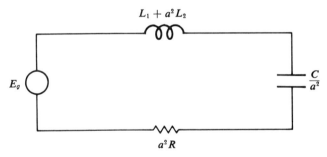

Fig. 6-5

Illustrative Example 6.5. Apply the node method to the electrical circuit illustrated in Fig. 6-6. There is no initial current in the inductor.

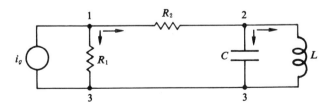

Fig. 6-6

Solution. Since there are three circuit nodes at locations 1, 2, and 3, we have two independent nodes. Let node 3 be the reference node, so that nodes 1 and 2 become the independent nodes. To establish current directions, assume that $E_1 > E_2 > E_3$, as shown in Fig. 6–6. Applying Kirchoff's current law to nodes 1 and 2, using the assumed current directions, we have

$$\frac{E_1 - E_3}{R_1} + \frac{E_1 - E_2}{R_2} = i_g$$

$$C\frac{d}{dt}(E_2 - E_3) + \frac{1}{L}\int_0^t (E_2 - E_3)\,dt = \frac{E_1 - E_2}{R_2}$$

Since the voltage E_3 at the reference node is treated as the reference voltage, we can rewrite these equations considering E_3 to be at zero voltage, as follows :

$$\frac{E_1}{R_1} + \frac{E_1 - E_2}{R_2} = i_g$$

$$\frac{E_2 - E_1}{R_2} + C\frac{dE_2}{dt} + \frac{1}{L}\int_0^t E_2\,dt = 0$$

If the second equation is differentiated with respect to time, we have the following two equations, where one is algebraic, so that we have only one differential equation to solve:

$$\frac{E_1}{R_1} + \frac{E_1 - E_2}{R_2} = i_g$$

$$C\ddot{E_2} + \frac{\dot{E_2}}{R_2} + \frac{E_2}{L} - \frac{\dot{E_1}}{R_2} = 0$$

Illustrative Example 6.6. For the electrical circuit illustrated in Fig. 6-7, apply the method that will result in the fewest number of equations. There is no initial current in the inductor.

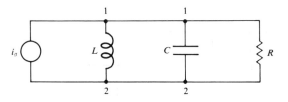

Fig. 6-7

Solution. This circuit has three independent loops. Since there are two nodes, the circuit has only one independent node. Thus, we shall use the node method. Let node 1 be the independent node, assume that E_1 is larger than E_2 in order to obtain the current flow directions, and let E denote the voltage drop ($E_1 - E_2$). Applying Kirchoff's current law to the independent node at 1, we have

$$C\frac{dE}{dt} + \frac{E}{R} + \frac{1}{L}\int_0^t E\,dt = i_g$$

Illustrative Example 6.7.   Show that the electrical circuits in Figs. 6-2 and 6-7 are dual circuits.

Solution.   From Ill. Exs. 6.1 and 6.6, we see that the loop equation for the first circuit and the node equation for the second circuit are

$$L\frac{di}{dt} + Ri + \frac{1}{C}\int_0^t i\,dt = E_g \qquad (6\text{-}7)$$

$$C\frac{dE}{dt} + \frac{E}{R} + \frac{1}{L}\int_0^t E\,dt = i_g \qquad (6\text{-}8)$$

Since the previous two equations are mathematically analogous to each other, the two specified circuits are dual circuits. For these two circuits to be exact dual circuits, the capacitance C in farads of the parallel-element circuit must numerically equal the inductance L in henries of the series-element circuit, the resistance of the parallel-element circuit must equal the inverse of the resistance of the series-element circuit, and the inductance L in henries of the parallel-element circuit must numerically equal the capacitance C in farads of the series-element circuit. Also the timewise variation of the current magnitude for the current source must equal the timewise voltage magnitude variation for the voltage source.

Illustrative Example 6.8.   Obtain the dual circuit of the circuit that is illustrated in Fig. 6-8a.

Solution.   Placing dots 1 and 2 inside the two independent loops of the original circuit, placing dot 3 outside this circuit, and going through the rest of the procedure as stated in this section, we obtain the dual circuit that is illustrated in Fig. 6-8b. It is wise to redraw dual circuits obtained this way to

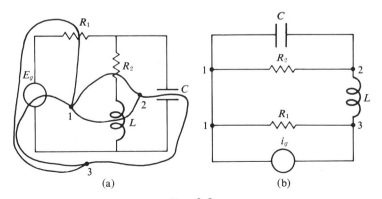

(a)   (b)

Fig. 6-8

a neater form. If we want an exact dual circuit, then numerically, $C = L$, $L = C$, $R_1 = 1/R_1$, $R_2 = 1/R_2$, and $i_g = E_g$ for the dual circuit element values in terms of those for the original circuit.

6.2   Electromechanical Analogies

Comparison of Eq. (2–4) in Sec. 2.3 with Eq. (6–7) rewritten in the following form, where $i = dq/dt$ and $E_g = E \sin \omega t$:

$$L\ddot{q} + R\dot{q} + \frac{1}{C}q = E \sin \omega t \qquad (6\text{–}9)$$

shows that the two systems shown in Figs. 2–3 and 6–2 are mathematically analogous if $E_g = E \sin \omega t$. Here, charge q is analogous to displacement x, and voltage is analogous to force. That is, the voltage source is analogous to the applied force, $E \sin \omega t$, and the three voltage drops are analogous to the inertia, damping, and spring forces. Thus, this analogy is called a *force-voltage analogy*. Section 2.4 and Table 2–1 show the analogy between rectilinear and rotational systems, where moment is analogous to force. Since Eqs. (2–14) and (6–9) are mathematically analogous, we have a *moment-voltage analogy*. Because of the existence of dual circuits, where current is analogous to voltage, it is possible to have a *force-current analogy*. Table 6–1 tabulates the analogous quantities for a force-voltage and moment-voltage mechanical analogy and for a force-current and moment-current mechanical analogy. The force-voltage and moment-voltage terms were obtained by taking the corresponding terms from Eqs. (2–4), (2–14), and (6–9), which are mathematically analogous. The force-current and moment-current analogy terms were obtained by making a current-voltage analogy, using Eqs. (6–7) and (6–8) in Ill. Ex. 6.7, which are mathematically analogous to each other.

TABLE 6-1

Mechanical Translational or Rotational System	Electrical System (Force-Voltage or Moment-Voltage Analogy)	Electrical System (Force-Current or Moment-Current Analogy)
W/g or I	L	C
c or c_t	R	$1/R$
k or k_t	$1/C$	$1/L$
F or M	E	i
x or θ	q	$\int_0^t E\,dt$
$\dot{x}$ or $\dot{\theta}$	$i = \dot{q}$	E
$\ddot{x}$ or $\ddot{\theta}$	$di/dt = \ddot{q}$	$\dot{E}$

To obtain the electrical analogy (force-voltage or force-current) for a mechanical translational system, first write the equations of motion for the mechanical system. For each mechanical element, identify the points to which it is connected to other elements and by which its displacement may be defined. List the force-voltage or force-current electrical elements analogous to each mechanical element and substitute these electrical terms in the differential equations of motion. Interpret the resulting differential circuit equations and make the proper connections between electrical elements to yield the analogous electrical network. This procedure is exemplified in Ill. Ex. 6.9 of this section. The procedure for obtaining a moment-voltage or a moment-current electrical analogy circuit for a mechanical rotational system is almost the same.

A similar procedure can be used to find a force-voltage or a force-current mechanical translational system analogy for an electrical circuit. That is, substitute force-voltage or force-current analogous mechanical system terms for each term in the circuit equations, and interpret the resulting equations to draw the analogous mechanical system. Besides being a more impractical experimental analogy, the resulting mechanical system can be an unusual one. If the original electrical circuit has no inductor, then its analogous force-voltage mechanical system will have no mass. Similar statements can be made about mechanical rotational system analogies of electrical circuits.

An ideal transformer, where $E_1/E_2 = N_1/N_2 = i_2/i_1$, may be used as an electrical analogy for such mechanical coupling devices as gears, friction wheels, and levers. This is a major reason for the use of ideal transformers in this text. For a pair of engaged gears

$$\frac{\theta_1}{\theta_2} = \frac{T_2}{T_1} \qquad (6\text{--}10)$$

where θ is the angle that the gear moves and T is the number of teeth on the gear. For a pair of non-slipping friction wheels

$$\frac{M_1}{M_2} = \frac{r_1}{r_2} = \frac{\theta_2}{\theta_1} \qquad (6\text{--}11)$$

where M is the torque exerted, r is the radius of the wheel, and θ is the angular displacement of the indicated wheel. For a lever

$$\frac{F_1}{F_2} = \frac{d_2}{d_1} \qquad (6\text{--}12)$$

where F is the vertical component of the force exerted at one end of the lever, and d is the distance along the lever between the fulcrum and the indicated applied force. The analogy between an ideal transformer and these three mechanical coupling devices is obvious.

The reader may wonder why we have discussed methods for obtaining the electrical analogy of mechanical systems. One reason, besides creating a better understanding between the two systems, is that electrical systems are often easier to study experimentally than mechanical systems. For example, it is easier to change a resistance than a damping constant or a coefficient of friction, and it is easier to measure the results. Thus, it can be advantageous to represent and study mechanical vibration systems by their equivalent electrical circuits. Very many actual mechanical systems are nonlinear, and assumptions must be made to obtain a linear system before a linear electrical analogy circuit may be obtained. Two simple examples are the small angle assumption for pendulum oscillations and viscous damping. The assumptions should not invalidate the solution. Some nonlinear mechanical terms can be represented by nonlinear circuit elements. Electrical analogies are also the basis of network analyzers, which are electric analog computers and which are discussed in Sec. 6.4. Analogies are also useful in the analysis of electromechanical systems (i.e., systems composed of both electrical and mechanical elements).

Illustrative Example 6.9.   Obtain the force-voltage electrical analogy circuit for the spring-mass system that is shown in Fig. 6-9.

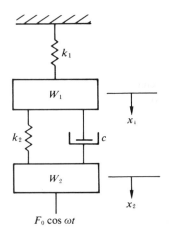

Fig. 6-9

Solution.   The equations of motion for this spring-mass system are

$$\frac{W_1}{g}\ddot{x}_1 = -k_1x_1 - k_2(x_1 - x_2) - c(\dot{x}_1 - \dot{x}_2)$$

$$\frac{W_2}{g}\ddot{x}_2 = F_0 \cos \omega t + k_2(x_1 - x_2) + c(\dot{x}_1 - \dot{x}_2)$$

which can be rewritten as

$$\frac{W_1}{g}\ddot{x}_1 + k_1 x_1 + k_2(x_1 - x_2) + c(\dot{x}_1 - \dot{x}_2) = 0$$

$$\frac{W_2}{g}\ddot{x}_2 + k_2(x_2 - x_1) + c(\dot{x}_2 - \dot{x}_1) = F_0 \cos \omega t$$

Using Table 6-1, we substitute force-voltage analogous circuit terms for the terms in the above two equations. Upon doing this, we obtain the following two circuit equations:

$$L_1 \ddot{q}_1 + \frac{1}{C_1} q_1 + \frac{1}{C_2}(q_1 - q_2) + R(\dot{q}_1 - \dot{q}_2) = 0$$

$$L_2 \ddot{q}_2 + \frac{1}{C_2}(q_2 - q_1) + R(\dot{q}_2 - \dot{q}_1) = E_0 \cos \omega t$$

To draw the resulting electrical circuit, the above equations must be interpreted. Since they contain q_1, q_2, and their derivatives, they are loop equations for this circuit. Since there are two equations, this circuit has two independent loops. The term $(q_1 - q_2)/C_2$ in the previous two equations tells us that capacitor C_2 is in both loops, while the term $R(\dot{q}_1 - \dot{q}_2)$ tells us that resistor R is in both loops. Since they all appear in the first equation, the first loop contains inductor L_1, capacitors C_1 and C_2, and resistor R. Since they all appear in the second equation, the second loop contains inductor L_2, capacitor C_2, resistor R, and the voltage source. The force-voltage electrical analogy circuit is shown in Fig. 6-10. For an exact electrical analogy, then numerically we make $L_1 = W_1/g$, $C_1 = 1/k_1$, $C_2 = 1/k_2$, $R = c$, $L_2 = W_2/g$, and $E_0 = \cos \omega t = F_0 \cos \omega t$.

A similar procedure can be used to obtain the force-current analogy of this spring-mass system. That is, substitute force-current electrical analogy terms in the equations of motion and draw the electrical circuit from the circuit equations that result. Another way to obtain the force-current analogy circuit for the given mechanical system is to draw the dual circuit of the force-voltage analogy circuit shown in Fig. 6-10.

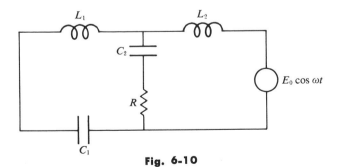

Fig. 6-10

Illustrative Example 6.10. Consider the pair of friction wheels illustrated in Fig. 6-11. The upper disk has a moment of inertia I_1, radius r_1, and an external moment M_E applied to it. The lower disk, which is mounted on a shaft whose shaft stiffness constant is k_t, has a moment of inertia I_2 and radius r_2.

The motion of the lower disk is retarded by both linear damping, whose coefficient is c_t, and shaft elasticity. Find the moment-voltage analogous circuit.

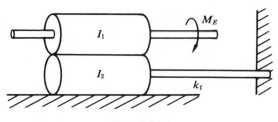

Fig. 6-11

Solution. Since this is a simple mechanical system, the moment-voltage analogous circuit will be drawn by inspection. The first loop will represent the $\sum M = I_1\ddot{\theta}_1$ equation for the top disk. This loop will contain an inductor L_1 to represent I_1, and a voltage source E_g to represent M_E. The second loop, which represents the $\sum M = I_2\ddot{\theta}_2$ equation for the bottom disk, has an inductor L_2 to reqresent I_2, a resistor R to represent c_t, and capacitor C to represent $1/k_t$. The friction wheel coupling is represented by a transformer. This analogous circuit is shown in Fig. 6-12. A one-loop equivalent circuit for the previous two-loop transformer circuit is shown in Fig. 6-13, where a is the transformer turns-ratio. A moment-current analogy circuit can be obtained by drawing the dual of the one-loop circuit in Fig. 6-13. The dual circuit will contain a current source, two capacitors, a resistor, and an inductor that are all connected in parallel to each other.

The moment-voltage analogous electrical circuit could have been drawn by first writing the equations of motion, instead of by inspection. The equations of motion are as follows, where M_1 and M_2 are the friction wheel torques:

$$I_1\ddot{\theta}_1 = M_E - M_1$$
$$I_2\ddot{\theta}_2 = M_2 - c_t\dot{\theta}_2 - k_t\theta_2$$
$$\frac{M_2}{M_1} = \frac{\theta_1}{\theta_2} = \frac{r_2}{r_1} = \frac{1}{a}$$

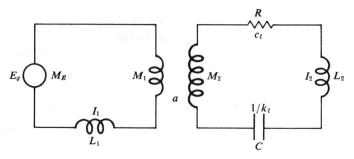

Fig. 6-12

If we substitute analogous moment-voltage electrical terms in the above equations, the resulting circuit equations will be the loop equations for the circuit illustrated in Fig. 6-12.

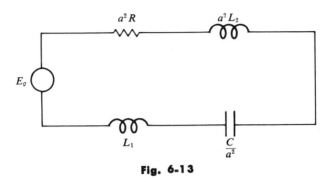

Fig. 6-13

6.3 Brief Comparison of Digital and Analog Computers

In general, computing devices may be classified as either analog or digital. A *digital computer* works with specific numerical quantities and performs its operations in a discrete, stepwise procedure. Examples are the abacus, adding machines, mechanical desk calculators, and the modern electronic digital computers. Most digital computers perform their operations by combinations of additions and subtractions (e.g., desk calculators and electronic digital computers perform a multiplication by successive additions). An *analog computer* performs its computing operations by manipulating *continuously-varying physical quantities* (such as voltage, electrical current, angular rotation of a shaft, position on a scale, etc.) that are used to represent numerical quantities. Some very simple analog computers are the slide rule, the automobile speedometer, and the planimeter. The accuracy of an analog computer is limited by the accuracy of its least-accurate components. Components of electronic analog computers that have an accuracy of one-tenth of one per cent are rather expensive, since they require careful design. Their accuracies usually vary between one-half and five per cent. Digital computers are more general in purpose and more accurate, but are less economical than analog computers. An analog computer has components that can generate such functions as e^{at}, sin bt, etc., while a digital computer must employ formulas that approximate these functions. As shown in Sec. 7.25, a digital computer performs an integration by using an approximate numerical technique, while certain types of analog computers have components that can perform an integration.

By the use of analog-to-digital converters and digital-to-analog converters, digital and analog computers have been combined into one computational system. Because of the timing problems involved, this combination is uneconomical, and is usually used only for special purposes. The *digital differential analyzer*, whose chief purpose is to solve ordinary differential equations, works with digital quantities and performs its operations on a discrete, digital basis. This special-purpose type of computer has a built-in numerical integration scheme and imitates an analog computer with regard to problem setup, but still has digital computer accuracy.

There are no absolute criteria that specify which types of problems should be solved on which types of computers. This depends upon too many factors, some of which could be the allotted time, number of parameters, accuracy required, type of input, most desirable output form for a clear interpetation of the results, availability, economic considerations, and, especially, the type of problem. The digital computer can handle a much wider range of problems. For example, an analog computer would not, in general, be used for matrix arithmetic calculations, or to find the roots of equations, or to apply the Holzer method. The d-c electronic differential analyzer, which is by far the most widely used type of analog computer, is very useful for solving linear and nonlinear differential equations whose coefficients do not have an overly complex mathematical form. This type of computer provides an economical means for obtaining plotted results of fair accuracy; and it allows the user an easy means for changing the input parameters, after a glance at the plotted results, for successive cases. For time-varying vibration problems where an oscillation is involved, there is no big error problem with integration on an analog computer, while a digital computer solution should employ an accurate type of numerical-integration technique, as discussed in Sec. 2.14, to solve this type of problem.

6.4   Classification of Analog Computers

Analog computers have been classified according to the type of physical quantity employed (e.g., mechanical, electrical, electronic, electromechanical, etc.), but are more generally classified as either *direct analog computers* or *differential analyzers*. The *direct analog computer* is a model of the physical system to be analyzed and, thus, does not directly perform mathematical operations (such as addition, multiplication, integration, etc.). The particular direct analog computer should, of course, be easier to build and modify than the physical system to be analyzed. Thus, an aerodynamic wind tunnel and other scale models are types of direct analog computers. Two other types are the field plotter, which measures the electrostatic potential distribution across a thin sheet of conducting paper, and a soap-film membrane of uniform tension, whose shape is to be measured and whose boundary is a bent wire. Both of these systems are used to represent physical systems that may be represented by Laplace's partial differential equation (e.g., steady-state temperature distributions, fluid flow streamlines, magnetic potential distributions, etc.) whose form is $\partial^2\phi/\partial x^2 + \partial^2\phi/\partial y^2 + \partial^2\phi/\partial z^2 = 0$. The most popular direct analog computer is the network analyzer which will be discussed next. There are many other examples of direct analog computers, but they are designed only for specialized applications.

The *network analyzer* consists of combinations of passive electrical components (resistors, inductors, and capacitors) whose magnitudes are both variable and wide in range and which may be flexibly interconnected. It is natural to use a direct analog computer consisting of passive electrical components, because they are much easier to build and modify than their physical counterparts, and because they may be used to represent many types of physical systems (e.g., mechanical vibrations, heat flow, fluid flow, acoustic vibrations, etc.). This latter point has already been shown in Sec. 6.2 for vibration problems. To indicate how such problems would be set up and solved on a network analyzer, we must first obtain the electrical network analogy of the physical system to be studied, connect the proper electrical components, and then adjust these components to the proper values. The *thermal analyzer* is a special network analyzer, consisting of variable resistors and capacitors, for studying heat flow problems, while the *transient analyzer* is a special type of network analyzer for studying transient phenomena. The more general types of network analyzers will have output-recording devices and provisions for generating various forcing functions. The network analyzer is an economical device, but is limited in application mainly to system analogies that may be expressed as ordinary differential equations with constant coefficients. If an analogy exists between two physical quantities (e.g., force F and voltage E), they are linearly related even if they vary with time. Since $E(T) = F(kt)$ for a force-voltage analogy, a fast timewise simulation can be obtained by setting the time scale-factor k to a value less than one. Scaling will be discussed in Sec. 6.5 in more detail. As a rather simple example, the displacement of the spring-mass system in Fig. 2-3 can be solved on a network analyzer by connecting a voltage source of output $E_g = E \sin \omega t$, an inductor, a resistor, and a capacitor in series as a one-loop circuit, as shown in Fig. 6–2. This is a solution circuit for a force-voltage analogy because the loop equation for this circuit is given by Eq. (6–9) of Sec. 6.2. This circuit equation is mathematically analogous to the equation of motion for the linear spring-mass system, which is given by Eq. (2–4) of Sec. 2.3. Comparison of these two equations shows that charge q is analogous to displacement x, inductance L is analogous to mass W/g, resistance R is analogous to damping coefficient c, capacitance C is analogous to $1/k$, and voltage E is analogous to force F, as shown in Table 6–1 for a force-voltage analogy. The initial displacement and initial velocity for this spring-mass system can be simulated on the network analyzer by an initial charge on the capacitor and by an initial loop current. The force-voltage-analogy, network-analyzer circuit that would be used to solve the two-degree-of-freedom spring-mass system in Fig. 3–1 would consist of two loops, and would include two inductors to simulate two masses and three capacitors to simulate the three linear springs. Section 6.2 furnishes other details and examples (e.g., Ill. Exs. 6.9 and 6.10) on electrical analogies of mechanical systems.

The *differential analyzer*, which is designed mainly to solve ordinary differential equations, performs mathematical operations on the varying, physical quantities that are employed. It consists of a number of different components, each of which performs a specific mathematical operation (e.g., addition, multiplication, integration, etc.), and when these components are appropriately interconnected they represent the mathematical equations to be solved (instead of a physical system). Analog differential analyzers may be physically classified as to being of the *mechanical, d-c electronic,* or *a-c electromechanical type*. By far the most popular type of analog computer is the *d-c electronic differential analyzer*—so much so, that analog computers are often inappropriately considered to be synonymous with d-c electronic differential analyzers. The *a-c electromechanical types* have been used primarily for special purpose applications (e. g., aircraft navigation and military fire control systems). The d-c voltages may be integrated by purely electronic devices, but integration of a-c voltages requires a mechanical or electromechanical device.

The usage of mechanical differential analyzers was never widespread, but they do have a historical importance. The variable quantities are represented on this type of computer by the angular positions of shafts from a referenced zero position. These shafts and other computer components (i.e., mechanical adders, multipliers, integrators, etc.) are connected in such a manner that a mathematical equation is represented. Addition of two quantities (i.e., angular displacements)

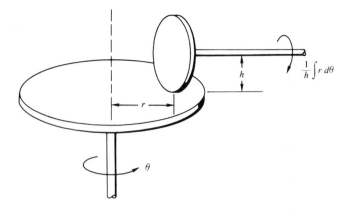

$$\frac{1}{h} \int r \, d\theta$$

Fig. 6-14

is performed mechanically by differential gearing similar to that in the rear axle of an automobile. Subtraction is performed in similar fashion, where the negative input is obtained by reversing the direction of its shaft rotation. Quantities are positive or negative in this computer according to the direction of its shaft rotation. If the gear ratio of two meshed gears is R (where R is the ratio of the number of driver teeth to the number of teeth driven), then the driven or output gear will move R times as fast as the driver or input gear. Thus, multiplication by a constant is obtained by setting the proper gear ratios between the shafts whose angular positions represent the physical variables. Arbitrary functions can be generated by using special cams. The output is plotted on an output table. Input tables may be used to provide input data in graphical form. Initial values are obtained by initially rotating the proper shafts.

Integration on a mechanical differential analyzer is obtained by means of a *Kelvin wheel-and-disk integrator*, which is illustrated in Fig. 6-14. In principle, the Kelvin integrator consists of a friction wheel that rotates without slipping on a disk, where the axes of the wheel and disk are normal to each other as shown in Fig. 6-14. If the disk moves an angle $d\theta$ when the wheel is at a distance r from the center of the disk, then the rim of the wheel moves a distance $r \, d\theta$, and the wheel itself has turned an angular distance $r \, d\theta/h$. By letting the distance r be a variable input quantity over a finite period when rotating the disk, the output shaft of the wheel rotates an angular displacement which is given by the integral $\int r \, d\theta/h$. In spite of its rather good accuracy, the use of the mechanical differential analyzer was never widespread because of its slowness of operation and because of the awkward paths that the represented variables must travel. The principles for setting up problems on the mechanical and electronic differential analyzers are similar, and this will be discussed in more detail in the next section.

6.5 The D-C Electronic Differential Analyzer

As mentioned previously in Sec. 6.4, a differential analyzer has components that perform specific mathematical operations (e.g., addition, multiplication, integration, etc.), and it uses a combination of these components to solve the mathematical equations for a given problem. Because a problem is solved on any differential analyzer type of analog computer by connecting components that perform specific mathematical operations, a simple schematic symbol to denote a specific component is very helpful. That is, if the diagram that shows the connections of the differential analyzer components for solving a specific problem is both simple and clear, and also shows enough detail so that it is complete, then there is a smaller chance that the components will be connected improperly during computer operation. Thus, we shall employ simple schematics to represent the different types of components for a d-c electronic differential analyzer. We shall not discuss how to connect these components manually, nor shall we discuss the detection and correction of component malfunctions (for example, amplifier drift) since this requires a rather good electrical background.

For *d-c electronic differential analyzers*, the variable quantities are represented by d-c voltages that may vary with time. This type of computer can solve both linear and nonlinear ordinary differential equations. The solutions can be recorded by special output devices (e.g., pen recorder or oscilloscope) that plot the timewise variations of the voltage-represented physical quantities. Thus, time is usually the independent variable, but we can solve problems where a displacement is the independent variable by letting the computer time represent this displacement. Just as the Kelvin wheel-and-disk integrator is the heart of the mechanical differential analyzer, the heart of the d-c electronic differential analyzer is the *operational amplifier*. These are d-c electronic amplifiers of extremely high gain and large impedance. Basically, they accept a d-c input voltage and generate a much larger voltage of opposite polarity. Output voltage E_0 is approximately equal to $-KE_i$, where gain $K \gg 1$ (e.g., 10^5 to 10^8 for modern transistorized, electronic differential analyzers). The operational amplifier is schematically represented in Fig. 6-15.

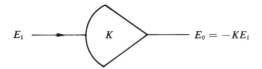

Fig. 6-15

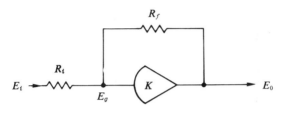

Fig. 6-16

Now let us consider an operational amplifier having an input resistor R_1, and a feedback resistor R_f, as shown in Fig. 6-16, where all voltages are referenced to ground. Since the input voltage to the amplifier is E_g, $E_0 = -KE_g$. Since the amplifier impedance is so high, most of the input current goes through the feedback resistor R_f. Thus, we have the approximation

$$\frac{E_i - E_g}{R_i} = \frac{E_g - E_0}{R_f}$$

Substitution of $E_0 = -KE_g$ in the previous equation gives

$$\frac{1}{R_i}\left(E_i + \frac{E_0}{K}\right) = -\frac{1}{R_f}\left(\frac{E_0}{K} + E_0\right)$$

which can be rewritten as

$$E_0\left[\frac{1}{KR_i} + \frac{1}{R_f}\left(1 + \frac{1}{K}\right)\right] = -\frac{E_i}{R_i}$$

Thus, the ratio of the output voltage E_0 to the input voltage E_i is given by

$$\frac{E_0}{E_i} = -\frac{R_f}{R_i}\left[\frac{1}{1 + (1/K)(1 + R_f/R_i)}\right] \qquad (6\text{-}13)$$

Since the amplifier gain K is extremely large, we may use the approximation

$$\frac{E_0}{E_i} = -\frac{R_f}{R_i} \qquad (6\text{-}14)$$

so that this operational amplifier circuit just changes the sign and magnitude of the input voltage. If we modified the operational amplifier circuit so that there are two input resistors R_1 and R_2, as shown in Fig. 6-17, it can be shown in an analogous manner that

$$E_0 = -\frac{R_f}{R_1}E_1 - \frac{R_f}{R_2}E_2$$

Thus, if we have n input resistances and a feedback resistance of value R_f, then

$$E_0 = -\sum_{k=1}^{n} \frac{R_f}{R_k}E_k \qquad (6\text{-}15)$$

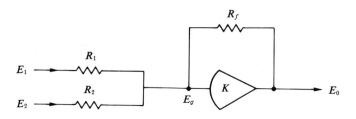

Fig. 6-17

Such an amplifier circuit is called a *summer*. For many analog computers, the resistance ratios are controlled to be either 1, 4, or 10. Schematically, the summer can be represented by the diagram shown in Fig. 6-18. By paralleling the input resistors, values other than 1, 4, or 10 can be obtained.

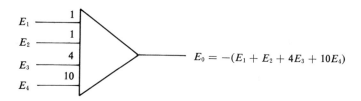

Fig. 6-18

If in Fig. 6-16, we replaced the feedback resistor R_f by a capacitor of capacitance C, then because of the high amplifier impedance, the current through the capacitor almost equals the input current, as before. Since the feedback loop has a capacitor, this gives

$$\frac{E_i - E_g}{R_i} = C\frac{d(E_g - E_0)}{dt}$$

Substitution of $E_g = -E_0/K$, and rearranging gives

$$R_i C \frac{dE_0}{dt} = -E_i - \frac{1}{K}\left(E_0 + R_i C \frac{dE_0}{dt}\right)$$

Since the amplifier gain K is very large, we may use the approximation

$$\frac{dE_0}{dt} = -\frac{1}{R_i C}E_i$$

Thus, we have

$$E_0 = -\frac{1}{R_i C}\int_0^t E_i\, dt + E_0(0) \tag{6-16}$$

where $E_0(0)$ is the initial voltage on the capacitor. This is how operational amplifier circuits are used to perform integrations in a d-c differential analyzer. Initial values for a problem are set in a d-c differential analyzer by charging the feedback capacitor in each *integrator* (which has just been described) to the required voltage value, before the switch to start computer computation is thrown. By using an input capacitance of value C, and a feedback resistance of value R, an operational amplifier can be made to perform differentiations. The resulting equation is $E_0 = -RC(dE_i/dt)$. Because differentiation amplifies the unavoidable input noise signals, which might cause amplifier saturation and which completely mask the signal of interest, the problem equations should be rearranged so that differentiation is avoided.

An *integrator* is an amplifier circuit that contains one feedback capacitor and one or more input resistors. For many large analog computer installations, the input resistors and the feedback capacitor are internally connected and fixed in value. The schematic in Fig. 6-19

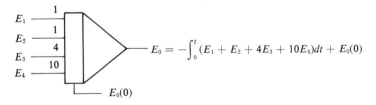

$$E_0 = -\int_0^t (E_1 + E_2 + 4E_3 + 10E_4)dt + E_0(0)$$

Fig. 6-19

can be used to represent such an integrator, where multiplication of an input voltage by 1 or 4 or 10 can be obtained by connecting a patchboard cord to the proper input jack. This fixed-value type of analog computer uses standard resistors of 10.0×10^5, 2.5×10^5, and 1.0×10^5 ohms and a standard capacitor of 1.0×10^{-6} farad to achieve these $1/RC$ gains of 1, 4, and 10. These values can be changed by putting two resistors in parallel or series by jack connections, if desired. Other types of analog computers, however, have plug-in resistors and capacitors. Thus, for this type of analog computer, which is not uncommon at colleges, the schematics in Figs. 6-18 and 6-19 for a summer and an integrator should be modified to contain input resistors and a feedback resistor or capacitor of specified magnitudes. This should be done because these resistors and capacitors must be physically inserted before the computer is run to solve a specific problem. We shall use the fixed-value type of summer and integrator in this text; the reader should be able to convert to the plug-in type if needed, since the gains for a summer or an integrator are specified by its R_f/R_i or $1/R_iC$ values.

Multiplication of voltages by constants less than unity are generally done by using *potentiometers*. That is, if the total resistance of the potentiometer is R_p and if the total resistance from the point of contact by the potentiometer arm to the ground is R_a, then the output voltage E_0 equals $(R_a/R_0)E_i$, where E_i is the input voltage. Since the divided resistance R_a is less than the total resistance R_p, then we can use simple potentiometers to multiply voltage quantities by constants whose values are positive and less than one. The coefficient setting P (i.e., R_a/R_p) is usually read on a potentiometer dial whose scale is often linear. Thus, multiplication by a constant may be represented schematically as shown in Fig. 6-20, where the constant P, which equals R_a/R_p, must be less than one. The schematic in Fig. 6-21 illustrates the use of a summer and a potentiometer to multiply voltages by constants whose values are greater than unity.

Since we have mentioned output recorders, summers, integrators, and potentiometers in this section, we have covered the components

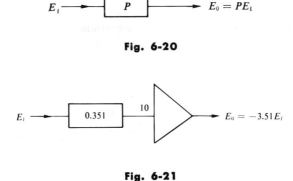

$$E_i \longrightarrow \boxed{P} \longrightarrow E_0 = PE_i$$

Fig. 6-20

$$E_i \longrightarrow \boxed{0.351} \longrightarrow^{10} \triangleright \longrightarrow E_0 = -3.51E_i$$

Fig. 6-21

that are necessary to solve homogeneous, linear differential equations with constant coefficients, using a d-c differential analyzer. These components can be used to solve some nonhomogeneous differential equations because they can also be used to generate certain time-varying functions. To furnish some examples, the generation of the functions e^{-at}, sin at, and cos at is shown in Ill. Ex. 6.11. The *first step* in solving an engineering problem on a differential analyzer type of analog computer is, of course, to determine the mathematical equations that describe this physical problem and to write them in a proper form. Assuming that these equations contain one or more differential equations, each differential equation should be written in a form that solves for its highest derivative. That is, we want a set of equations that solves for the highest derivative for each variable. The equation for the lower-order derivatives and the algebraic equations should be written after the particular differential equation that they affect. This step of the setup procedure is exemplified in Ill. Exs. 6.12 and 6.13. The *second step* is to scale the problem, and this is described in the next paragraph. The *third step* is to draw a diagram which shows the component magnitudes and settings, where each component is represented by a simple schematic, and which shows how these components are connected. This block diagram represents the problem equations in scaled form. The *fourth step* is to connect these components in proper fashion and to run the problem. A static check should be made after the problem is patched on the analog computer. The potentiometers, initial values, etc. must be set before the problem is run. At some computer installations, the setting of these devices is done automatically instead of manually. The *fifth, and last, step* is to interpret the results, where scale factors must be taken into account.

In a d-c electronic differential analyzer, physical quantities are represented by voltages. *Magnitude scale factors* indicate the specific relations between voltages and these physical quantities. Thus, if $\dot{x}$ is velocity and x is displacement, they are represented by $X = S_1 x$ and $\dot{X} = S_2 \dot{x}$, where $\dot{X}$ and X are voltages, and constants S_1 and S_2, which linearly relate the terms, are magnitude scale factors. For accuracy considerations, to obtain as much significance as possible in the results, the scale factors should be as large as possible. That is, if the maximum displacement for a vibration problem is six inches, we would not want to have one volt represent a one-inch displacement in such a problem, since the plotted results would be relatively insignificant (i.e., a six-volt maximum output is much less than the allowed maximum). Also, if the voltage were low for a long interval of time, the per cent accuracy of the solution would be decreased by drift, noise, and other effects. For efficient vacuum tube and computer component operation, however, the voltages of the operationl amplifiers for most d-c electronic differential analyzers must not exceed the range of $\pm$ 100 volts, and some have smaller maximums. Outside these allowed voltage ranges, the amplifiers become nonlinear and saturation occurs. We shall assume a $\pm$ 100-volt computer in this text. Thus, since the problem variables will have many different values, the magnitude of a scale factor S is set to equal 100 volts divided by the estimated maximum value for the particular physical quantity. If we allow a margin of safety for our estimates of the maximum values by letting the maximum voltage value be 90 volts, then our scale factor S_2 for $\dot{x}$ is $90/\dot{x}_{max}$ volt/(in./sec), and scale factor S_1 for x is $90/x_{max}$ volt/in. This gives a constant safety factor of $10/9$. It may be preferable to have variable safety factors by basing the calculations on 100 volts and by using displacement and velocity values (for x_{max} and $\dot{x}_{max}$) that are reasonably larger than their estimated maximums.

It may not be desirable to solve our problems on this type of computer with a one-to-one time scale. Too fast a time can cause amplifier instability and servo unreliability problems; but fast running times are desirable in order to minimize amplifier drift, capacitor leakage, and other accumulative errors, and to obtain results more quickly. Time scale is also affected by the type and capabilities of the recording device (e.g.. pen recorder or oscilloscope). Because of inertia and friction, the pen of a pen recorder can follow voltage changes only within a certain velocity range. For the physical equation that is transformed to a voltage, computer-time equation, we let computer time T be related to physical time t by the equation $T = kt$, where k is the *time scale factor* and where the physical problem is slowed down by the computer if $k > 1$. It can be shown that the nth order derivative $d^n x/dt^n$ can be time-scaled by the equation $d^n x/dt^n = k^n d^n x/dT^n$, where k is the time scale factor. After the magnitude and the time scaling is completed, we then draw a computer block diagram, using schematic representations, to show how the computer components must be connected in order to solve these transformed voltage equations. Time scaling should precede magnitude scaling, if they are done separately, because the derivatives with respect to computer time T, instead of physical time t, are the terms that should be magnitude scaled. The scaling of a physical problem is exemplified in Ill. Exs. 6.13 and 6.14. The time and amplitude scaling methods shown in these two examples are probably the most obvious and most-easy-to-explain methods, but it is preferable in professional practice to use quicker scaling methods. For this reason, the reader is referred to the analog computer references in Appendix A-3 (such as the one by Leon Levine) if he wishes to learn about scaling shortcuts.

Illustrative Example 6.11. Show how we may generate the functions e^{-at}, sin at, and cos at on a d-c electronic differential analyzer, where $a < 1$.

Solution. We know that e^{-at} is the solution of the differential equation $\dot{y} + ay = 0$ when $y(0) = 1$. Thus, the function e^{-at} can be generated by the method shown in Fig. 6-22, which simulates the equations

$$\dot{y} = -ay$$
$$y = -\int (ay)\, dt$$

We know that sin at is the solution of the differential equation $\ddot{y} + a^2y = 0$ when $y(0) = 0$ and $\dot{y}(0) = a$, and that cos at is the solution of the same differential equation when $y(0) = 1$ and $\dot{y}(0) = 0$. Thus, we can generate either function using the same circuit but different initial values. The function sin at can be generated by the method shown in Fig. 6-23, which simulates the equations that follow this paragraph. It may be noted that this method also generates $-a$ cos at, since $\dot{y} = a$ cos at. The generation of functions of the dependent variable (e.g., sin x and cos x) are somewhat more involved, as shown in Sec. 6.6, because this type of computer integrates with respect to time.

$$\ddot{y} = -a^2y$$
$$\dot{y} = -\int (a^2y)\, dt$$
$$y = \int \dot{y}\, dt$$

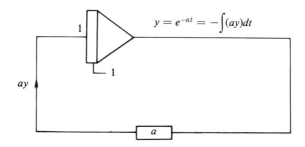

Fig. 6-22

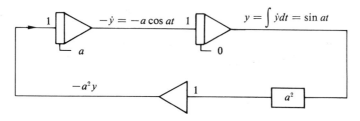

Fig. 6-23

Illustrative Example 6.12.   Consider a one-degree-of-freedom, spring-mass-damper system consisting of a mass that weighs 77.2 lb, a linear spring whose spring constant k equals 1.60 lb/in., and a linear damper whose damping coefficient c equals 0.140 lb-sec/in. This spring-mass-damper system is illustrated in Fig. 1-19 of Sec. 1.7. Draw the block diagram that will show how to solve, on a d-c electronic differential analyzer, for the timewise displacement and velocity variations for this linear spring-mass-damper system. Use unit scale factors throughout.

Solution.   As stated previously, the first step is to obtain the system equation. The equation of motion for this spring-mass-damper system is the following linear, homogeneous differential equation with constant coefficients:

$$\frac{W}{g}\ddot{x} + c\dot{x} + kx = 0$$

Substitution of the given numerical values for $W, c,$ and $k,$ along with $g = 386$ in./sec², gives

$$0.200\ddot{x} + 0.140\dot{x} + 1.60x = 0$$

Solving this equation for its highest derivative $\ddot{x}$ gives

$$\ddot{x} = -0.700\dot{x} - 8.00x$$

Since $\dot{x} = \int \ddot{x}\, dt,$ we want the computer to simulate the following equations:

$$\dot{x} = \int (-0.700\dot{x} - 8.00x)\, dt$$

$$x = \int \dot{x}\, dt$$

as is done by the block diagram in Fig. 6-24. The first step in drawing this block diagram is to draw the integrator that integrates the highest derivative $\ddot{x}$, where the input $-(0.700\dot{x} + 8.00x)$ and the output $-\dot{x}$ are both shown. Note that the integrator also changes the sign. This represents our first equation. Then we draw our second integrator which integrates $-x$ to give us x. Now we can draw the summer and the two recorders that measure $\dot{x}$ and x. The last step is to draw the connections and the potentiometers that will furnish the inputs, $-0.700\dot{x}$ and $8.00x$, for the first integrator (the one that integrates $\ddot{x}$ to obtain $-\dot{x}$).

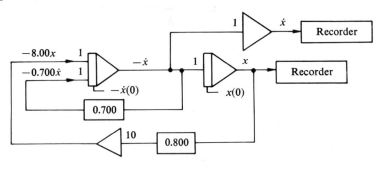

Fig. 6-24

Illustrative Example 6.13.   Scale the problem in Ill. Ex. 6.12 if the estimate for maximum displacement, velocity, and acceleration are 9 in., 45 in./sec, and 90 in. /sec^2, respectively. We also want the computer to run at the same speed as the physical or real time.

Solution.   To allow a tolerance or safety margin in case we miscalculated the maximum values, we shall let the maximum voltages that represent the velocity, displacement, and acceleration all equal 90 volts. Thus, the scale factor for displacement x is (90 volts)/(9 in.) or 10 volts/in., and $X = 10x$, where X is the measured voltage that represents displacement x. The scale factor for velocity $\dot{x}$ is 90 volts/(45 in./sec) or 2 volt-sec/in., and $\dot{X} = 2\dot{x}$, where $\dot{X}$ is the measured voltage that represents velocity $\dot{x}$. Since the scale factor for acceleration $\ddot{x}$ is 90/90 or 1.00 volt-sec^2/in., $\ddot{X} = \ddot{x}$. The problem equation that was to be solved in Ill. Ex. 6.12 was

$$\ddot{x} = -0.700\dot{x} - 8.00x$$

Since that example had unit scale factors, it had the relationships $\ddot{X} = \ddot{x}$, $\dot{X} = \dot{x}$, and $X = x$, so that the previous equation could be used to represent a voltage-time equation for that problem. Substitution of $\ddot{x} = \ddot{X}$, $\dot{x} = 0.500X$, and $x = 0.100X$ in the previous equation gives

$$\ddot{X} = -0.350\dot{X} - 0.800X$$

Since $\dot{X} = \int \ddot{X}\, dt$, we want the computer to simulate the following voltage-time equations:

$$\dot{X} = \int (-0.350\dot{X} - 0.800X)\, dt$$

$$X = \int \dot{X}\, dt$$

Thus, we now draw a block diagram for a d-c electronic differential analyzer solution of the variation of voltages $\dot{X}$ and X with time t. The procedure for doing this is given in Ill. Ex. 6.12, and the diagram will be similar to that in Fig. 6-24, except for the potentiometer settings. That is, we want the input for the first integrator to be $-(0.350\dot{X} + 0.800X)$ and the input for the second integrator to be $-\dot{X}$. The initial values for the two integrators would equal $2\dot{x}(0)$ and $10x(0)$ volts. The scale factors must be used to convert the measured voltages $\dot{X}$ and X to the physical quantities $\dot{x}$ and x, when the output data is interpreted. That is, 1.00 volt for X equals 0.100 in. for displacement x, and 1.00 volt for $\dot{X}$ equals 0.500 in./sec for velocity $\dot{x}$.

Illustrative Example 6.14.   Change the time scale of the problem in Ill. Ex. 6.12 so that the analog computer solution will be speeded up by a factor 2. Let the initial conditions be $x(0) = 2$ in. and $\dot{x}(0) = -5$ in./sec.

Solution.   Since $T = t/2$, we substitute $dx/dt = 0.5\ dx/dT$ and $d^2x/dt^2 = 0.25\ d^2x/dT^2$ into

$$0.200\ddot{x} + 0.140\dot{x} + 1.60x = 0$$

to obtain

$$0.050\frac{d^2x}{dT^2} + 0.070\frac{dx}{dT} + 1.60x = 0$$

or

$$\frac{d^2x}{dT^2} = -1.40\frac{dx}{dT} - 32.0x$$

and the corresponding initial conditions are $x(0) = 2$ and $dx(0)/dT = (-5)/(0.5) = -10.0$. Since the time scaling is completed, we can now do the magnitude scaling if needed. The details for magnitude scaling are shown in Ill. Ex. 6.13. The scaled equation of motion would have the same mathematical form as the unscaled equation in Ill. Ex. 6.12, so the procedure and the block diagram for the d-c electronic differential analyzer solution of this scaled equation is similar, except for the potentiometer settings, to the solution in Ill. Ex. 6.12. For this solution, real time t equals $2T$ (i.e., twice the measured time T).

To show how magnitude and time scaling might be done together in a more general fashion, let us scale a second-order linear, homogeneous differential equation, with constant coefficients, that is written in general form as follows:

$$\ddot{x} + a\dot{x} + bx = 0$$

Substitution of $\ddot{x} = k^2 d^2x/dT^2$ and $\dot{x} = k\, dx/dT$, where k is the time scale factor, in the previous equation gives

$$k^2\frac{d^2x}{dT^2} + ak\frac{dx}{dT} + bx = 0$$

The next step is to calculate the three magnitude scale factors S_1, S_2, and S_3 from x_{max}, dx_{max}/dT, and d^2x_{max}/dT^2, so that X_{max}, dX_{max}/dT, and d^2X_{max}/dT^2 all equal about 90 volts. Substitution of $x = X/S_1$, $dx/dT = (dX/dT)/S_2$, and $d^2x/dT^2 = (d^2X/dT^2)/S_3$ in the previous equation gives

$$\frac{k^2}{S_3}\frac{d^2X}{dT^2} + \frac{ka}{S_2}\frac{dX}{dT} + \frac{b}{S_1}X = 0$$

Solving the previous equation for its highest derivative gives

$$\frac{d^2X}{dT^2} = -\frac{aS_3}{kS_2}\frac{dX}{dT} - \frac{b}{k^2}\frac{S_3}{S_1}X$$

For the problem in this example, $k = 1/2$, $S_1 = S_2 = S_3 = 1$, $a = 0.700$, and $b = 8.00$. Substitution of these values in the previous equation gives

$$\frac{d^2X}{dt^2} = -1.40\frac{dX}{dt} - 32.0X$$

as before. For the problem in Ill. Ex. 6.13, $k = 1$, $S_1 = 10$, $S_2 = 2$, $S_3 = 1$, $a = 0.700$, and $b = 8.00$. Substitution of these values in the d^2X/dT^2 equation gives

$$\frac{d^2X}{dT^2} = -0.350\frac{dX}{dT} - 0.800X$$

which agrees with the result that is given in Ill. Ex. 6.13, since computer time T equals physical time t for that problem. It should be noted that there are many scaling shortcuts, whose details may be found in textbooks on analog computers.

Illustrative Example 6.15.   Draw a block diagram to show how to solve, on a d-c electronic differential analyzer, for the timewise displacement variations of the masses in the two-degree-of-freedom system shown in Fig. 3-1. Ignore scaling details.

Solution.   The equations of motion are given by Eqs. (3-1a, b) in Sec. 3.2. Solving these two equations for $\ddot{x}_1$ and $\ddot{x}_2$, we have

$$\ddot{x}_1 = ax_1 + bx_2$$

$$\ddot{x}_2 = ex_1 + dx_2$$

where $a = -(k_1 + k)/m_1$, $b = k/m_1$, $e = k/m_2$, and $d = -(k_2 + k)/m_2$. To ease the details of this example, assume that a, b, d, and e are all less than unity. To obtain the displacement x_1, we apply the $\ddot{x}_1$-equation and use two integrators, where the input for the first integrator is $ax_1 + bx_2$, the input for the second integrator is $-\dot{x}_1$ (which is the output of the first integrator), and the output of the second integrator is x_1. To obtain displacement x_2, we apply the $\ddot{x}_2$-equation and use two integrators, where the input for the first integrator is $ex_1 + dx_2$. The next step is to draw the recorders that measure x_1 and x_2 and the four potentiometers that are needed to compute ax_1, bx_2, ex_1, and dx_2. The block diagram for this d-c electronic differential analyzer solution is given in Fig. 6-25.

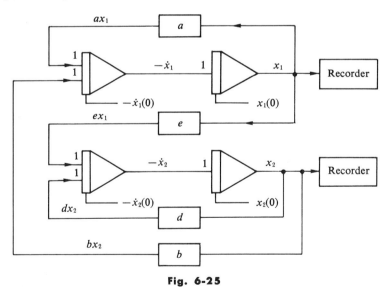

Fig. 6-25

6.6   Other Components for the D-C Electronic Differential Analyzer

In order to solve nonlinear differential equations that represent nonlinear vibration problems, linear differential equations with variable coefficients, and nonhomogeneous differential equations that represent forced vibration problems, other components than those mentioned in the previous section are necessary. The discussion on the components in this section will be very brief, because complete descriptions would take up much space. The reader is referred to analog computer texts which give much more detail about these components. These components are specialized, and if one expects to use the d-c electronic differential analyzer frequently, then he should also gain a thorough understanding of the equipment, including the electronic components, that are involved. Methods for multiplying two variable quantities (i.e., two voltages) are of two types. A *servo-multiplier* performs such a multiplication by electromechanical means. The *electronic multiplier* uses electronic means.

A *servo-multiplier* consists of a servo system, whose output shaft drives the arms of two (or more) potentiometers. The ends of the first potentiometer are connected to $+100$ and -100 volts, while the ends of the second potentiometer are connected to $+E_2$ and $-E_2$ volts. The servo system is used to cause the output of the first potentiometer to equal the input voltage to the servo system which we shall call E_1. Thus, the servo system moves the arm of the first (and also the second) potentiometer so that the potentiometer setting P for the first (and the second) potentiometer equals $E_1/100$. Since the input for the second potentiometer is E_2, its output equals PE_2 or $E_1E_2/100$. Thus, we have achieved a means for multiplying voltages E_1 and E_2. The division of E_1E_2 by 100 helps to keep the output of this multiplier from exceeding 100 volts in absolute value. The servo system portion consists of an amplifier connected to a motor whose output shaft turns the potentiometer arms. The output voltage of the motor is fed back and subtracted from the input voltage. This differential voltage is fed into the amplifier which amplifies this magnitude so that it can operate the motor. When this differential voltage is non-zero, the motor is driven, thus turning the potentiometer arms, and the motor will stop when the system is balanced so that the output voltage of the motor equals the input voltage for the servo system. The frequency-response characteristics of a servo-multiplier are limited, so that *electronic multipliers* must be used for higher-frequency applications.

The *time-division multiplier* and the *quarter-square multiplier* are two commonly used kinds of electronic multipliers. Some electronic multipliers reverse the sign (i.e., their output is $-E_1E_2/100$). The schematic in Fig. 6-26 will be used to represent a multiplier (i.e., either a servo-multiplier or an electronic multiplier). Analog computer texts show how a multiplier can be combined with an amplifier in different ways to perform a division, a square root, or a cube root operation.

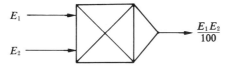

E_1

E_2

$\dfrac{E_1 E_2}{100}$

Fig. 6-26

A special device or component that is devised to calculate sines and cosines is called a *resolver*, and such a device can calculate $E_1 \sin \theta$ and $E_1 \cos \theta$, when E_1 is an input voltage and θ is an input angle. A *polar resolver*, whose voltage inputs are E_1 and E_2, can be used to compute angle θ, where $\theta = \tan^{-1}(E_2/E_1)$, and magnitude R, where $R = \sqrt{E_1^2 + E_2^2}$. Vacuum tube and transistor diodes are convenient components for use in a d-c differential analyzer. A diode basically conducts only when plate voltage E_1 exceeds cathode voltage E_2, and hence acts as a one-way check valve. That is, the output voltage E_0 equals E_1 if $E_1 > E_2$, and E_0 equals zero if $E_1 < E_2$. A relay, which is an electrical switch that can go to one of two positions depending upon whether a coil is energized or unenergized, can be used to simulate physical discontinuities, which are common in vibration problems. One example of a discontinuity is a rectangular pulse, where the applied force equals F_0 for $t < T$ and where this force equals zero for $t > T$, so that there is a sudden force discontinuity when time t equals T.

E_i

$E_0 = f(E_i)$

Fig. 6-27

A very important class of components are *function generators*, which can be represented by the schematic shown in Fig. 6-27. Two kinds of function generators are of the curve-follower type. The *x-y plotter* uses a magnetic coil which follows a curve that is drawn with an ink of high electrical conductivity. An alternating current is passed through this line to produce a magnetic field, which can be sensed by this coil so that it follows the curve in the *y*-direction, as the coil is driven in the *x*-direction. The *photoelectric curve-follower* consists of a cathode ray tube, upon which the curve is drawn and where the portion above this curve is transparent and the portion below is opaque. This can be done by placing an opaque mask over the lower portion of the cathode ray tube, where the upper boundary of the mask, which can be followed by an electron beam from an electron gun, is the shape of the given function.

Other kinds of function generators approximate a curve by a series of straight-line segments, as shown in Fig. 7-6 in Sec. 7.23. This type of approximation can be done electromechanically by a *tapped potentiometer* in which a servo system moves the potentiometer arm to various positions according to the input voltage E_i. The various portions of the potentiometer have different resistances according to the slope of the given straight-line segment. Another means is to use a *group of relays*, which can be used to approximate both continuous and discontinuous functions by a series of straight-line segments, where each relay would switch on at a different end-point for a straight-line segment. A rectangular-wave force or displacement excitation, which has many discontinuities, can be easily handled by such a device. The most widely used type of function generator is the *biased-diode function generator*, which is flexible and can be run on a rather fast time scale. It is an electronic device that consists of several diode bridge circuits, where each circuit corresponds to a straight-line segment. Each diode bridge circuit also has controls by which the length and slope of the corresponding straight-line segment can be set. As mentioned before, when the applied voltage exceeds the cathode voltage, a diode tube conducts, thus effectively closing this electronic switch. Thus, the voltage at which a specific diode tube begins to conduct is set to correspond to the x-value for which its simulated straight-line segment begins.

The generation of certain specific time-varying functions was discussed in Sec. 6.5. By use of multipliers, integrators, etc., we can also generate functions of dependent variables, which might be coefficients or a term in an equation of motion as in the nonlinear pendulum oscillation problem. For example, since $d(e^x)/dt = (e^x)(\dot{x})$, we can use the equation $e^x = \int \dot{x}e^x \, dt$ to generate the function e^x. That is, the circuit would have a multiplier, whose inputs are $\dot{x}$ and e^x and whose output is $\dot{x}e^x/100$, and an integrator, whose input is $-\dot{x}e^x$ and whose output e^x is also a multiplier input. We can also generate the functions $\cos x$ and $\sin x$ by a circuit consisting of two multipliers and two integrators. Since $d(\cos x)/dt = -\dot{x}\sin x$ and $d(\sin x)/dt = \dot{x}\cos x$, we can apply the equations $\cos x = -\int \dot{x}\sin x \, dt$ and $\sin x = \int \dot{x}\cos x \, dt$, so that the two integrators can generate $\cos x$ and $\sin x$ by these integral formulas. The integrands can be obtained from the multiplier outputs, where the inputs for each multiplier are $\dot{x}$ and an integrator output. We can also generate the function $\log_e x$ by applying the equation $\log_e x = \int \dot{x}(1/x)\,dt$, and by use of a division circuit to generate $1/x$.

Illustrative Example 6.16. Draw the block diagram that will show how to solve, on a d-c electronic differential analyzer, for the variables x and y in the following two simultaneous, differential equations. Use unit scale factors throughout. Function $F(t)$ can be approximated by a series of straight-line segments.

$$\dot{x} - 0.402xy = 0$$
$$\dot{y} - 0.581x = F(t)$$

Solution. These two differential equations can be solved by a circuit consisting of two integrators, two potentiometers, two summers, a function generator, and a multiplier, as shown in Fig. 6-28. The first integrator calculates x from the equation $x = \int (0.402xy)\,dt$ while the second integrator calculates y from the equation $y = \int (F(t) + 0.581x)\,dt$.

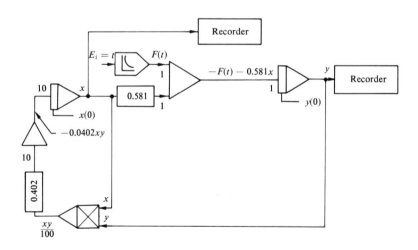

Fig. 6-28

PROBLEMS

6.1. Write the current-variation equations for the electrical circuits illustrated in Figs. 6-29 and 6-30 using the loop method. Draw their dual circuits and list the circuit element values that are necessary to make them exact dual circuits.

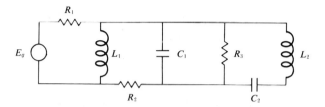

Fig. 6-29

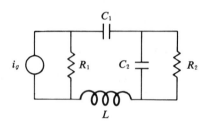

Fig. 6-30

6.2. Write the current-variation equations for the two transformer electrical circuits illustrated in Figs. 6-31 and 6-32 using the loop method. For each circuit, draw an equivalent analogous circuit that has no transformer and label the values of its circuit elements.

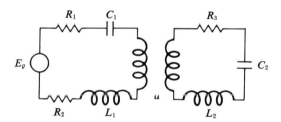

Fig. 6-31

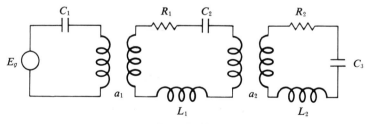

Fig. 6-32

6.3. Write the voltage-variation equations for the electrical circuits illustrated in Figs. 6-29 and 6-30 using the node method.

6.4. Draw the force-voltage electrical analogy circuit for each of the following spring-mass systems. Obtain the force-current analogy circuit by the method of exact duals.

(a) The spring-mass system in Fig. 3-1 with an external force F_E applied to mass m_1.

(b) The spring-mass sytem in Fig. 3-1 with a linear damper c_1 between mass m_1 and the upper surface and a linear damper c between masses m_1 and m_2.

(c) The spring-mass system in Fig. 3-3 with a linear damper c_1 between masses m_1 and m_2 and a linear damper c_2 between masses m_2 and m_3.

(d) Same spring-mass system as in Prob. 6.4c, except replace spring k_4 with an external force $F \sin \omega t$.

(e) The branched spring-mass system in Fig. 3-13.

(f) The spring-mass system in Fig. 4-1 with a linear damper c between masses m_1 and m_2 and an external force $F_1 \sin \omega_1 t$ applied to mass m_1 and an external force $F_2 \sin \omega_2 t$ applied to mass m_3.

6.5. Write the equations of motion for the following mechanical systems. Draw the force-voltage analogy circuit for each system.

(a) The spring-mass system in Fig. 6-33.

(b) The spring-mass system in Fig. 6-34.

(c) The spring-mass system in Fig. 6-35.

(d) The spring-mass-lever system in Fig. 6-36.

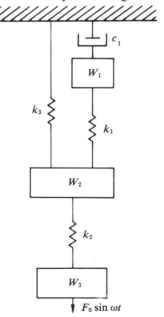

Fig. 6-33

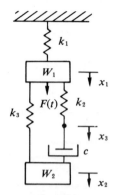

Fig. 6-34

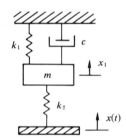

Fig. 6-35

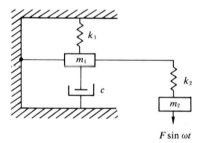

$F \sin \omega t$

Fig. 6-36

(e) A lever system supporting a weight W_1 at one end, which is d_1 ft
from the fulcrum, and a weight W_2 at the other end, which is d_2
ft from the fulcrum. The motion of weight W_2 is resisted by a
damper of damping coefficient c, that is vertically mounted between the weight and a rigid surface that is above the weight.
The motion of weight W_1 is resisted by a spring of spring constant
k, that is vertically mounted between W_1 and the rigid floor below
this weight.

6.6. Write the equations of motion for the following disk-shaft systems. Draw the spring-mass analogy system and the moment-voltage electrical analogy circuit.

(a) The disk-shaft system in Fig. 3-4 with an external torque M_E applied to disk I_3.

(b) The disk-shaft system in Fig. 6-37, where the friction wheel diameters are D_2 and D_3. Use a lever in the spring-mass system analogy.

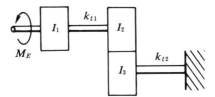

Fig. 6-37

6.7. Draw the force-voltage mechanical analogy system for the same electrical circuit as that in Fig. 6-3, except that inductor L_1 is replaced by a capacitor whose capacitance is C.

6.8. Using only integrators, summers, and potentiometers, show how we may generate the functions t^3, e^{at}, sinh at, cosh at, and $\log_e t$ on a d-c electronic differential analyzer. To generate sinh at and cosh at, use the fact that A cosh $at + B$ sinh at is the solution of the differential equation $\ddot{y} - a^2 y = 0$.

6.9. Draw the schematic block diagram for Ill. Ex. 6.13. Draw another schematic block diagram for the case when we desire to slow down the computer solution of the problem by a factor of 3.20. Show your scaling details first.

6.10. Repeat Ill. Ex. 6.13 for the case when the maximum displacement, velocity, and acceleration are 5 in., 60 in./sec, and 110 in./sec^2, respectively.

6.11. Draw the block diagram for the solution of Ill. Ex. 3.1 on a d-c electronic differential analyzer. Use the results of Ill. Ex. 3.1 to magnitude scale this problem.

6.12. Same as Prob. 6.11, but also time scale this problem to slow it down by a factor of two.

6.13. Scale Ill. Ex. 6.16 and redraw the block diagram circuit for the case where the maximum values of x, $\dot{x}$, y, $\dot{y}$, and $F(t)$ are 20, 180, 12, 80, and 45, respectively, and where we want to slow down the computer solution of the problem by a factor of 2.50.

6.14. Draw block diagrams to show the solutions of the six spring-mass systems in Prob. 6.4 by a d-c electronic differential analyzer. Record all of the time-varying displacements and velocities, and use unit scale factors throughout. Use your own symbol to denote a resolver when needed, and use a function generator when an arbitrary external force F_E is applied.

6.15. Draw block diagrams to show the solutions of the following systems by use of a d-c electronic differential analyzer. Record all time-varying displacements and velocities, and use unit scale factors throughout.

(a) The spring-mass system in Prob. 4.37 with a linear damper, whose damping coefficient is 0.0812 lb-sec/in., between mass m_1 and mass m_2.

(b) The spring-mass system in Prob. 4.41.

(c) The spring-mass system in Fig. 3-1 with a nonlinear damper, whose resistance force is $0.00162\dot{x}^2$, between mass m_1 and the upper surface.

(d) The spring-mass system in Ill. Ex. 6.12, except that the linear damper is replaced by a nonlinear damper, whose resistance force is $0.00807\dot{x}^2$.

6.16. Repeat Prob. 6.15b, but magnitude scale this problem assuming that all accelerations have a maximum value of 90 in./sec², all velocities have a maximum value of 65 in./sec, displacement x_1 has a maximum value of 7 in., and displacements x_2 and x_3 have a maximum value of 4 in.

6.17. Draw the block diagrams which show how a d-c electronic differential analyzer can generate the functions e^x, $\sin x$, $\cos x$, and $\log_e x$. Use the methods suggested in Sec. 6.6, and use your own symbol to denote a division circuit.

6.18. Draw the block diagram schematics that will show how to solve, on a d-c electronic differential analyzer, the following set of simultaneous differential equations. Function $F(t)$ can be approximated by a series of straight-line segments. Use unit amplitude scale factors throughout.

$$\ddot{x} + 9\dot{x}\dot{z} + 14x = F(t)$$
$$\ddot{y} + 6\dot{x}\dot{y} - 17y = 0$$
$$\ddot{z} + 4\dot{x}\dot{z} - 8yz = 0$$

7 Digital Computer Solution Methods

7.1 Introduction

Because of the ever-growing use of digital computers for solving vibration and other engineering problems that arise in professional practice, it becomes increasingly important for the engineering student to learn how to use these types of computers. In the field of vibration analysis, digital computers are especially useful when it is desired to compute the natural frequencies for a large-degree-of-freedom system by such tedious calculation techniques as the Holzer, Myklestad, and matrix-iteration methods. This chapter contains a discussion of digital computer programming using Fortran, and it contains and discusses digital computer programs for solving vibration problems of various selected types. That is, this chapter uses various selected examples to show how vibration problems of different types can be solved by use of a digital computer. If the reader has had a previous course in Fortran, which is becoming a trend, the sections in this chapter that deal with Fortran programming (e.g., Secs. 7. 3 to 7. 13) can be omitted or used for review purposes. Because digital computer solutions of engineering problems often require the use of numerical methods, roots of polynomials, matrix inversion, and other numerical solution techniques have been discussed throughout Chaps. 2 to 5.

A digital computer solves problems by executing a series of previously designated *instructions*. The series of instructions that a computer executes must be expressed in a form or language that the computer can understand, and for this reason they are called *machine-language instructions*. Each machine-language instruction is a number (usually consisting of 7 to 10 digits), and it performs only one elementary operation, such as an addition or the shifting of a number.

The trend today is to program engineering problems, not in the computer's machine language, but by using a system whose language is oriented to that of the user. *Fortran* is a *progamming language* that is very similar to the mathematical language of the scientist and engineer. The computer must translate a Fortran program into its own machine-language before it can perform or execute the various steps of the program. That is, the computer itself first converts a Fortran program into a machine-language program, and then executes the latter program in order to solve the particular problem. In this way, the user does not have to be concerned with the basic language of the computer or its details of operation. The original Fortran program is called a *source program*. The translated, machine-language program is called an *object program*. A Fortran instruction consists of symbols, numbers, etc., that are easily understood by the engineer (instead of being a pure number) and usually performs several computer operations. Thus, a Fortran program is shorter and easier to interpret than a machine-language program. The Fortran language can be used on many different types of digital computers, which means that a problem that is programmed in Fortran may be solved on various types of computers. This is especially useful when a company purchases a new type of digital computer.

Before we can write a digital computer program, we must first learn the rules for using a programming language. This chapter contains an introduction to the Fortran language and discusses the rules for using this language. There are several versions of Fortran, and each version has its differences. This text uses the *Fortran II version* that is designed for the *IBM 1620 computer*. Deviations to be made when *Fortran I* is used for the IBM 1620 computer are also mentioned in this text. This text does not attempt to be a reference for all of the rules for these two versions because the details for these rules can be found elsewhere and because it makes it less confusing for the beginner to learn Fortran programming. The details and complete set of rules for a specific Fortran version can be found in pamphlets available at the computer center or which are supplied by the computer manufacturer. The reader will probably use a Fortran version for a larger computer than the IBM 1620 (and therefore use a slightly different set of rules) when he enters professional practice.

7.2   Steps in Setting Up and Solving an Engineering Problem on a Digital Computer

We can arbitrarily divide the procedure for solving a typical engineering problem on a digital computer into the following eight steps:

1. Definition of the problem and formulation of the physical model, based on a thorough understanding of the problem and its details.
2. Formulation of the mathematical equations, using the physical model as basis.
3. Determination of the numerical techniques to use, if they are necessary.
4. Determination of the problem inputs and the desired outputs.
5. Writing the calculation sequence and possibly drawing a block diagram.
6. Programming the problem solution.
7. Running the program on the digital computer for both program checkout and problem solution.
8. Interpretation of the results.

Note that knowledge of a programming language is very helpful, but not necessary, for the engineer to do six of the eight steps (i.e., the first five steps and the eighth step) of this procedure. Even if in professional practice the engineer has a professional programmer write the programs for solving his problems on digital computers, he should be able to set up the problem for digital computer solution. He should also write the calculation sequence and be able to specify or discuss the numerical solution technique. For a more effective communication with the programmer, the engineer should know something about programming languages and should be capable of programming the solution of small engineering problems. Much expensive computer time has been wasted in the engineering profession, during the checkout of a program, because the engineer did not know how to communicate properly with a programmer. The checkout debugging of a computer program requires knowledge of both the programming language and of the details of the problem. The engineer may have to decide whether an erroneous result is due to a programming error or to an error in his analysis or specification of the problem. The engineer should know the theory of analysis, the problem assumptions and compromises, and the computer program's limitations, whenever he utilizes a computer program, so that he can properly interpret the printed results.

As stated before, unlike analytical methods, numerical methods are often approximate, repetitive step-by-step methods, which can be tedious to apply when done by hand calculation. They are usually easy to set up on a digital computer, which can quickly do this labor. This text will arbitrarily classify numerical methods in the twelve categories listed as follows. This classification is arbitrary, because some of these categories are interrelated. The names and details of some specific numerical methods within some of these categories are discussed throughout this text. Again, most complex engineering problems require the use of one or more numerical solution techniques if they are to be solved on a digital computer. For example, the analysis of random vibrations often requires the use of statistical numerical methods.

1. Evaluation of functions and expressions, summation of series, etc.
2. Solution and roots of single equations.
3. Solution of systems of linear and nonlinear, simultaneous equations.
4. Solution of ordinary differential equations.
5. Solution of partial differential equations.
6. Solution of integral equations.
7. Interpolation of tabulated data.
8. Integration and differentiation of tabulated data.
9. Smoothing and curve- and equation-fitting tabulated data.
10. Matrix methods and linear algebra.
11. Statistical analyses.
12. Simulation and Monte Carlo methods.

7.3 Example of a Fortran Program

Figure 7–1 shows a Fortran program for solving a very elementary problem. The reader will not understand all of the details of this program until he goes further in this chapter, but he will get an idea of what a Fortran program looks like. As stated before, a Fortran program consists of a series of statements or instructions that are written in the Fortran language.

C FOR COMMENT

STATEMENT NUMBER (1–5)	Cont. (6)	FORTRAN STATEMENT (7–)
1		READ 4, A, B
4		FØRMAT (F3.2, F5.1)
		C = 6.541 * B
		D = A + B
		E = C - A
		F = E / D
		G = F ** 3.58
		PRINT 5, A, B, C, D, E, F, G
5		FØRMAT (7F10.4)
		GØ TØ 1
		END

Fig. 7-1

This program calculates the terms C, D, E, F, and G from the equations

$$
\begin{aligned}
C &= 6.541\, B \\
D &= A + B \\
E &= C - A \\
F &= E/D \\
G &= F^{3.58}
\end{aligned}
$$

for specific values of the algebraic quantities A and B. The reader can easily recognize where these equations are used in the Fortran program given by Fig. 7–1.

Figure 7–1 shows the Fortran program as it would appear if it were written on a standard Fortran programming sheet. These programming sheets are usully available at the computer installation. Columns 2 to 5 are used for writing the *statement number*, which is a means of identifying a particular Fortran statement. Only statements that are referred to by another Fortran statement need to have a statement number. In Fig. 7–1, the statement numbers (i.e., 1, 4, and 5) were put in column

5, but they also could be put in column 2, 3, or 4. That is, a statement number may be placed anywhere in columns 2 to 5.

The *Fortran statement* itself is written in columns 7 to 72. Thus, a Fortran statement can contain 65 symbolic characters on one line. A Fortran statement is unaffected if it contains blank spaces (i.e., blanks anywhere in columns 7 to 72). Thus, the individual characters in a Fortran statement may be spaced as desired in order to improve the readibility of that statement. This was done in the GO TO 1 statement in Fig. 7–1. GOTO1 has the same meaning. Also, the statement does not have to start in column 7. If more than 65 symbolic characters are required for a Fortran statement, then the statement can be continued on the following line of the Fortran programming sheet if a numerical digit (i.e., 1 to 9) is placed in column 6 of the *continued statement*. That is, if a digit appears in column 6, then the contents of that line is a continuation of the statement on the previous line. Fortran I does not permit this continuation feature (i.e., each statement must be written on one line). A Fortran II statement must be written within a maximum of five lines, where the first line must contain a blank or zero in column 6, and each of the four continuation lines must contain a numerical digit (i.e., 1 to 9) in column 6. The rules for writing Fortran statements themselves are covered in the rest of this chapter.

The instuctions that are written on a Fortran programming sheet are then punched on *IBM cards*. One line of information on a Fortran programming sheet is punched on an IBM card, and this information must be punched within the first 72 columns of the IBM card. Columns 73 to 80 on an IBM card are usually used for identification purposes. The IBM cards containing the Fortran instructions are called the *source program deck*. The source program deck is loaded into the digital computer, and the Fortran instructions are translated into machine-language instructions. The computer then executes the machine-language instructions.

The Fortran program of Fig. 7–1 requires that numerical values of the quantities A and B be supplied. This is done by punching these numerical values on an IBM card that is placed behind the source program deck. The Fortran statement numbered 1 reads these values from the IBM card and loads them in the computer. Figure 7–2 contains an IBM data card for the case where $A = 1.21$ and $B = 321.4$. The statement numbered 4 tells where to locate the decimal points on this IBM input data card. The eighth statement of the program in Fig. 7–1 prints the values of the quantities $A, B, C, D, E, F,$ and G on one line as ouput.

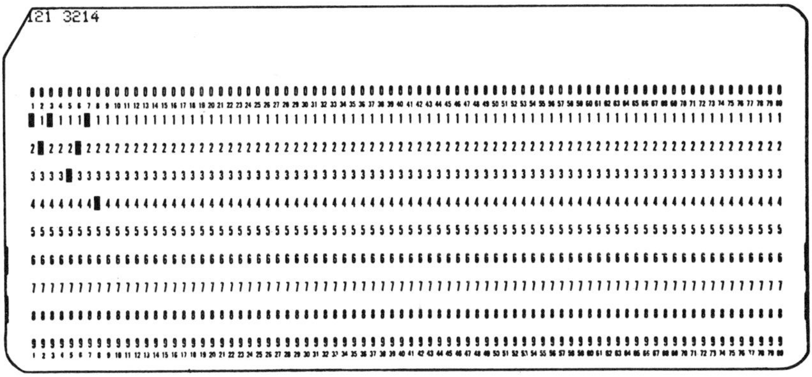

Fig. 7-2

7.4   Fortran Constants

The numbers used by Fortran are classified into two types: *fixed-point* and *floating-point*. A Fortran *constant* is a *particular value* of one of these numbers (i.e., a fixed-point constant or a floating-point constant), and its value cannot be changed during program execution. A *fixed-point number* is an integer (i.e., whole number) used for specific purposes. A *fixed-point constant* is a fixed-point number of specific value which cannot be changed during program execution, and is written *without* a decimal point. For positive numbers, a preceding + sign is optional. For the IBM 1620 computer, the maximum value of a fixed-point number is variable (up to ten digits), but most installations set the maximum value to be 99999 (i.e., a five-digit integer). This maximum will be assumed in this text. The first three of the following numbers are acceptable fixed-point constants. The last three numbers are *not* acceptable because one contains a decimal point, the next one contains more than five digits, and the last one contains a comma.

$$+62$$
$$-1104$$
$$78$$
$$6.89$$
$$184619$$
$$5,148$$

To simplify matters at the present time, consider a *floating-point number* to be merely a real decimal number that can be an integer, or a fraction, or can have an integral and fractional part. A *floating-point constant* is a floating-point number of specific value, and is always written with a decimal point. A preceding + sign is optional if the number is positive. Most numbers in a Fortran program are written in floating-point form. The maximum number of significant digits for a floating-point number is variable (up to 28 digits) for the IBM 1620 computer, but most installations set a limit of eight significant digits. This maximum will be assumed in this text. The numbers -79.7 and 0.00108 are acceptable floating-point constants, and the number 48 is unacceptable because of the absence of a decimal point.

7.5   Fortran Variables

A *Fortran variable* is a term or quantity whose magnitude can be varied as the Fortran program is executed. As stated before, the magnitude of a Fortran constant does not change during program execution. A Fortran variable is referred to in a program by a symbolic name consisting of one to six upper-case alphabetic and numeric characters. The first character must be alphabetic. For Fortran I, the maximum is five alphabetic and numeric characters. A constant, on the other hand, is written as a number. As with Fortran constants, a Fortran variable is in either a fixed-point or a floating-point form.

A *fixed-point variable* is a fixed-point number that can assume various magnitude values during program execution. Its symbolic name must begin with one of the following letters: I, J, K, L, M, or N. The most commonly used type of variable is the *floating-point variable*, which is a floating-point number that may assume different magnitude values during program execution. Its symbolic name may begin with any alphabetic letter except I, J, K, L, M, or N.

Illustrative Example 7.1.   State whether the following Fortran terms are fixed-point variables or floating-point variables: B5R, T$M, N2R, 3PR, H4.9, and MINIMUM.

Solution.   The term B5R is a properly written floating-point variable and the term N2R is a properly written fixed-point variable, but all of the other terms are unacceptable. The terms T$M and H4.9 contain an improper character that is neither alphabetic nor numeric (i.e., a dollar sign and a period), the term 3PR does not start with an alphabetic character, and the term MINIMUM contains more than six characters.

7.6   Fortran Operations and Expressions

Five mathematical *operations* are provided by Fortran : *addition, subtraction, multiplication, division,* and *exponentiation.* Addition is represented by the $+$ symbol. Thus, the expression $A + B$ adds the two floating-point variables A and B. Subtraction is represented by the $-$ symbol. The expression $C - A$ subtracts A from C. Multiplication is represented by the $*$ symbol. Thus, the expression $6.541 * B$ multiplies the floating-point variable B by the floating-point constant 6.541. Division is represented by the $/$ symbol. The expression E/D divides E by D. Exponentiation is represented by the symbol $**$. Note that $**$ is considered to be one symbol. This is the only case where two operation symbols may be written side by side (i.e., next to each other). For example, the expression $E + (-F)$ is permissible, but not $E + -F$. The expression $F ** 3.58$ calculates the term $F^{3.58}$. The ambiguous expression $E ** F ** G$ must be written either as $E ** (F ** G)$ or as $(E ** F) ** G$, depending upon which was originally intended. The exponent may be a fixed-point or a floating-point quantity, but only floating-point quantities may have floating-point exponents. Thus, $MAX ** F$ is not permissible.

A *Fortran expression* is a combination of constants, variables, and/ or functions separated by *operation symbols, commas,* and/or *parentheses,* where certain rules must be fulfilled. In Fortran expressions, *parentheses* can be used to indicate *groupings of terms* just as is done in algebra. As examples, the expression $X + Y ** 4$ differs from the expression $(X + Y) ** 4$, and the expression $X - Y + Z$ differs from the expression $X - (Y + Z)$. The expression $X (Y + Z)$ is an improper one and does not produce the product of X and $(Y + Z)$. The expression should be written as $X * (Y + Z)$, since the parentheses here are only used to group terms. It may also be mentioned that XY represents only one term and not the product of X and Y.

An expression can consist of all floating-point quantities or all fixed-point quantities, but it must not contain both types of quantities. The only exceptions are that fixed-point quantities may appear in floating-point expressions as exponents and as subscripts (which will be described later). Thus, in the following expressions, the first one is proper, while the rest are improper because they are *mixed expressions* that contain both fixed-point and floating-point quantities in an improper manner.

$$(A + B) ** (I + 2)$$
$$E ** (J + 3.2)$$

$$F + R - K$$
$$E * F * (R + 7)$$

As for the order of performing the mathematical operations in a Fortran expression, the order of the calculations is from left to right if there are no parentheses and if the operations are either all additions and subtractions or all multiplications and divisions. Thus, the order of calculations in the expression $X + Y - Z$ is

$$X + Y$$
$$(X + Y) - Z$$

and the order of calculations in the expression $X * Y/Z$ is

$$X * Y$$
$$(X * Y)/Z$$

Now let us consider any expression without parentheses. The exponentiations are performed first (from left to right), then the multiplications and divisions (from left to right), and finally the additions and subtractions (from left to right). Thus, the order of the calculations in the expression $A + B * C ** D$ is

$$C^D$$
$$B \times (C^D)$$
$$A + (BC^D)$$

Parentheses are used in a Fortran expression to specify the order of the calculations. Operations within the innermost parentheses are performed first. The expressions $(X/Y) - (C ** D)$ and $X/Y - C ** D$ are equivalent expressions, since the exponentiation and the division operations are performed before the subtraction is done in either expression.

Illustrative Example 7.2.   Write the corresponding mathematical expressions for the following Fortran expressions: $(F + D/E) * R$, $X * B * (K + 4)$, $E ** 5/A + 1.62$, $5 * B ** A$, and $A * B ** (4 * K)$.

Solution.   The mathematical expressions for the first, third, and fifth Fortran expressions are

$$\left(F + \frac{D}{E}\right) R$$
$$\frac{E^5}{A} + 1.62$$
$$AB^{4K}$$

The second and fourth expressions are improper because they are mixed expressions that contain floating-point and fixed-point quantities in an improper manner.

7.7   Arithmetic Statements

The *arithmetic statement* is generally the most commonly used type of statement in a Fortran program, because it is the statement that performs mathematical calculations. Its general form is $a = b$, where a is a Fortran variable (fixed- or floating-point) and b is an expression. The expression b must be properly written according to the rules specified in Sec. 7.6. Thus, the following:

$$C = 6.541 * B$$
$$D = A + B$$
$$E = C - A$$
$$F = E/D$$
$$G = F ** 3.58$$

are proper Fortran arithmetic statements. It may be noted that these five statements were used in the program given by Fig. 7–1. The following statements are improper because the left-hand side does not consist only of a single Fortran variable.

$$-A = E ** F$$
$$R * S = (A + B) ** 1.82$$
$$3.8 = R * T/Y$$

In an arithmetic statement, the equal sign specifies a replacement or substitution, rather than an equivalence (as it does in algebra). An arithmetic statement $a = b$ replaces the value of the variable a with the computed value of the expression b. Thus, the statement $N = 7$ replaces the value of N with the fixed-point number 7. If N had a previous value, it is now destroyed. The statement $D = X * Y - R$ first calculates the value of the expression $X * Y - R$, and then sets the new value of D to equal the computed value of the expression. The Fortran algebraic statement $Y = Y + D$ does not state that D equals zero because it replaces the old value of Y by the sum of its old value plus D. Effectively, this statement increments the value of Y by the amount D. Similarly, the statement $N = N + 4 * L$ increments the value of N by 4L. Since the rules for writing an expression must be fulfilled, a statement such as

$$X = 2 * A + B(E + F)$$

is an incorrect algebraic statement. This is because the expression is mixed (i.e., we should use 2. instead of 2), and because there is no multiplication symbol (i.e., $*$) between B and $(E + F)$.

Something might be said here about fixed-point arithmetic calculations. If a fixed-point calculation gives a result that is not a whole number, the fraction is dropped (i. e., the result is *truncated* instead of rounded). Thus, the result of the fixed-point operation 11/4 is 2 and not 2.75. Since fixed-point additions, subtractions, and multiplications give correct whole-number results, we only have to be cautious about fixed-point divisions. This truncation feature is often a useful tool for programmers. As another example, the expression 11/4 ∗ 3 has a value of (2)(3) or 6, while the expression 11 ∗ 3/4 has a value of 33/4 or 8. This difference occurs because the operations are performed from left to right.

Usually in an arithmetic statement, $a = b$, the variable a and the expression b are both in floating-point form, or a and b are both in fixed-point form. If the expression b is in floating-point form while the variable a is in fixed-point form, then the computed result of the expression b is truncated (i.e., the fraction is dropped after the complete floating-point calculation is performed). This truncated result is converted to a fixed-point number and becomes the new value of the variable a. This feature can be used to *convert* a floating-point quantity to a fixed-point quantity. For example, the statement J = A truncates A to an integer and converts the result to fixed-point form before this result becomes the new value of J. If the variable a, in the algebraic statement $a = b$, is in floating-point form while the expression b is in fixed-point form, then the result of the fixed-point computation is converted to floating-point form before it becomes the new value of the variable a. For example, the value of the fixed-point variable I may be obtained in a floating-point form by use of the statement AI = I. This statement converts I to floating-point form before it becomes the new value of the variable AI.

Illustrative Example 7.3.　　Write the corresponding mathematical equations for the following Fortran arithmetic statements. If any of these statements are improper, state why.

$$A = P * R ** (5 * M)$$
$$9 = I * J - L$$
$$R = M * K ** (L-4)$$
$$C + D = R * T - A$$
$$M = A(B + D)$$

Solution.　　The second and fourth arithmetic statements are improper because 9 and C + D are not Fortran variables. The last arithmetic statement is improper because there is no multiplication symbol (i.e., ∗) between A and (B + D). The mathematical equations for the first and third arithmetic statements are $A = PR^{5M}$ and $R = MK^{L-4}$.

7.8   READ, PRINT, and PUNCH Statements

Punched IBM cards, magnetic tape, and paper tape are the most common means by which input data are received by a digital computer. We shall consider only punched IBM-card inputs. The *card reader* is a device that reads the data from punched cards, by scanning the holes in these cards, and transmits this data to the digital computer, to be stored. The Fortran statement

READ 9, A, B, C, D, E

causes the computer to take an IBM card, containing five quantities, from the card reader and store these quantities in the storage unit. The first quantity is assigned the symbol A, the second B, the third C, the fourth D, and the fifth E. The number 9 is the statement number of a FORMAT *statement* that describes the arrangement of these five quantities on the card. FORMAT statements are discussed in Sec. 7.9. Suppose that quantities A, B, and C are on one IBM card and that D and E are punched on another IBM card. The READ *statement* will cause the program to read card after card from the card reader until *all* of the quantities *specified in that statement* have been read and stored. Thus, several cards may be read by one READ statement.

Punched IBM cards, magnetic tape, paper tape, and the printed page (by either a typewriter or a line-printer) are the most common means by which data or problem results from a computer are transmitted as output. So that the output data may be visualized by the engineer, data obtained on magnetic tape, paper tape, or punched cards are later taken to a line-printer, in order to be printed. Magnetic and paper tape and punched cards are used for output, because the direct use of a printer slows down the operation of a digital computer. That is, the computer can write on magnetic tape or punch an IBM card faster than it can print the data, thus saving expensive computer time. In this text, the only output means we shall consider are the printing of data by a typewriter during computer operation and the punching of output data on IBM cards. The punched-card output is then later taken to a line-printer, where these output data are printed. The following Fortran statement, which is a PRINT *statement*,

PRINT 8, A, B, C, D, E, X, Y, Z, AVG

causes the digital computer to print the specified quantities (i.e., A to AVG) by typewriter on one or several lines. Line after line is printed in an arrangement specified by statement 8 (which must be a FORMAT statement) until all of the specified quantities have been printed. If instead, we desired to punch this same output data on an IBM card for eventual printout, the following Fortran statement, which is a PUNCH *statement*, should be used

PUNCH 8, A, B, C, D, E, X, Y, Z, AVG

Note that the only change was to substitute the word PUNCH for the word PRINT.

Illustrative Example 7.4.   Write Fortran statements to read quantities K, X, Y, Z, and R from an IBM card and to print these quantities using an on-line typewriter.   Let statement 11 specify the arrangement of the quantities on the IBM card and let statement 12 specify the arrangement of the printed quantities.

Solution.   The Fortran statements are as follows:

READ 11, K, X, Y, Z, R
PRINT 12, K, X, Y, Z, R

7.9   FORMAT Statements

The FORMAT *statement* describes the arrangement of input or output data on an IBM card or the manner in which output data are printed. That is, this statement tells how many card columns should be allocated for each quantity, where to locate the decimal point, whether the number is fixed or floating point, etc. Since FORMAT statements are not executed statements, they may be placed anywhere in a Fortran program, but one should never transfer to a FORMAT statement. The FORMAT statement has many flexibilities, and we shall utilize a simple example before we discuss the details of this type of statement.

Consider the following two statements from the Fortran program of Fig. 7–1:

1   READ 4, A, B
4   FORMAT (F3.2, F5.1)

along with the IBM card given in Fig. 7–2. This IBM card contains the values of the input data (i.e., A and B) for the problem given in Sec. 7.3. In this section, it was given that these values were $A = 1.21$ and $B = 321.4$. These two statements cause a card reader to read these two values from the IBM card and to transmit them to the computer. In statement 4 (i.e., the FORMAT statement), F3.2 specifies the arrangement for quantity A, and F5.1 specifies the arrangement for quantity B. Also note that these arrangement terms (i.e., F3.2 and F5.1) are separated by a comma. It is a requirement that these arrangement specification terms in a FORMAT statement must be separated by a comma. The field description term F3.2 specifies that quantity A must occupy three card columns and that the decimal point must be located such that two digits lie to the right of this decimal point. Note in Fig. 7–2 that quantity A occupies three card columns (i.e., columns 1 to 3). The field description term F5.1 specifies that quantity B must occupy five card columns and that one digit is to lie to the right of the decimal point. In Fig. 7–2, quantity B occupies five card columns (i.e., columns 4 to 8), where the blank in column 4 is assigned to quantity B. It may be noted that the decimal points can be punched in the IBM cards for the input quantities. If the decimal point location on the IBM card *contradicts* the location specified by the corresponding FORMAT statement, then the Fortran program will use the decimal point location that is punched on the IBM card. This flexibility permits the handling of *special-magnitude inputs*. A maximum of 72 characters may be punched on an IBM card.

Also in the program of Fig. 7–1 were the two statements

PRINT 5, A, B, C, D, E, F, G
5   FORMAT (7F10.4)

The first of these statements causes a typewriter to print the values of A, B, C, D, E, F, and G on one line in an arrangement specified by statement 5. The FORMAT statement (i.e., statement 5) specifies that each of these 7 printed quantities must occupy 10 spaces and that each quantity must be printed such that 4 digits lie to the right of the decimal point. The specification 7F10.4 is valid for Fortran II, but when Fortran I is used, one must write F10.4 seven times in the FORMAT statement. It should be cautioned when writing a FORMAT statement for output terms that one should specify enough spaces for each quantity to allow for the *digits*, a possible *minus sign*, a *decimal point*, and some *blank characters* to separate the output term from the adjacent one. The printed quantities are not rounded by the IBM 1620 Fortran. It may be noted that all seven of these printed quantities were floating-point quantities. A maximum of 87 characters may be printed on one line by the typewriter. Thus, a FORMAT statement for typewritten output should not ask to print more than 87 characters per line (including blank spaces).

Now we shall discuss FORMAT statements in more detail. The printing or reading of a fixed-point quantity is represented in a FORMAT statement by the form I*w*, where *w* is the number of card columns or print spaces to use for this integral quantity. To print or read a floating-point quantity without expressing it in an exponent form, we represent this quantity in a FORMAT statement by the form F$w.d$, where w is the number of card columns or print spaces used to represent this quantity and d is the number of digits to the right of the decimal point. Thus, the number -15.1687 would print as -15.168 if an F7.3 specification is used; and improperly as 5.1687 if an F6.4 specification is used. This type of specification has been applied previously in this section. If d equals zero, no decimal point is printed. As mentioned before, when F$w.d$ is used for printing purposes, the number w must be large enough to allow room for a sign, decimal point, digits, and blank spaces (which appear at the left) to separate printed quantities.

To print or read a floating-point quantity in an *exponential form*, we represent this quantity by the form E$w.d$ in a FORMAT statement, where w is the number of card columns or print spaces used to represent this quantity and d is the number of digits to the right of the decimal point. When the form E$w.d$ is used on the IBM 1620 for output printing or punching, the number of significant digits printed or punched is often, but not always, equal to $(w - 6)$. This is to allow for the printing of a sign, two-digit exponent, exponent sign, decimal point, etc. If an E10.2 specification is used, the number -8.114 would be printed as -81.14 E -01 when the Fortran II version is used. When the E$w.d$ form is used to represent an input quantity, the quantity punched on the IBM card need not have a decimal point. Also, the first digit of the exponent may be omitted if it is a zero. If either the quantity sign or the exponent sign is positive, it may be omitted. Also, the term E may be omitted if the exponent sign is included. Thus, for floating-point input the following exponent forms are equivalent: E + 04, E04, E4, E | 4, | 04, and | 4.

As a summarizing example for this section, the following two Fortran statements:

PRINT 4, A, B, J
4   FORMAT (E12.4, F10.5, I4)

would print the quantities A $= -86.7211$, B $= 37.13811$, and J $= 31$ as follows:

$$-86.7211E+00bb37.13811bb31$$

where the symbol b indicates a blank space in this example.

Illustrative Example 7.5.   Write FORMAT statements for the READ and PRINT statements in Ill. Ex. 7.4. Let each input quantity occupy five card columns, and let each output quantity occupy eight print spaces. Let each floating-point quantity be read and printed in non-exponential form with two decimal places.

Solution. The two FORMAT statements can be written as follows if Fortran II is used.

$$11 \quad \text{FORMAT (I5, 4F5.2)}$$
$$12 \quad \text{FORMAT (I8, 4F8.2)}$$

If Fortran I is used, the two FORMAT statements must be written as follows

$$11 \quad \text{FORMAT (I5, F5.2, F5.2, F5.2, F5.2)}$$
$$12 \quad \text{FORMAT (I8, F8.2, F8.2, F8.2, F8.2)}$$

7.10   Use of FORMAT Statements for Printing Alphameric Information and for Output Spacing

By using the / symbol to indicate the beginning of a new IBM card or a new printed line, one FORMAT statement may be used to read or punch several cards or print several lines, where the quantities on each card or line have a different arrangement. For example, the statements

$$\text{PRINT 10, I, A, B, C}$$
$$10 \quad \text{FORMAT (I5/F5.1, F5.2, F4.1)}$$

may cause the following printout

$$bb512$$
$$b-1. 2b6.81b9.2$$

where the symbol b indicates a blank space in this example. Thus, the / symbol is used to *separate* two different one-line or one-card formats. In the previous example, statement 10 specifies a two-line arrangement where the first line has format I5 and the second line has format F5.1, F5.2, F4.1. Blank lines can be obtained by putting consecutive / symbols in a FORMAT statement. N consecutive / symbols cause $(N-1)$ blank lines. For example, the following FORMAT statement will cause three blank lines to occur between a printed line of format F10.3 and a printed line of format F8.6, F10.1.

$$12 \quad \text{FORMAT (F10.3////F8.6, F10.1)}$$

It may be mentioned again that *all* of the quantities listed in a READ, PRINT, or PUNCH statement are read, printed, or punched by that statement. The FORMAT statement specifies the arrangement of these quantities on an IBM card or a printed line. The / symbol and the closing parenthesis each indicate the termination of an IBM card or a printed line. For example, the Fortran statements

PRINT 10, A, B, C
10   FORMAT (F5.1, F5.2)

may cause the following printout:

$$-16.8b7.41$$
$$312.2$$

where the value of C is printed on the second line with an F5.1 specification. That is, the first statement causes the printing of three quantities (i. e., A, B, and C), while statement 10 allows no more than two
quantities to be printed per line.

It is also possible to specify one arrangement for the first (or more)
cards or lines and a different arrangement for *all* of the cards or lines
that follow. That is, one FORMAT statement can be used to specify
one arrangement for the first card or line and a second arrangement for
all of the successive cards or lines. This is done by enclosing the *repeating* specification in *parentheses*. For example, the statements

PRINT 8, K, A, B, C, D
8   FORMAT (I5/(F5.1, F5.2))

may cause the following printout

$$bb601$$
$$-21.7b6.85$$
$$434.2b3.08$$

In this example, statement 8 is equivalent to the following statement:

8   FORMAT (I5/F5.1, F5.2/F5.1, F5.2)

It is often desired to print, read, or punch n successive quantities
in the same manner on a single line or an IBM card. As was previously
stated in Sec. 7.9 and exemplified in Ill. Ex. 7.5, this may be simplified
when using Fortran II, but not Fortran I, by putting this number n
before the letter I, F, or E in the format specification. Statement 5 in
the program of Fig. 7–1 used this feature. Thus, 3I5 is equivalent to
I5, I5, I5, and 2E11.3 is equivalent to E11.3, E11.3.

The printed output becomes much more readable if we put alphabetic titles above the numerical values. When it is desired to print
alphameric (i.e., alphabetic and numeric) information, use nH followed
by the alphameric characters to be printed in the FORMAT statement.
The number n denotes the *number* of alphameric characters to be
printed. *Blanks* are considered to be alphameric characters and must
be accounted for by the number n. This is the *only* case where a blank
space in a FORTRAN statement is not ignored. The following two

statements print the two words INPUT DATA, starting at the left-hand margin of the page:

PRINT 15

15   FORMAT (10HINPUT DATA)

A means of providing *successive blank characters* in a printed output is to use nX in the FORMAT statement, where n is the *number* of successive blank characters. Thus, the statements

PRINT 9, C

9   FORMAT (4X1HC/F5.2)

would result in the following printout if C $= -6.45$:

C

-6.45

That is, the first line contains 4 blank spaces (that are caused by the 4X in statement 9) followed by the letter C (caused by the 1HC in statement 9). The second line contains the value of C printed according to an F5.2 specification. If the ninth statement in Fig. 7-1 were changed to be

5   FORMAT (///9X1HA, 9X1HB, 9X1HC, 9X1HD,
        9X1HE, 9X1HF, 9X1HG/7F10.4)

then the letters A, B, C, D, E, F, and G would be printed above the numerical values that these symbols represent in that program.

Illustrative Example 7.6.   Write Fortran statements to read quantities C, D, E, and F from one IBM card and quantities G and H from a second IBM card. Let 21 be the number of the FORMAT statement, do not use any / symbols in this statement, and read these quantities in non-exponential form. Let each quantity occupy eight card columns with three digits after the decimal point.

Solution.   This can be done using two Fortran statements as follows. Note that data from two IBM cards are read using only one READ and one FORMAT statement because the closing parenthesis indicates the termination of an IBM card after each set of four quantities is read.

READ 21, C, D, E, F, G, H

21   FORMAT (4F8.3)

Illustrative Example 7.7.   Write Fortran statements to read the quantity KODE from one IBM card, quantities A and B from the second card, quantities C and D from the third card, and quantities E and F from the fourth card. Use an I5 format specification for each fixed-point quantity, and use an F8.4 format specification for each floating-point quantity. Use only one

READ statement and one FORMAT statement. Choose 17 to be the number of this FORMAT statement, and do not use more than one / symbol in this statement.

Solution. This can be done using two Fortran statements as follows. Note the use of parentheses within parentheses to specify the format arrangement for the last three IBM cards.

READ 17, KODE, A, B, C, D, E, F

17   FORMAT (I5/(2F8.4))

Illustrative Example 7.8. Write Fortran statements to print quantity K on one line using an I6 specification; skip two lines; print quantities V, X, and Y on the next line using an F9.4 specification for all three quantities; and print quantities R and T on the last line using an F9.1 and an F9.7 specification respectively. Do not use more than two Fortran statements.

Solution. This can be done as follows, where three consecutive slash symbols were used to cause the skipping of two lines:

PRINT 51, K, V, X, Y, R, T

51   FORMAT (I6///3F9.4/F9.1, F9.7)

7.11   STOP, PAUSE, and END Statements

The STOP *statement* tells the computer, during program execution, to stop calculating. This is one of the two ways in which to prevent the digital computer from making any additional calculations during the execution of a given program. The other way is to cycle the program such that it keeps executing a specific READ statement. When there are no more data to read and store (i.e., no more cards in the card reader), the computer will stop the computation for that program automatically.

The PAUSE *statement* permits a temporary stop in the execution of a program. When the *start button* on the *computer console* is pressed, the program execution is continued, starting with the first statement *after* this PAUSE statement. This feature allows the engineer to inspect the printed output, stop the program if the results are satisfactory, or put in new input values and do further calculations.

The last statement of every Fortran program must be an END *statement.* This statement has no effect during program execution. It *separates* the Fortran program from the input data (if there are any), and this statement takes effect when the Fortran source program is converted to machine language. It tells the computer that there are no more Fortran instructions in the program to be translated into the language of the computer.

7.12   GO TO and IF Statements

Fortran statements are executed in the same sequence as they are written on the programming sheet, unless a different sequence is specified. The GO TO and IF *statements* are two types of statements that can cause a *different sequence*, and, hence, are commonly called *control statements*. These two types of statements tell the program to go to a *specific* statement in the program. As stated previously, statements that are referred to by another statement (e.g., by a control statement or by an input-output statement) in the program must have a statement number. These are the *only* statements that need to be numbered, and no two statements may have the same number. There is no requirement that the statements must be numbered in sequence. That is, the first statement can be numbered 32, the second statement numbered 4, etc.

The GO TO *statement* takes the form GO TO *n*, where *n* is a *statement number*. It causes a transfer of control to the statement whose number is *n*. The program then continues from statement *n*. Thus, the GO TO statement provides a means for transferring control to any executable statement in the program. The GO TO statement is utilized in the program of Fig. 7–1, and it is the last executable statement in this program. It causes the program to go to statement 1, thus reading a new set of values for quantities A and B. Since we have now discussed all of the types of statements utilized in this program, we can now examine the program more thoroughly. The sequence of operations (which we shall call a *calculation sequence*) for this program is as follows:

1) Read the values of A and B from the card reader.
2) Calculate the values of C, D, E, F, and G using the following equations:

$$C = 6.541B$$
$$D = A + B$$
$$E = C - A$$
$$F = E/D$$
$$G = F^{3.58}$$

3) Print the values of A, B, C, D, E, F, and G on a typewriter.
4) Go to (1) for the next case (or set of calculations), using new values for quantities A and B.

The IF *statement* is another statement that allows a transfer of control to some statement other than the next one is sequence. Its form is IF (*d*) *n*1, *n*2, *n*3, where *d* is a Fortran *expression* and *n*1, *n*2, and *n*3 are *statement numbers*. This statement transfers control to statement

$n1$ if quantity d (which can be algebraic) is negative. If quantity d equals zero, statement $n2$ is the next statement to be executed. This IF statement causes the program to skip to statement $n3$ if expression d is positive (i.e., statement $n3$ is executed next if $d > 0$). Thus, the IF statement is a *conditional transfer statement* that depends upon the sign of expression d. As an example, suppose that we wished to calculate the quantity F, where

$$
\begin{aligned}
F &= 0 && \text{if } A < D \\
F &= A^2 && \text{if } A = D \\
F &= 5.2D && \text{if } A > D
\end{aligned}
$$

This calculation can be programmed in FORTRAN as follows:

```
        IF(A − D)2, 3, 4
    2   F = 0.0
        GO TO 5
    3   F = A ** 2
        GO TO 5
    4   F = 5.2 * D
    5
```

where the expression in the IF statement is $(A - D)$. If $(A - D)$ is negative, then $A < D$ and $F = 0$. If $(A - D) = 0$, then $A = D$ and $F = A^2$. If $(A - D) > 0$, then $A > D$ and $F = 5.2D$. The two GO TO 5 statements are used to prevent a recalculation of the term F when $(A - D) \leq 0$. Statement 5 is the next statement to be executed in the program after quantity F is calculated.

7.13 Library Functions and Arithmetic Statement Functions

A digital computer performs arithmetic operations, and like a desk calculator, it does not automatically produce the square root, sine, logarithm, etc., of a number. From calculus, we learned that terms such as $\sin x$, $\cos x$, and e^x may be approximated by a finite number of terms in an infinite series and that a square root may be computed by use of a Newton-Raphson iteration. Equations that closely approximate the curves for $\sin x$, $\cos x$, $\log_e x$, and e^x are also available. Thus, several Fortran statements are usually required to calculate such functions as $\sin x$, $\cos x$, $\log_e x$, e^x, $\sqrt{x}$, etc. Very fortunately, we do *not* have to program or write a calculation subroutine when we wish to calculate $\sqrt{x}$, $\sin x$, etc., when Fortran is used. To do this, we can use *library functions*, which in actuality consist of a set of Fortran statements, since such functions must be numerically calculated or approximated. For the Fortran II system for the IBM 1620 computer, there are seven available library functions. For these seven library functions, use

1. SINF(X)   to compute sin x
2. COSF(X)   to compute cos x
3. EXPF(X)   to compute e^x
4. SQRTF(X)   to compute $\sqrt{x}$
5. ABSF(X)   to compute $|x|$
6. LOGF(X)   to compute $\log_e x$
7. ATANF(X)   to compute $\tan^{-1} x$

The IBM 1620 Fortran I system has all of the previous library functions except ABSF(X). A library function is handled like any other variable in a Fortran arithmetic statement, and a Fortran expression may be contained within the parentheses. Thus, Fortran provides a very easy means for programming the use of the more common functions. For example, $A\sqrt{BC}$ may be computed by the following Fortran arithmetic statement:

$$Y = A * SQRTF(B * C)$$

Suppose that we wished to have a means to compute tan x in the same manner that a library function does. Since we already have library functions that compute sin x and cos x, this can be done by use of the following *arithmetic statement function*:

$$TANF(X) = SINF(X)/COSF(X)$$

This type of statement must be located before the first executable statement of a program, since it only defines a function (i.e., it does not cause any calculations to result). In the rest of the program, the function TANF(X) is used in the same manner that a library function is, where X is a dummy variable for later specification of the argument (or angle in this example) for the given function. For example, the following Fortran statement computes (tan A + tan B):

$$D = TANF(A) + TANF(B)$$

The *arithmetic statement function* is limited in application to functions that have only one value and which can be expressed by only one statement. Its form is $e(g) = f$, where e is the *function name*, g is the *argument*, and f is a *Fortran expression*. The name is arbitrary except it must not contain more than six characters, the first character must be alphabetic, and the last character must be F. The first letter must be X if the function is fixed-point in value. Argument g follows name e and is enclosed in parentheses. If there are several arguments, they must be separated by commas. Expression f may use other functions if they have been previously defined by other arithmetic function statements or are library functions. This had been previously done in this section to calculate tan $x = \sin x / \cos x$.

Illustrative Example 7.9. Write two Fortran statements to calculate $y = xe^{a+b}$ and $t = \sqrt{a + \log_e b}$.

Solution. This can be done, using the EXPF(X), SQRTF(X), and LOGF(X) library functions, as follows:

$$Y = X * EXPF(A + B)$$
$$T = SQRTF(A + LOGF(B))$$

Illustrative Example 7.10. Obtain the functions $\log_{10} x$, tanh x, sinh x, cosh x, and $\sin^{-1} x$ using arithmetic statement functions.

Solution. Using the following formulas:

$$\log_{10} x = \frac{\log_e x}{2.3026}$$

$$\tanh x = \frac{e^x - e^{-x}}{e^x + e^{-x}}$$

$$\sinh x = \tfrac{1}{2}(e^x - e^{-x})$$

$$\cosh x = \tfrac{1}{2}(e^x + e^{-x})$$

$$\sin^{-1} x = \tan^{-1}\left(\frac{x}{\sqrt{1 - x^2}}\right)$$

and using Fortran library functions, we can obtain these five functions, for later use in a Fortran program, as follows:

```
LOGT(X) = LOGF(X)/2.3026
TANH(X) = (EXPF(X) − EXPF(−X))/(EXPF(X) + EXPF(−X))
SINH(X) = 0.50 * (EXPF(X) − EXPF(−X))
COSH(X) = 0.50 * (EXPF(X) + EXPF(−X))
ASIN(X) = ATANF(X/SQRTF(1.0 − X ** 2))
```

7.14   Exact Solution of a Forced Vibration

Even though we may know the analytical solution for a problem, the computation of values for a given set of numerical inputs can be eased by use of a digital computer. This is especially true when we wish to compute the timewise variation of displacement for a vibration problem whose analytical solution is known, since such calculations can be very lengthy. We shall program the computation of the timewise displacement variation for a free and forced harmonic vibration of a single-degree-of-freedom system, in order to furnish an exact solution example. This is a relatively easy problem to program. The details for programming other time-varying vibration problems whose analytical solutions are known (including damped vibrations and multiple-degree-of-freedom systems) will be similar to the details for this program.

From Eq. (2–3) in Sec. 2.2, we see that the exact solution for this problem, when the applied force is $F \sin \omega t$ as shown in Fig. 2–1, is given by

$$x = x_0 \cos \omega_n t + \left(\frac{v_0}{\omega_n} - \frac{F/k}{1 - r^2} r \right) \sin \omega_n t + \left(\frac{F/k}{1 - r^2} \right) \sin \omega t$$

where $r = \omega/\omega_n$, $\omega_n = \sqrt{kg/W}$, the last term is the steady-state solution x_p, and the sum of the first two terms is the transient solution x_h. Since this is an exact solution, no numerical techniques are necessary. Our inputs for this problem, and hence also the computer program, are weight W, spring constant k, force magnitude F, force frequency ω, initial displacement x_0, initial velocity v_0, time interval Δt, and maximum time $t_{\max}$. The outputs, that we wish to be printed are time t, displacement x of the mass, transient displacement x_h, and steady-state displacement x_p. These three displacements will be printed columnwise (i.e., like a table) versus time t, from $t = 0$ to time $t_{\max}$ at an interval Δt. We have now completed the first four steps of the digital-computer solution procedure that is given in Sec. 7.2. The next step is to write a calculation sequence, which is given below.

(1) Load inputs: W, k, F, ω, x_0, v_0, Δt, and $t_{\max}$.
(2) Print the inputs.
(3) $t = 0$.
(4) $\omega_n = \sqrt{kg/W}$, where $g = 386$ in./sec^2.
(5) $r = \omega/\omega_n$.
(6) $A = x_0$.
(7) $C = (F/k)/(1 - r^2)$.
(8) $B = (v_0/\omega_n) - Cr$.
(9) $x_h = A \cos \omega_n t + B \sin \omega_n t$.
(10) $x_p = C \sin \omega t$.
(11) $x = x_h + x_p$.
(12) Print t, x, x_h, and x_p.
(13) New value of time $t = $ old $t + \Delta t$.
(14) Test $(t - t_{\max})$: If $t > t_{\max}$, go to (1) for the next case.
      If $t \leq t_{\max}$, go to (9) for the next time cycle.

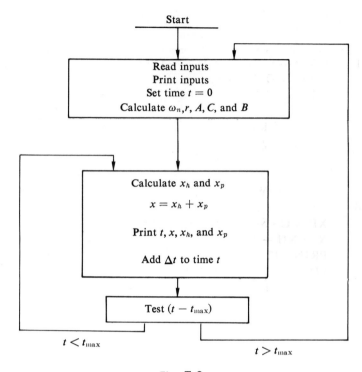

Fig. 7-3

The previous calculation sequence is the basis of the Fortran program, which solves for the variation of displacement x with time t for the spring-mass system shown in Fig. 2–1, that follows. Figure 7–3 is a block diagram for both the calculation sequence and the Fortran program. In the calculation sequence or the block diagram, note that all of the inputs were printed. This was done so that the user can determine whether the proper inputs were used for a specific problem, and whether the input keypunch forms were properly filled in. Input errors due to carelessness are sometimes too common. Note that the constant terms ω_n, r, A, C, and B are computed first, so that they will not be recalculated during every Δt-time-cycle. Also, note that we need initial values for time t and displacement x and that we need a final value t_{max} for time t, so that we can end the problem (i.e., not go through an endless number of cycles). The Fortran symbols W, CON, F, FREQ, X, DX, DT, TMAX, T, WN, R, XH, and XP are used to represent the physical terms W, k, F, ω, x (and also x_0), v_0, Δt, t_{max}, t, ω_n, r, x_h, and x_p, respectively.

```
 1      READ 10, W, CON, F, FREQ, X, DX, DT, TMAX
10      FORMAT (8F8.4)
        PRINT 11, W, CON, F, FREQ, X, DX, DT, TMAX
11      FORMAT (///6HINPUTS/4F12.4/4F12.4////8X4HTIME,
           2X12HDISPLACEMENT, 5X9HTRANSIENT,
           2X12HSTEADY STATE)
        T = 0.0
        WN = SQRTF (386.0 * CON/W)
        R = FREQ/WN
        A = X
        C = (F/CON)/(1.0 — R ** 2)
        B = (DX/WN) — (C * R)
 4      XH = A * COSF (WN * T) + B * SINF (WN * T)
        XP = C * SINF (FREQ * T)
        X = XH + XP
        PRINT 12, T, X, XH, XP
12      FORMAT (F12.4, 3F14.6)
        T = T + DT
        IF (T — TMAX) 4, 4, 1
        END
```

If the inputs for this program were $W = 40$ lb, $k = 50$ lb/in., $F = 60$ lb, $\omega = 2$ rad/sec, $x_0 = 0.100$ in., $v_0 = 0.200$ in./sec, $\Delta t = 0.050$ sec, and $t_{\max} = 0.3000$ sec, the inputs would be printed under the title "INPUTS," and this would be followed by the following printout for this program:

TIME	DISPLACEMENT	TRANSIENT	STEADY STATE
0.0000	.100000	.100000	0.000000
.0500	.076318	−.044483	.120801
.1000	.099906	−.140490	.240396
.1500	.274193	−.083395	.357588
.2000	.535789	.064581	.471208
.2500	.722298	.142178	.580119
.3000	.748069	.064834	.683235

7.15   Numerical Solution of a Nonlinear Vibration

In Sec. 2.13, we discussed the numerical solution of the forced vibration of the damped, one-degree-of-freedom spring-mass system shown in Fig. 2–3 for the case of nonlinear damping. As stated before, we are unable to obtain an analytical solution for this nonlinear vibration problem, and Sec. 2.13 shows how to solve this problem numerically by the Euler and trapezoidal methods [(i.e., Eqs. (2–90) and (2–93)]. Thus, Sec. 2.13 contains the first three steps of the digital-computer solution procedure (as given in Sec. 7.2) for a Fortran program to solve this nonlinear vibration problem. Since Sec. 2.13 also furnishes the equations to compute $\ddot{x}_t$, $\dot{x}_{t+\Delta t}$, and $x_{t+\Delta t}$, as given by Eqs. (2–94) to (2–96); that section also contains a major part of the calculation sequence. The inputs for this nonlinear problem are KODE, W, a, b, k, F, ω, t_0, x_0, $\dot{x}_0$, Δt, t_{max}, and $(\Delta t)_{print}$. The term KODE is an identification number for each case that is run, a and b are constants in the equation for the nonlinear damping force F_d as given by Eq. (1–32) in Sec. 1.6, t_0 is the initial value of time, x_0 is the displacement at time t_0, $\dot{x}_0$ is the velocity at time t_0, Δt is the time-interval for applying the Euler and trapezoidal numerical methods, and $(\Delta t)_{print}$ is the time-interval at which the output is printed. Since the purpose of this problem is to determine the timewise variation of the displacement of the mass, the desired output terms are time t and displacement x.

Something will be said here about program checkout, because of its importance. The inputs to a problem can be easily checked if the computer program prints these terms. To do a complete job of checking out this nonlinear vibration program, it is wise to calculate by hand (i.e., by desk calculator and mathematics tables) each of the terms computed by this program to the same accuracy as that done by the digital computer (i.e., eight significant digits). If this degree of accuracy is used in the hand calculations, then it should be sufficient to do these calculations for the first three or four values of time t. Now we want to devise a means for printing out all of the terms computed by the program so that they may be checked and compared with those calculated by hand. If two corresponding terms do not agree, then the computer- or the hand-calculated term is in error. Because a computer program may be modified later on and because sometimes a program, especially a large one, is not completely checked out at first, it is useful to have a *permanent checkout feature* in a computer program. We shall use the input term KODE for this purpose. If KODE ≥ 0, we shall print the normal engineering output, which are t and x. If the KODE term is a negative input number, we shall print all of the calculated terms for checkout purposes. Thus, if KODE < 0, we shall print the values of displacement x, velocity $\dot{x}$, acceleration $\ddot{x}$, external force F_E, damping force F_d, and spring force F_s at time t. It should also be mentioned that all new and modified Fortran programs should be checked over visually before they are ever run on a digital computer.

A good way to do this is to write the calculation sequence from the Fortran program itself, and then check to determine whether it matches the original calculation sequence.

Since we have now completed steps 1 to 4 of the computer setup-and-solution procedure given in Sec. 7.2, we are now ready to write a calculation sequence, which is as follows. The subscript t refers to the value at time t, while the subscript $t + \Delta t$ refers to the value at time $t + \Delta t$. The terms F_{E_t}, F_{d_t}, and F_{s_t} are the values of the external force, damping force, and spring force, respectively, at time t. The background for the equations for $\ddot{x}_t$, $\dot{x}_{t+\Delta t}$, and $x_{t+\Delta t}$ is furnished in Sec. 2.13.

(1) Load the inputs: KODE, W, a, b, k, F, ω, t_0, x_0, $\dot{x}_0$, Δt, $t_{\max}$, and $(\Delta t)_{\text{print}}$.

(2) Print the inputs.

(3) Set the initial value of print-time, t_{print}, equal to t_0.

(4) $F_{E_t} = F \sin{(\omega t_t)}$.

(5) $F_{d_t} = a|\dot{x}_t|^b \dot{x}_t$.

(6) $F_{s_t} = k x_t$.

(7) $\ddot{x}_t = (F_{E_t} - F_{d_t} - F_{s_t})(g/W)$, where $g = 386$ in./sec².

(8) Test $(t_t - t_{\text{print}})$: If ≥ 0, go to (9).
  If < 0, go to (14).

(9) New value of print-time $t_{\text{print}} = $ old value $ + (\Delta t)_{\text{print}}$.

(10) Test (KODE): If ≥ 0, go to (13).
  If < 0, go to (11).

(11) Print t_t, x_t, $\dot{x}_t$, $\ddot{x}_t$, F_{E_t}, F_{d_t}, and F_{s_t}.

(12) Go to (14).

(13) Print t_t and x_t.

(14) $t_{t+\Delta t} = t_t + \Delta t$.

(15) Save the value of $\dot{x}_t$.

(16) $\dot{x}_{t+\Delta t} = \dot{x}_t + (\Delta t)\ddot{x}_t$.

(17) $x_{t+\Delta t} = x_t + (\Delta t/2)(\dot{x}_t + \dot{x}_{t+\Delta t})$.

(18) Test $(t_{t+\Delta t} - t_{\max})$: If > 0, go to (1) for next case.
  If ≤ 0, go to (4) for next Δt-time-cycle.

As the *first* step in writing the computer program, we shall assign symbolic Fortran names to the various terms in the preceding calculation sequence. The symbols KODE, W, A, B, CON, F, FREQ, T, X, DX, DT, TMAX, DTPRT, TPRT, FE, FD, FS, D2X, and DXB are used to represent the terms KODE, W, a, b, k, F, ω, t_t (also t_0 and $t_{t+\Delta t}$), x_t (also x_0 and $x_{t+\Delta t}$), $\dot{x}_t$ (also $\dot{x}_0$ and $\dot{x}_{t+\Delta t}$), Δt, $t_{\max}$, $(\Delta t)_{\text{print}}$, t_{print}, F_{E_t}, F_{d_t}, F_{s_t}, $\ddot{x}_t$, and $\dot{x}_t$ respectively. It should be noted that input x_0 is the initial value of x_t and that $x_{t+\Delta t}$ is the next value of x_t. Thus, we can use the same Fortran name to represent x_0, x_t, and $x_{t+\Delta t}$. For a similar reason, we shall use the same Fortran name to represent t_0, t_t, and $t_{t+\Delta t}$ and the same Fortran name to represent $\dot{x}_0$, $\dot{x}_t$, and $\dot{x}_{t+\Delta t}$. The following is a Fortran program that is written from the preceding calculation sequence and which may be used to solve this vibration problem. Thus, we have now completed the first six steps of the computer setup-and-solution procedure given in Sec. 7.2

```
1      READ 11, KODE, W, A, B, CON, F, FREQ, T, X, DX, DT,
          TMAX, DTPRT
11     FORMAT (I5/(6F8.4))
       PRINT 12, KODE, W, A, B, CON, F, FREQ, T, X, DX, DT,
          TMAX, DTPRT
12     FORMAT (////6HINPUTS/I5/6F12.4/6F12.4///
          6X4HTIME, 3X9HDISPLMENT)
       TPRT = T
5      FE = F * SINF (FREQ * T)
       FD = A * (ABSF (DX) ** B) * DX
       FS = CON * X
       D2X = (FE — FD — FS) * 386.0/W
       IF (T — TPRT) 4,6,6
6      TPRT = TPRT + DTPRT
       IF (KODE) 2,3,3
2      PRINT 13, T, X, DX, D2X, FE, FD, FS
13     FORMAT (F10.5, F12.5, 5F10.4)
       GO TO 4
3      PRINT 13, T, X
4      T = T + DT
       DXB = DX
       DX = DX + DT * D2X
       X = X + (DT/2.0) * (DX + DXB)
       IF (T — TMAX) 5, 5, 1
       END
```

Let us now discuss why we are using a different time-interval for printout than the Δt-value used in the Euler and trapezoidal numerical-solution techniques. As stated before, the Euler method is a relatively inaccurate method to use for the solution of this vibration problem, and the required value of Δt may be so small that the printout may be excessive, unless we printed the output only when a specified printout interval $(\Delta t)_p$ is attained. That is, we want to prevent an excessive amount of unimportant, printed data, which takes extra computer time and may hamper the readability of the output. It will also be easier to apply the interval-halving technique (which is discussed in Sec. 2.14) to determine the proper value of Δt, when the output for *all* cases is printed at a specified interval $(\Delta t)_p$, since this makes *comparison* of results easier. In this Fortran program, we have used the symbol DTPRT for $(\Delta t)_p$ and the symbol TPRT for t_{print}, where t_{print} is the next value of time at which we wish to print output data. That is, we shall print the output *only* whenever $t_t = t_{print}$. The statement TPRT = T (i.e., the fifth statement in the Fortran program) causes the output to be first printed at time t_0. The tenth and eleventh program statements, which are

$$\text{IF (T} - \text{TPRT) 4, 6, 6}$$
$$6 \quad \text{TPRT} = \text{TPRT} + \text{DTPRT}$$

cause the output to be printed at an interval of $(\Delta t)_p$ thereafter. The first of these two statements tells the program to skip to statement 4 (i.e., jump over the printout statements) if $t_t < t_{print}$. If $t_t \geq t_{print}$, the program goes to statement 6, after which it prints the output according to the sign of the KODE term. Statement 6 calculates the new value of t_{print} (i.e., the next value of time at which to print the output) by adding $(\Delta t)_p$ to the old value of t_{print}. Thus, the first value of t_{print} is t_0, the second value is $t_0 + (\Delta t)_p$, the third value is $t_0 + 2(\Delta t)_p$, etc. Figure 7–4 is a block diagram for the Fortran program for the solution of this nonlinear vibration problem, where this print-interval feature is included. As mentioned previously in Sec. 2.14, we could use the midpoint slope or Adams method, which are given by Eqs. (2–99) and (2–101), instead of the Euler method in order to more accurately solve this problem. An even more accurate solution procedure would be to use the Runge-Kutta method, which is discussed in Sec. 2.14, to compute both $\dot{x}_{t+\Delta t}$ and $x_{t+\Delta t}$. As was shown in Sec. 3.11, the numerical solution, and hence the correspoding Fortran program, of this single-degree-of-freedom problem is similar to the procedure for numerically solving linear and nonlinear, multiple-degree-of-freedom vibration problems, unless the Runge-Kutta method is used. Thus, we could modify this Fortran program (by adding inputs and Fortran statements) to make it solve a two-degree-of-freedom nonlinear vibration problem by the Euler and trapezoidal methods, where the calculation sequence in Sec. 3.11, or its basis, might be employed.

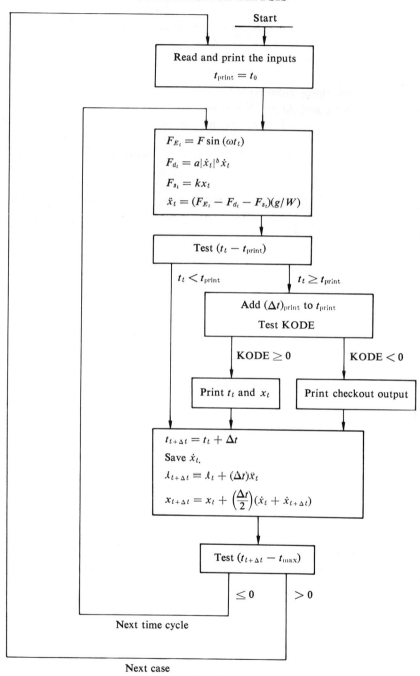

Fig. 7-4

7.16   Arrays and Subscripted Variables and DIMENSION Statements

Suppose we had quantities $A_1, A_2, A_3, \ldots, A_9,$ and A_{10} as data. We can represent these subscripted quantities in Fortran by A(1), A(2), A(3), ..., A(9), and A(10). Note that the specific term A_3 is represented by A(3) and that the term inside the parentheses is a *fixed-point constant*. We can also represent the arbitrarily subscripted quantity A_i by A(I), and it should be noted that the term inside the parentheses is a *fixed-point variable*. If a Fortran instruction refers to A(I) at a time when I = 3, then the instruction refers to A(3) at that time. Here we have used a *subscripted variable* (with only the name A) to represent the ten quantities A(1) to A(10). The complete set of ten quantities is called an *array*, and the name of this array is A. An individual quantity in an array, such as A(4), is called an *element* of that array. The first element is A(1), the second element is A(2), etc.

Now suppose that we had the following matrix consisting of two rows and four columns:

$$B_{11} \; B_{12} \; B_{13} \; B_{14}$$
$$B_{21} \; B_{22} \; B_{23} \; B_{24}$$

The arbitrarily subscripted term B_{ij}, where i is the row number and j is the column number, can be represented in Fortran by B(I, J). Thus, for the term B(I, J), the first subscript I is the row number and the second subscript J is the column number. Note that the subscripts are separated by commas within the parentheses. We can represent the first row of the above matrix in Fortran by B(1, 1), B(1, 2), B(1, 3), and B(1, 4); and the second row can be represented by B(2, 1), B(2, 2), B(2, 3), and B(2, 4).

The name of a subscripted variable must follow the same rules as those previously given for an unsubscripted Fortran variable. Thus, the name of an array consisting of fixed-point elements must start with the letter I, J, K, L, M, or N. An array connot consist of both fixed- and floating-point elements. A subscripted variable may have one, two, or three subscripts, and we shall call these variables one-, two-, and three-dimensional arrays, respectively. Fortran I does not permit more than two subscripts. A *subscript*, which must always be positive and a fixed-point quantity, may be written in the following forms:

$$k$$
$$m$$
$$m + k$$

$$m - k$$
$$k * m$$
$$k * m + k$$
$$k * m - k$$

where k means any positive fixed-point constant, and m means any positive fixed-point variable. Fortran I permits only the first four of the previous subscript-forms. Thus, Fortran II permits the following notations for the elements of subscripted variables:

$$AB(I - 4)$$
$$XR(4 * J, K, 4)$$
$$MS(6 * K + 2)$$

but does not permit the following notations:

$$AB(4 - I)$$
$$RX(J * 4)$$
$$TM(4 * B - 1)$$
$$SK(I + J + K)$$

because the subscripts do not fulfill any of the seven permissible forms.

The DIMENSION *statement* is an unexecuted statement that specifies the maximum number of elements in an array and, thus, tells the computer, during machine-language translation, how many storages to allocate for that array. All subscripted variables in a program must appear in a DIMENSION statement, and this DIMENSION statement must appear before any other appearance of this variable in the program. For this reason, it is wise to locate the DIMENSION statements at the beginning of a program. The following statements may be used to specify the maximum size of arrays A, X, Y, and Z:

DIMENSION A(15, 10, 12)
DIMENSION X(20), Y(30), Z(11, 8)

The same statement states that array X has at most 20 elements, array Y has at most 30 elements, and array Z has at most 11 rows and 8 columns (i.e., 88 elements). Note that several arrays may be mentioned in a DIMENSION statement and that the maximum size is specified by a positive fixed-point number within the parentheses.

If a READ or PRINT statement contains specifically given elements of an array, the elements will be read or printed according to the sequence listed in the Fortran statement. If this data is input, it must be punched on the IBM card in the corresponding sequence (i.e., the first data field on the card goes with the first variable name, etc.), but this sequence is arbitrary if these elements are *explicitly stated*. Thus, the following Fortran statement is a valid one:

$$\text{READ } 11, B(9), B(7), B(21), A(3, 8), R, B(2), F$$

For Fortran II, a one-dimensional array may be read into the computer as follows:

$$\text{READ } 18, (A(K), K = 1, N)$$
$$18 \qquad \text{FORMAT (8F8.3)}$$

which means that the first N elements of array A are read from IBM cards and that each card contains eight elements. The first quantity read is A(1), the second is A(2), the third is A(3), etc. Thus, the first card should contain A(1) to A(8), the second card (if necessary) should contain A(9) to (A16), etc. Note that the array's variable name and the indexing information are all enclosed in parentheses. Similar statements may be made when a one-dimensional array is printed or punched. The same type of indexing can be used to read or print a row or a column of a matrix and other types of two-dimensional arrays. Thus, the statement

$$\text{PRINT } 14, (B(I, 4), I = 1, 8)$$

prints eight elements from column 4 of matrix B, and the statement

$$\text{PRINT } 15, (C(6, J), J = 1, 9)$$

prints nine elements from the sixth row of matrix C. The reading and printing of complete two-dimensional arrays are discussed in Sec. 7.17.

Illustrative Example 7.11.   Write Fortran instructions to compute the following equations:

$$A_{11}X_1 + A_{12}X_2 + A_{13}X_3 = R_1$$
$$A_{21}X_1 + A_{22}X_2 + A_{23}X_3 = R_2$$

Solution.   This may be done as follows, where use was made of arrays A, R, and X:

$$R(1) = A(1, 1) * X(1) + A(1, 2) * X(2) + A(1, 3) * X(3)$$
$$R(2) = A(2, 1) * X(1) + A(2, 2) * X(2) + A(2, 3) * X(3)$$

These Fortran instructions should be preceded in the program by a DIMENSION statement such as

$$\text{DIMENSION } R(2), X(3), A(2, 3)$$

Illustrative Example 7.12.   The following points were taken from a graph of y versus x: $(0, 6.1), (1, 8.7), (3, 3.2), (7, -1.1)$, and $(12, -7.3)$. Show how to represent this graph in a Fortran program.

Solution. We can represent the x-points by array X, where $X(1) = 0.0$, $X(2) = 1.0, X(3) = 3.0, X(4) = 7.0$, and $X(5) = 12.0$. We can represent the y-points by array Y, where $Y(1) = 6.1, Y(2) = 8.7, Y(3) = 3.2, Y(4) = -1.1$, and $Y(5) = -7.3$. Since $X(3) = 3.0$ and $Y(3) = 3.2$, we know that $y = 3.2$ when $x = 3.0$. Thus, we have shown how an array may be used to represent points on a graph and other types of tabulated data.

Illustrative Example 7.13. For a certain nonlinear spring, the values of its spring constant were tabulated for 20 different values of displacement. Show how to represent this tabulated data in a Fortran program.

Solution. We can represent the displacement values by an array called DISPL and the spring constant values by an array called SCON. Thus, SCON(12) is the value of the spring constant at displacement DISPL(12). The DIMENSION statement for this problem can be written as:

DIMENSION DISPL(20), SCON(20)

Illustrative Example 7.14. Suppose that we have quantities $A_1, A_2, A_3, \cdots,$ A_n as data. Write Fortran instructions to calculate the average value of these n quantities.

Solution. Using N to represent n, array A to represent A_1 to A_n, and the following equation to calculate the average value:

$$A_{av} = \frac{1}{n} \sum_{j=1}^{n} A_j$$

we obtain the following portion of a Fortran program:

```
        J = 0
        SUM = 0.0
   15   J = J + 1
        SUM = SUM + A(J)
        IF(J − N)15, 16, 16
   16   AN = N
        AVG = SUM/AN
```

The above program portion represents A_j by A(J), $\sum A_j$ by SUM, and A_{av} by AVG. The first two statements set both j and $\sum A_j$ equal to zero, while the third, fourth, and fifth statements are executed n times (because of the IF statement) to calculate $\sum A_j$. During the first trip or cycle through the loop, $j = 1$, A_1 is added to SUM, and the IF statement causes the program to go to statement 15 (which sets $j = 2$ and starts the second trip through the loop). During the nth trip or cycle, $j = n$, A_n is added to SUM, and control goes from the IF statement to statement 16 (since $j = n$) which obtain n in floating-point form. The last statement calculates the average value.

7.17   DO and CONTINUE Statements

In Ill. Ex. 7.14, we had calculated the average value of the elements in array A (i.e., A(1) to A(N)). This could have been done more easily using a DO statement (instead of an IF statement) as follows:

$$\text{SUM} = 0.0$$
$$\text{DO } 17 \text{ J} = 1, \text{N}$$
$$17 \quad \text{SUM} = \text{SUM} + \text{A(J)}$$
$$\text{AN} = \text{N}$$
$$\text{AVG} = \text{SUM}/\text{AN}$$

First, note that the program-portion in Ill. Ex. 7.14 has two more statements than the previous one. The first statement of the above program-portion sets $\sum A_j$ (i.e., SUM) to initially equal zero. The last two statements obtain n as a floating-point number and calculate the average value using the equation

$$A_{av} = \frac{1}{n} \sum_{j=1}^{n} A_j$$

The DO statement and statement 17 calculate the term $\sum_{j=1}^{n} A_j$. These two statements are executed n times (i.e., the DO statement causes the program to go through n loops or cycles). During the first trip through the loop, J = 1 and A(1) is added to SUM (i.e., A_1 is added to $\sum A_j$), during the second trip or cycle J = 2 and A(2) is added to the SUM term, and during the last cycle J = N and A(N) is added to the SUM term. After the Nth cycle is completed, the SUM term will equal A(1) + A(2) + $\cdots$ + A(N), and the program will go to the AN = N statement (instead of going from statement 17 to the DO statement as was done during the first N − 1 cycles). Thus, we have shown that the DO statement is a rather powerful one.

Let us now discuss the features of a DO statement more thoroughly. Note in the previous example that the DO statement caused certain consecutive statements to be repeated several times and that a fixed-point variable was increased in value between repetitions. The DO *statement* can be written in one of the following two forms:

$$\text{DO } s \ j = n1, n2$$
$$\text{DO } s \ j = n1, n2, n3$$

where s is a *statement number*, j is a *fixed-point variable* which is called the *index*, and $n1$, $n2$, and $n3$ may be *fixed-point constants* or *variables*. If $n3$ is not stated in a DO statement, then $n3 = 1$. The DO statement instructs the computer to repeatedly execute the statements that follow the DO statement, up to and including the statement numbered s. These repeated statements are called the *range* of the DO statement. The first time these range statements are executed with index j equal to $n1$, and the second time with $j = n1 + n3$. That is, for each succeeding execution of the range statements, j is increased by $n3$. After the range statements have been executed with j equal to the highest value that does not exceed $n2$, the program stops the repetition and goes to the statement located after statement s. The reader should now look again at the previous Fortran example and see how this paragraph applies to the DO statement and statement 17, which calculates $\sum A_j$. As stated before, whenever enough loops are executed so that the next value of index j will exceed $n2$, then the program will go to the statement that follows statement s (which is the last statement in the DO range) and, thus, transfer out of the DO loop. Thus, for the statement DO 19 $J = 1,17$, the program will transfer out of the DO range (to the statement following statement 19) after 17 loops or cycles are executed. An IF or a GO TO statement, inside the DO range, may be used to cause a transfer outside the DO range. This usage is illustrated in Ill. Ex. 7.16. A rather valuable feature is that the value of index j is available for use in computations whenever an IF or GO TO statement is used for transfer outside a DO loop. We are not allowed, however, to transfer into the range of a DO statement by use of a transfer statement located outside the range of that DO statement. Thus, the following program portion

$$\text{GO TO } 17$$
$$\text{DO } 17 \text{ } J = 1, \text{N}$$
$$17 \quad \text{SUM} = \text{SUM} + \text{A(J)}$$

is not permissible. Illustrative Example 7.16 shows a valid transfer from the inside to the outside of a DO range. The last statement in a DO range, however, cannot be a transfer statement (e.g., IF, GO TO, etc.). If this results, then this particular transfer or control statement should be followed by a CONTINUE statement. The CONTINUE *statement*, which does nothing (i.e., causes no action) when the Fortran program is executed, is used merely to satisfy the rule that the last statement in a DO range cannot be a transfer statement. This usage of the CONTINUE statement as the last statement in a DO range is exemplified in Ill. Ex. 7.16. Illustrative Example 7.15 shows another use of the CONTINUE statement as a dummy, no-operation instruction.

If a DO statement lies within the range of another DO statement, then *all* statements in the range of the inner DO statement must lie *within* the range of the outer DO statement. The range of the two DO loops can end with the same statement, as shown in Ill. Ex. 7.17, but the last statement of the inner DO loop must not be located after the last statement of the outer DO loop.

In Fortran II, it is also permissible to use indexing so that an entire two-dimensional or three-dimensional array may be read by one READ statement or printed by one PRINT statement. For example, matrix R can be printed as follows, where six elements are printed on each line:

$$\text{PRINT } 31,((R(K, L), L = 1, M), K = 1, 7)$$

31   FORMAT (6F11.4)

In principle, the inner and outer indexes of the above PRINT statement act like the inner and outer DO statements, respectively, of a double DO loop. Thus, matrix R will be printed row-wise in this example, and seven rows (i.e., $K = 1$ to 7) of this matrix will be printed. Note that if $M = 4$ at this instant, then the first two lines of printout will be as follows:

$$R_{11} \qquad R_{12} \qquad R_{13} \qquad R_{14} \qquad R_{21} \qquad R_{22}$$
$$R_{23} \qquad R_{24} \qquad R_{31} \qquad R_{32} \qquad R_{33} \qquad R_{34}$$

This printout would be more readable if the Fortran statements were written such that each line of printout will contain elements of only one row of R, no matter what the value of M is at that instant. This can be done as follows:

$$\text{DO } 7 \text{ K} = 1, 7$$

7   PRINT 31, $(R(K, L), L = 1, M)$

31   FORMAT (6F11.4)

Each cycle of the DO loop prints one row (i.e., M elements) of matrix R. If $M \leq 6$, this row will be printed on one line. Thus, if $M = 4$, the first line will contain R_{11} to R_{14}, the second line will contain R_{21} to R_{24}, etc.

Illustrative Example 7.15.   Suppose that we had quantities $A_1, A_2, A_3, \cdots,$ A_n as data. Write Fortran instructions to calculate the average of all the positive non-zero quantities in this array.

Solution.   We can accomplish this by adding an IF statement, a CONTINUE statement, and two other statements to the first Fortran example in this section. Thus, we have

$$\text{SUM} = 0.0$$
$$\text{NPOS} = 0$$
$$\text{DO } 17 \text{ J} = 1, \text{N}$$

```
        IF (A(J))17, 17, 18
   18   SUM = SUM + A(J)
        NPOS = NPOS + 1
   17   CONTINUE
        AN = NPOS
        AVG = SUM/AN
```

The CONTINUE statement was used, because statement 17 must be executed during each cycle of the DO loop. The term NPOS is a count of the number of positive terms.

Illustrative Example 7.16. Suppose that we had quantities $x_1, x_2, x_3, \cdots,$ x_n as data, where $x_1 < x_2 < \cdots < x_n$. Write Fortran statements to determine between which x_i-values the value of a quantity p lies.

Solution. In the following Fortran statements array X is used to represent the x_i-terms, N represents n, and P represents p:

```
        DO 23 I = 1, N
        IF (P − X(I))24, 24, 23
   23   CONTINUE
   24
```

Since the X(I) (i.e., x_i) elements are arranged in an increasing-value order, the statements in the DO range will be repeated (where index I is increased by one after each repetition) until the IF statement finds the first X(I)-term in the array that equals or exceeds quantity P. When this happens, the IF statement causes a transfer to statement 24 which is outside the DO loop. The value of index I is also available for use in a computation. When statement 24 is reached, we know that the value of quantity P either equals X(I) or it lies between the values of X(I − 1) and X(I). Later in this chapter, we shall utilize Fortran statements similar to those in this example as the first part of a linear or a polynomial interpolation routine for application to tables and graphs.

Illustrative Example 7.17. Write Fortran statements to compute

$$\sum_{j=1}^{21} \sum_{i=1}^{12} A_{ij} B_i + C_i$$

Solution. This computation may be performed by the following Fortran statements, where it should be noted that the inner DO loop is contained within the range of the outer DO loop:

```
        SUM = 0. 0
        DO 37 J = 1, 21
        DO 37 I = 1, 12
   37   SUM = SUM + A(I, J) * B(I) + C(I)
```

Illustrative Example 7.18.   Write a complete program to do the calculations of Ill. Ex. 7.17.

Solution.   One possible Fortran program to do this is as follows:

```
      DIMENSION B(12), A(12, 21)
  1   READ 11, (B(I), I = 1, 12)
      READ 11, (C(I), I = 1, 12)
 11   FORMAT (6F8.2)
      DO 7 I = 1, 12
  7   READ 11, (A(I, J), J = 1, 21)
      SUM = 0.0
      DO 37 J = 1, 12
      DO 37 I = 1, 12
 37   SUM = SUM + A(I, J) * B(I) + C(I)
      PRINT 12, SUM
 12   FORMAT (22HDOUBLE SUM OF PRODUCTS/F15.2)
      GO TO 1
      END
```

This program first reads arrays B and C and two-dimensional array A; then it performs the calculations of Ill. Ex. 7.17, prints the result, and goes to statement 1 for the next case.

7.18   Calculation of Magnification Factors and Phase Angles

Figures 2–4 and 2–5 furnish a plot of magnification factor M.F. versus frequency ratio r and damping ratio ζ and a plot of phase angle ϕ versus r and ζ. These two figures are based on Eqs. (2–12) and (2–13) in Sec. 2.3, which are

$$\text{M.F.} = X/X_0 = 1/\sqrt{(1 - r^2)^2 + (2\zeta r)^2}$$
$$\phi = \tan^{-1} [2\zeta r/(1 - r^2)]$$

We may want a tabulation of M.F.-and ϕ-values that have more than three-digit accuracy and which have a smaller ζ-interval than that shown in Figs. 2–4 and 2–5. We shall write a Fortran program that will give us such a tabulation of M.F.-and ϕ-values as a function of r and ζ, for any desired Δr-and $\Delta \zeta$-intervals. The calculation sequence for this program is as follows:

(1)   Read and print NR (the number of r-values), ND (the number of ζ-values), Δr, and $\Delta \zeta$.

(2)   Set $\zeta = 0$.

(3)   DO (11) I = 1,ND.

(4) Set $r = 0$.

(5) DO (10) J $= 1$,ND.

(6) M.F. $= 1/\sqrt{(1 - r^2)^2 + (2\zeta r)^2}$.

(7) ϕ in degrees $= 57.29578 \tan^{-1}[2\zeta r/(1 - r^2)]$.

(8) Add $180°$ to ϕ if $r > 1$.

(9) Print r, ζ, M.F., and ϕ.

(10) Add Δr to r.

(11) Add $\Delta \zeta$ to ζ.

(12) Go to step (1) for another case.

Erroneous output or error statements might arise during the calculation of ϕ when $r = 1$ (due to a division by zero) at some computer installations. If this happens, this program should be modified to set $\phi = 90°$ if $r = 1$, and it should set M.F. equal to a very large number if $r = 1$ and $\zeta = 0$. These two additions can be included in step (8) of the previous calculation sequence. Using the symbols DELR, DELD, D, R, FM, and PH to represent Δr, $\Delta \zeta$, ζ, r, M.F., and ϕ, respectively, the Fortran program (as written from the previous calculation sequence) is as follows:

```
1      READ 10, NR, ND, DELR, DELD
10     FORMAT (2I5, 2F8.3)
       PRINT 11, NR, ND, DELR, DELD
11     FORMAT(///6HINPUTS/2I6, 2F12.3///4X10HFREQ
           RATIO, 4X10HDAMP RATIO, 3X11HMAGN
           FACTOR, 3X11HPHASE ANGLE)
       D = 0.0
       DO 7 I = 1, ND
       R = 0.0
       DO 6 J = 1, NR
       FM = 1.0/SQRTF((1.0 − R ** 2) ** 2 + (2.0 * D * R) ** 2)
       PH = 57.29578 * ATANF(2.0 * D * R/(1.0 − R ** 2))
       IF(R − 1.0)5, 5, 4
4      PH = 180.0 + PH
5      PRINT 12, R, D, FM, PH
12     FORMAT (4F14.3)
6      R = R + DELR
7      D = D + DELD
       GO TO 1
       END
```

If we used $\Delta r = \Delta \zeta = 0.500$ as the inputs for a chekout case, the inputs would be printed, and the next twelve lines of printout would be as follows:

FREQ RATIO	DAMP RATIO	MAGN FACTOR	PHASE ANGLE
0.000	0.000	1.000	0.000
.500	0.000	1.333	0.000
1.000	0.000	.999999E+99	89.999
1.500	0.000	.800	180.000
2.000	0.000	.333	180.000
0.000	.500	1.000	0.000
.500	.500	1.109	33.690
1.000	.500	1.000	89.999
1.500	.500	.512	129.805
2.000	.500	.277	146.309
0.000	1.000	1.000	0.000

In the previous output, note how the two successive DO statements in the Fortran program caused a calculation of M.F. and ϕ for all possible r-and ζ-combinations. In many situations, an engineering designer must determine the combination of values for the parameters of a system that will cause a desired output for the system. Often many combinations of these parameter values will do the desired job for this system. As a large-scale example, consider the design of an airplane or a missile to perform a specific military mission. There are many combinations of wing and body shapes, fuel tank locations and sizes, wing and body dimensions, etc., that will enable the aircraft or missile to perform its desired mission. Often the designs of the competing aircraft manufacturers are quite different, and the preferable design is based on such factors as initial cost, ease of maintenance, etc. If we had four input parameters for a vibration design problem, where there are several possible values for each parameter, then we could easily obtain a computer printout of the design calculations for all combinations of the values for these four parameters by use of four successive DO statements.

In Secs. 2.9 and 2.10, we found that the amplitude ratio for a seismic vibration-measuring instrument equals r^2(M.F.) and that the amplitude ratios for a rotating and reciprocating unbalance and for a rotating shaft during whirl also equal r^2(M.F.). The phase angle ϕ is given by Eq. (2–13) for these three problems. Thus, the data for Figs. 2–11 and 2–4, which plot r^2(M.F.) and ϕ versus r and ζ, can be computed by modifying the program in this section to compute r^2(M.F.), instead of M.F. Similarly, we can also modify this program so that it will compute the values of both transmissibility TR and the phase angle for vibration isolation versus r and ζ, where the necessary TR and phase angle equations are given in Sec. 2.8.

7.19 A Holzer-Method Program

In Sec. 4.3, we had discussed the Holzer method for calculating natural frequencies. A natural frequency is assumed for the spring-mass (or a disk-shaft) system, and the displacements and inertia forces (or moments) are calculated for the system. The system either consists of masses connected only by linear springs or of disks connected only by linear elastic shafts. If both ends of the system are free, the sum of the inertia forces or moments will equal zero if the assumed frequency is a natural frequency. The Fortran program in this section will also handle systems where one end is ridigly fixed and the other end is free. For this case, we start at the free end, and we will know that we have assumed a natural frequency if the calculated displacement at the fixed end equals zero. The calculation sequence for the Holzer-method Fortran program in this section is as follows, where a negative KODE input causes a calculation for a system where both ends are free. If KODE ≥ 0, the calculations are for a system with one end fixed and one end free. For both cases, mass m_1 is the mass at the free end, and mass m_n, for an n-degree-of-freedom system, can be either free (i.e., unsupported) or be a fixed surface (as denoted by the sign of the KODE input).

(1) Read and print the KODE input.
(2) Read the NF assumed natural frequencies, ω_1 to ω_{NF}.
(3) Read the values of the NW masses in the system, (i.e., m_1 to m_{NW}).
(4) Read the spring constant values for the NS springs in the system (i.e., k_1 to k_{NS}).
(5) DO (16), I = 1,NF.
(6) Set SUM = 0.
(7) $x(1) = 1.00$.
(8) Print the value of $\omega(I)$.
(9) DO (13), J = 1,NW − 1.
(10) $F(J) = [\omega(I)]^2 [m(J)][x(J)]$.
(11) Add F(J) to SUM, where SUM = $\sum F(J)$.
(12) Print J, $m(J)$, $k(J)$, $x(J)$, and F(J).
(13) $X(J + 1) = X(J) − [\sum F(J)]/k(J)$.

(14) Print NW, m(NW), and x(NW).

(15) If the KODE input is negative, add F(NW) to SUM to obtain $\sum_{J=1}^{NW} F(J)$ and print this SUM term.

(16) CONTINUE.

(17) Go to step (1) for the next case.

The Fortran program that follows is based on this calculation sequence. Arrays FREQ, WATE, and SCON denote the assumed natural frequencies ω_1 to ω_{NF}, the system masses m_1 to m_{NW}, and the spring constants k_1 to k_{NS}, respectively. The Fortran symbols X, FORCE, and SUM are used to denote displacement x(J) for mass m(J), inertia force F(J), and $\sum F$(J), respectively.

```
       DIMENSION FREQ(20), WATE(30), SCON(30)
1      READ 10, KODE
10     FORMAT(I5)
2      PRINT 15, KODE
15     FORMAT (/////9HCASE CODE/I9)
       READ 11, NF, (FREQ(I), I = 1, NF)
       READ 11, NW, (WATE(J), J = 1, NW)
       READ 11, NS, (SCON (K), K = 1, NS)
11     FORMAT (I5/(8F8.4))
       NWA = NW - 1
       DO 9 I = 1, NF
       SUM = 0.0
       X = 1.00
       PRINT 12, FREQ(I)
12     FORMAT (////2X9HFREQUENCY/F12.4/3X3HNO.,
       8X4HMASS, 3X15HSPRING CONSTANT,
       3X12HDISPLACEMENT, 7X5HFORCE)
       DO 8 J = 1, NWA
       FORCE = (FREQ(I) ** 2) * WATE(J) * X
       SUM = SUM + FORCE
       PRINT 13, J, WATE(J), SCON(J), X, FORCE
13     FORMAT (I6, F12.3, F18.3, F15.8, F12.4)
8      X = X - SUM/SCON(J)
       PRINT 14, NW, WATE(NW), X
14     FORMAT (I6, F12.3, F33.8)
       IF (KODE) 3, 9, 9
3      SUM = SUM + (FREQ(I) ** 2) * WATE(NW) * X
```

```
        PRINT 16, SUM
16      FORMAT (///12H TOTAL FORCE/F12.4
9       CONTINUE
        GO TO 1
        END
```

If the inputs are KODE = 1010, $m_1 = 4$, $m_2 = 3$, $m_3 = 2$, $m_4 = 1$, $m_5 = 9999.999$, $k_1 = 1$, $k_2 = 2$, $k_3 = 3$, and $k_4 = 4$, the printout for an assumed natural frequency of 0.800 rad/sec would be as follows, where DISPLACEMENT is the title for $x(J)$, the displacement of mass $m(J)$, and FORCE is the title for inertia force $F(J)$:

FREQUENCY
.8000

NO.	MASS	SPRING CONSTANT	DISPLACEMENT	FORCE
1	4.000	1.000	1.00000000	2.5600
2	3.000	2.000	−1.56000000	−2.9952
3	2.000	3.000	−1.34240000	−1.7182
4	1.000	4.000	−.62457600	−.3997
5	9999.999		.01372415	

Since $x_5 \approx 0$, we know that one natural frequency ≈ 0.800 rad/sec. If the inputs are KODE = −2020, $m_1 = m_2 = m_3 = 1$ and $k_1 = k_2 = 1$, the printout for an assumed natural frequency of 1.00 rad/sec would be as follows:

FREQUENCY
1.0000

NO.	MASS	SPRING CONSTANT	DISPLACEMENT	FORCE
1	1.000	1.000	1.00000000	1.0000
2	1.000	1.000	0.00000000	0.0000
3	1.000		−1.00000000	

TOTAL FORCE
0.0000

Since the total inertia force, $\sum_{i=1}^{3} \omega^2 m_i x_i$, equals zero, we know that one natural frequency is 1.00 rad/sec. This Fortran program will handle a disk-shaft system by letting the input arrays WATE and SCON represent the I_j-and k_{t_k}-inputs, respectively. This program can be easily modified to handle a spring-mass or disk-shaft system that is rigidly fixed at both ends, where the new equation to compute $x(J)$ is given in

Sec. 4.3. We can also program other tabular methods for calculating natural frequencies (e.g., the Myklestad and Stodda methods) in a somewhat similar fashion as this Fortran program.

For convenience of notation for an n-degree-of-freedom system, let y denote displacement x_n for a system that is fixed at one end and free at the other end, and let y denote $\sum_{j=1}^{n} F_j$ for a system that is free at both ends. Thus, for both types of systems, we will have assumed a natural frequency when the computed value of y equals zero. Thus, since the Holzer method is not an iterative technique, we have a trial-and-error problem where we want to assume a value of ω such that $y(\omega) = 0$. The Fortran program in this section computes several values of y for several corresponding, assumed values of ω. If $y(\omega_i)$ and $y(\omega_{i+1})$ have opposite signs, we know that one natural frequency lies between the values of ω_i and ω_{i+1}. We can determine this in a Fortran program by using an IF statement to test the sign of the product $(\omega_i)(\omega_{i+1})$. That is, ω_i and ω_{i+1} have opposite signs if $(\omega_i)(\omega_{i+1}) < 0$, and they have the same sign if $(\omega_i)(\omega_{i+1}) > 0$. In Sec. 4.3, we saw that we could draw a curve through the computed (y, ω)-points to find the next values of ω to assume for a Holzer-method calculation.

At this point, we want to consider how we can add statements to this Fortran program so that, after we have obtained a set of (y, ω)-values by the Holzer method for the assumed set of ω-values, we can iterate to find the natural frequencies. To do this, we shall consider some numerical methods that are classified as being roots-of-equations techniques. For this iteration, the Fortran program must first use an IF statement to locate all of the pair of successive (ω_i, ω_{i+1})-values that have opposite signs, where the nature of this IF statement was previously explained. For such a pair, let the Fortran symbols W1 and W2 represent ω_i and ω_{i+1}, respectively, and let Y1 and Y2 represent y_i and y_{i+1}. We could use the second-degree Lagrange interpolation formula to fit a second-degree polynomial through the three points (y_i, ω_i), (y_{i+1}, ω_{i+1}), and (y_{i+2}, ω_{i+2}), and thus approximate, by interpolation of this quadratic polynomial curve, the value of ω for which $y = 0$. That is, we can use Eq. (7–4), where the new trial value of ω to assume is the value of $f(x)$ for which $x = 0$, $f(x_0) = \omega_i$, $x_0 = f(\omega_i)$, etc. To simplify matters, we shall also consider simpler methods to compute a trial value of ω, which is also between ω_i and ω_{i+1} (i.e., W1 and W2) in value, and we shall assign the Fortran symbol WN to this newly calculated frequency. The *bisection method* computes WN to be the midpoint between W1 and W2, and this method converges with surprising rapidity, since the (W1, W2) interval is halved after each iteration. Thus, for this method

$WN = (W1 + W2)/2$. The *linear-slope* or *false-position method* draws a straight line between the points (Y1, W1) and (Y2, W2), and WN is determined by the intersection of this straight line with the $y = 0$ axis. Thus, for this method

$$WN = W1 - \left(\frac{W2 - W1}{Y2 - Y1}\right) Y1$$

since the slope of the straight line is $(Y2 - Y1)/(W2 - W1)$, as shown in Fig. 7–5.

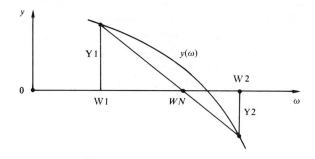

Fig. 7-5

To continue this iteration, for either the bisection or false-position method, we must now use the Holzer method to compute the value of y for which $\omega = WN$, and we shall use the Fortran symbol YN to represent this computed y-value. Suppose that $Y1 > 0$, $Y2 < 0$, and $YN > 0$. For the next iteration, we would let the new value of W1 equal WN (since Y1 and YN are both positive), and we would use the same value for W2 (since W2 is negative). We are closer to a solution since the interval (WN, W2) is smaller than the previous (W1, W2) interval, as shown in Fig. 7–5. A solution for natural frequency ω results when the interval is so small that $W2 \approx W1$. A simpler method for finding the new (W1, W2) interval is to use an IF statement to test the sign of the product (Y1)(YN) or (YN)(Y2). That is, the next interval is (W1, WN) if $(Y1)(YN) < 0$, since Y1 and YN have opposite signs; and the next interval is (WN, W2) if $(YN)(Y2) < 0$ or if $(Y1)(YN) > 0$.

7.20   SUBROUTINE Subprograms

A SUBROUTINE subprogram, which uses SUBROUTINE and RETURN types of Fortran statements and which is not available in

Fortran I, can be advantageously used in several ways to obtain a more efficient program. One of the chief advantages of subprograms (and these new statements) for engineers is that a matrix inversion (or an evaluation of the roots of a polynomial, etc.) can be programmed as a SUBROUTINE subprogram by someone with a competent mathematical background. This subprogram can be utilized by engineers in any of their programs that require the type of computation that is performed in that subprogram. Computer installations generally have SUBROUTINE subprograms that will preform interpolations, numerical integrations, numerical differentiations, matrix arithmetic, matrix inversions, curve-fit tabulated data, calculate roots of polynomials, determine eigenvalues, evaluate determinants, etc., for use in other programs. The form of a SUBROUTINE subprogram to perform the matrix multiplication $[A][B] = [C]$ could be as follows where matrix $[A]$ contains MA rows and NA columns and matrix $[B]$ contains MB rows and NB columns:

SUBROUTINE MATMUL (A, B, C, MA, NA, MB, NB)

.
.
.

RETURN
END

and where the Fortran statements between the SUBROUTINE and the RETURN statements perform the matrix multiplication $[A][B]$, as shown in Ill. Ex. 7.22. The general form of a SUBROUTINE *statement* (which is the first statement in a SUBROUTINE subprogram) is

SUBROUTINE Name (A1, A2, A3, ...)

where Name is the *symbolic name* of the subprogram (e.g., MATMUL) and A1, A2, etc. are the *arguments*. The subprogram name can consist of one to six alphabetic or numeric characters, but the first character must alphabetic. The subprogram arguments are the names of nonsubscripted, fixed- or floating-point variables, and these arguments can be considered to be inputs and outputs for the subprogram. The RETURN *statement* is the last executed statement when a SUBROUTINE subprogram is utilized or executed, and, thus, it usually precedes the END *statement*. This RETURN statement terminates a subprogram and returns control to the program that asked for use of this subprogram. Thus, a RETURN statement is used to state at what point in the subprogram to return to the main program (or the calling subprogram). There can be several RETURN statements or exits in a SUBROUTINE subprogram.

Suppose that we wished to use this matrix multiplication SUBROU-TINE subprogram to compute $[G] = [D][E][F]$, where matrices $[D], [E]$, and $[F]$ all have ten rows and ten columns. This could be done by the following Fortran program:

```
        DIMENSION D(10,10), E(10,10), F(10,10), G(10,10), H(10,10)
1       READ 10, (D(I, J), J = 1, 10), I = 1, 10)
        READ 10, (E(I, J), J = 1, 10), I = 1, 10)
        READ 10, (F(I,J), J = 1, 10), I = 1, 10)
10      FORMAT (5F15.5)
        CALL MATMUL (D, E, H, 10, 10, 10, 10)
        CALL MATMUL (H, F, G, 10, 10, 10, 10)
        PRINT 10, ((G(I, J), J = 1, 10), I = 1, 10)
        GO TO 1
        END
```

The second, third, and fourth statements read matrices $[D], [E]$, and $[F]$ into the computer, and the eighth statement prints the computed matrix $[G]$. The sixth and seventh statements use a CALL statement that uses the matrix multiplication subroutine to compute $[H] = [D][E]$ and $[G] = [H][F] = [D][E][F]$. This example program shows that a SUBROUTINE subprogram can be used as an easy means for handling any repeated calculations (e.g., matrix multiplications) that may occur in a problem or in several problems, since it only has to be programmed once in a SUBROUTINE subprogram. As shown here, a SUBROUTINE subprogram is utilized by the main program or another subprogram by the use of a CALL statement. The CALL *statement*, when executed, transfers control to the subprogram, and it specifies the name of the SUBROUTINE and its arguments. Its general form is as follows:

CALL Name (B1, B2, B3, ...)

The dummy arguments (A1, A2, etc.) of the SUBROUTINE statements are replaced by the arguments (B1, B2, etc.) given in the CALL statement when this subprogram is executed. Thus, the arguments in the CALL statement must agree in *number*, *order*, and *mode* with the corresponding arguments of the SUBROUTINE statement. The arguments in the CALL statement may be one of the following types: fixed- and floating-point constants, fixed- and floating-point variables (which may have subscripts), and arithmetic expressions. If an argument in a CALL statement is the name of an array, then the dimensions assigned to this array must be the same as the size of the corresponding array given in a DIMENSION statement in the called SUBROUTINE.

Illustrative Example 7.19.   Illustrative Example 7.12 in Sec. 7.16 shows how we can represent a graph of y versus x by tabulating N points and representing these points by arrays X and Y, where each array consists of N elements. Write a Fortran subroutine to read N points from a y-versus-x curve into a computer and to also print these values.

Solution.   The subroutine only has to read array X and array Y into the computer, and to print these values. To keep the points together, we have chosen the order of data reading to be: X(1), Y(1), X(2), Y(2), etc. A subroutine to do this is as follows:

```
            SUBROUTINE CURVRD (X, Y, N)
            DIMENSION X(20), Y(20)
            READ 10, N, (X(I), Y(I), I = 1, N)
     10     FORMAT (I5/(8F8.3))
            PRINT 11, N, (X(I), Y(I), I = 1, N)
     11     FORMAT (///11HINPUT CURVE/I6/(6F12.3))
            RETURN
            END
```

Illustrative Example 7.20.   Suppose that a computer installation had SUBROUTINE subprograms to invert a matrix, to evaluate a determinant, and to determine the eigenvalues of a matrix. The first statements for each of these three subprograms are as follows:

```
            SUBROUTINE MATINV (A, B, N)
            SUBROUTINE MATDET (A, VAL, N)
            SUBROUTINE MATEIG (A, EG, N)
```

where N is the size of input square matrix $[A]$, $[B] = [A]^{-1}$, VAL $= |A|$, and EG is a one-dimensional array that contains the N eigenvalues. Write Fortran statements to calculate $[R] = [T]^{-1}$, $D = |T|$, and the eigenvalues for square matrix $[T]$, whose size is 16, in a main Fortran program that uses these three subprograms.

Solution.   We shall utilize the following three CALL statements to do the three desired calculations by use of these three SUBROUTINE subprograms, where CHAR is the array that contains the eigenvalues.

```
            CALL MATINV (T, R, 16)
            CALL MATDET (T, D, 16)
            CALL MATEIG (T, CHAR, 16)
```

Illustrative Example 7.21.   Write SUBROUTINE subprograms to perform matrix addition, matrix subtraction, and the multiplication of a matrix by a constant.

Solution.   The following subroutine is for matrix addition. It performs the operation $[A] + [B] = [C]$, where NROW is the number of rows in $[A]$ and $[B]$ and NCOL is the number of columns in $[A]$ and $[B]$. Each element of $[C]$ is obtained using the equation $C_{ij} = A_{ij} + B_{ij}$. The second DO statement causes the calculation of C_{ij} for all column elements in a row. The first DO statement causes all of the rows to be included. Thus, all elements of $[C]$ are calculated.

```
          SUBROUTINE MATADD (A, B, C, NROW, NCOL)
          DIMENSION A(20, 20), B(20, 20), C(20, 20)
          DC 4 I = 1, NROW
          DO 4 J = 1, NCOL
     4    C(I, J) = A(I, J) + B(I, J)
          RETURN
          END
```

If the previous subprogram was modified so that the word MATSUB was substituted for the word MATADD in the first statement and if a minus sign is substituted for the plus sign in statement 4, then we would have a matrix subtraction subprogram. The following subroutine multiplies matrix $[A]$ by the scalar CNST, and the result is denoted by matrix $[C]$. Thus, $C_{ij} = (CNST)A_{ij}$ for all values of i and j.

```
          SUBROUTINE MATCON (A, C, CNST, NROW, NCOL)
          DIMENSION A(20, 20), C(20, 20)
          DO 4 I = 1, NROW
          DO 4 J = 1, NCOL
     4    C(I, J) = CNST * A(I, J)
          RETURN
          END
```

Illustrative Example 7.22.   Write a matrix multiplication SUBROUTINE subprogram.

Solution.   The following subroutine performs the operation $[A][B] = [C]$, where MA and NA are the number of rows and columns, respectively, of matrix $[A]$ and MB and NB are the number of rows and columns, respectively, of matrix $[B]$. If NA does not equal MB, the subroutine prints a statement to say that this matrix operation is not possible. The first two DO statements cause the element C_{ij} to be calculated for every row and column, respectively, since matrix $[C]$ will consist of MA rows and NB columns. The $C(I, J) = 0.0$ statement, the last DO statement, and statement 5 calculate each element C_{ij} according to the following equation:

$$C_{ij} = \sum_{k=1}^{NA} A_{ik} B_{kj}$$

```
          SUBROUTINE MATMUL (A, B, C, MA, NA, MB, NB)
          DIMENSION A(10, 10), B(10, 10), C(10, 10)
```

```
         IF (NA − MB) 3, 4, 3
3        PRINT 11
11       FORMAT (35HAN IMPOSSIBLE MATRIX MULTIPLICATION)
         GO TO 6
4        DO 5 I = 1, MA
         DO 5 J = 1, NB
         C(I, J) = 0.0
         DO 5 K = 1, NA
5        C(I, J) = C(I, J) + A(I, K) * B(K, J)
6        RETURN
         END
```

Illustrative Example 7.23.   Write a SUBROUTINE subprogram to read a matrix into the computer and, then, print this input matrix.

Solution.   The following subroutine reads matrix [A] into the computer and then prints this input matrix and its size which is NROW by NCOL, under the title "INPUT MATRIX." One row of matrix [A] is read and printed during each cycle of the DO loop.

```
         SUBROUTINE MATRED (A, NROW, NCOL)
         DIMENSION A(10, 10)
         READ 10, NROW, NCOL
10       FORMAT (2I5)
         PRINT 11, NROW, NCOL
11       FORMAT (/////12HINPUT MATRIX/2I5)
12       FORMAT (5F8.4)
13       FORMAT (5F12.4)
         DO 5 I = 1, NROW
         READ 12, (A(I, J), J = 1, NCOL)
5        PRINT 13, (A(I, J), J = 1, NCOL)
         RETURN
         END
```

7.21   A Matrix-Iteration Method Program

The matrix-iteration method for determining the natural frequencies of a vibrating system was explained in Sec. 4.9. That is, we can determine the natural frequency ω and the mode-shape column-matrix $\{x\}$ for a given vibration mode by iterating the following matrix equation (after first assuming a mode shape $\{x\}_1$) until $\{x\}_{i+1} = \{x\}_i$.

$$\{x\}_{i+1} = \omega^2[A][B]\{x\}_i \qquad (7\text{-}1)$$

To calculate the lowest natural frequency (i.e., the natural frequency of the first mode), let matrix $[A] = [\delta] = [k]^{-1}$, where $[\delta]$ is the influence coefficient matrix and $[k]$ is the stiffness or spring-constant matrix, and let $[B] = [m]$, where $[m]$ is the mass or inertia matrix. To calculate the reciprocal of the highest natural frequency [i.e., $\omega = 1/\omega_{max}$ in Eq. (7–1)], let $[A] = [m]^{-1}$ and $[B] = [k]$. To calculate the second lowest natural frequency, let $[A] = [\delta][m]$ and $[B] = [T_1]$, where $[T_1]$ is the first orthogonality transformation matrix, which is given in Sec. 4.9. The jth lowest natural frequency ω_j (where $j \geq 2$) can be calculated by iterating the equation $\{x\}_{i+1} = [D_j]\{x\}_i$, where $[D_j] = [D_{j-1}][T_{j-1}]$ and where $[T_{j-1}]$ is the $(j-1)$th orthogonality transformation matrix. That is, $[A] = [D_{j-1}]$ and $[B] = [T_{j-1}]$. Thus, we have shown that the mode number of the natural frequency and mode shape that is calculated by this matrix-iteration procedure depends upon the choice of the input matrices $[A]$ and $[B]$. The calculation sequence for the matrix-iteration-method computer program in this section is as follows, where the matrix-multiplication and matrix-read subroutines given in Ill. Exs. 7.22 and 7.23 are utilized:

(1) Read and print NE (the number of degrees of freedom and size of the matrices), ITMAX (the maximum number of iterations in case the iteration does not converge), and ϵ (the accuracy criteria for a solution).

(2) Read and print input matrices $[A]$, $[B]$, and $\{x\}_1$, where $\{x\}_1$ is the initially assumed mode shape, using the matrix-read subroutine.

(3) Calculate $[D] = [A][B]$ using the matrix-multiplication subroutine.

(4) Print matrix $[D]$.

(5) Set NITN $= 1$, where NITN is the count of the number of iterations.

(6) Calculate $\{E\} = [D]\{x\}$, which represents $\{x\}_{i+1} = [D]\{x\}_i$, using the matrix-multiplication subroutine.

(7) Normalize the first element of matrix $\{E\}$ by $\{F\} = \{E\}/E(1)$. Note that $\{X\} = \{x\}_i$ and $\{F\} = \{x\}_{i+1}$.

(8) Print iteration count NITN and matrix $\{x\}_{i+1}$.

(9) Test whether $|X(I) - F(I)| < \epsilon$ for all values of I. If so, go to step (10) since the iteration has converged. If not, go to step (13) to start the next iteration.

(10) Calculate natural frequency ω by $\omega = \sqrt{1/E(1)}$.

(11) Print the value of ω.

(12) Go to step (1) for the next case.

(13) Add unity to iteration count NITN.

(14) Set $\{x\}_i = \{x\}_{i+1}$ by $\{X\} = \{F\}$.

(15) Test (NITN − ITMAX) to see if the number of iterations has exceeded the specified maximum, ITMAX. If > 0, go to step (1) for the next case. If ≤ 0, go to step (6) to continue the iteration.

A Fortran program that is written from this calculation sequence is as follows, where EPS is a number that specifies the iteration accuracy and where the terms NITN and ITMAX are used to count the number of iterations and to prevent both a waste of expensive computer time and an endless number of cycles in case the iteration does not converge to a solution. It may be mentioned that the natural frequencies of a vibrating system may also be obtained in matrix fashion on a digital computer by use of one of the eigenvalue SUBROUTINE subprograms that are available to engineers and by a program that calculates the elements of the matrix $[m]^{-1}[k]$ whose n eigenvalues are the squares of the n natural frequencies.

```
       DIMENSION A(10, 10), B(10, 10), D(10, 10), X(10),
         E(10), F(10)
  1    READ 10, NE, ITMAX, EPS
 10    FORMAT (2I5, F8.5)
       PRINT 11, NE, ITMAX, EPS
 11    FORMAT (///6HINPUTS/2I6, F12.5)
       CALL MATRED (A, NE, NE)
       CALL MATRED (B, NE, NE)
       CALL MATRED (X, NE, 1)
       CALL MATMUL (A, B, D, NE, NE, NE, NE)
       PRINT 14
 14    FORMAT (////15HOUTPUT MATRIX D)
       DO 2 I = 1, NE
  2    PRINT 17, (D(I, J), J = 1, NE)
 17    FORMAT (5F12.7)
       NITN = 1
       PRINT 12
 12    FORMAT (////20HMODE SHAPE ITERATION/
         2 X 3 HNO.)
  7    CALL MATMUL (D, X, E, NE, NE, NE, 1)
       DO 3 I = 1, NE
  3    F(I) = E(I)/E(1)
       PRINT 15, NITN, (F(I), I = 1, NE)
```

```
15      FORMAT (I5, 5F12.7/(F17.7, 4F12.7))
        DO 4 I = 1, NE
        IF (ABSF (X(I) − F(I)) − EPS) 4, 5, 5
4       CONTINUE
        FREQ = SQRTF (1.0/E(1))
        PRINT 16, FREQ
16      FORMAT (///17HNATURAL FREQUENCY/F17.8)
        GO TO 1
5       NITN = NITN + 1
        DO 6 I = 1, NE
6       X(I) = F(I)
        IF (NITN − ITMAX) 7, 7, 1
        END
```

As for output, this program first prints the input terms NE, ITMAX, and ϵ and then prints the input matrices $[A]$, $[B]$, and $\{x\}_1$. For a case where NE = 3, ϵ = 0.00010, and

$$[A] = \begin{bmatrix} 0.3333 & 0.3333 & 0.3333 \\ 0.3333 & 1.3333 & 1.3333 \\ 0.3333 & 1.3333 & 2.3333 \end{bmatrix}$$

$$[B] = \begin{bmatrix} 4 & 0 & 0 \\ 0 & 2 & 0 \\ 0 & 0 & 1 \end{bmatrix}, \quad \{x\}_1 = \begin{Bmatrix} 1.000 \\ 2.000 \\ 4.000 \end{Bmatrix}$$

the rest of the output would be as follows:

MODE NO.	SHAPE	ITERATION	
1	1.0000000	3.0002000	4.0003000
2	1.0000000	3.1431143	4.0003643
3	1.0000000	3.1602676	4.0003765
4	1.0000000	3.1622807	4.0003788
5	1.0000000	3.1625163	4.0003793
6	1.0000000	3.1625439	4.0003793

NATURAL FREQUENCY
.45764505

Illustrative Example 7.24. Suppose that we had a matrix inversion

SUBROUTINE subprogram whose title is MATINV and an eigenvalue SUBROUTINE subprogram whose title is MATEIG. Use these two subprograms to compute the natural frequencies of a nine-degree-of-freedom system whose inertia and stiffness matrices, [m] and [k], are given.

Solution. The following Fortran program first uses the matrix-read subprogram in Ill. Ex. 7.23 to read and print matrices [m] and [k]. After inverting matrix [m] to obtain [m]$^{-1}$, this program uses the matrix-multiplication subprogram in Ill. Ex. 7.22 to compute $[H] = [m]^{-1}[k]$. The program uses the eigenvalue subprogram to determine the nine eigenvalues of matrix [H], which is represented by {GAM}. The program prints the nine natural frequency values, which were computed as the square roots of these eigenvalues.

```
      DIMENSION AM(9, 9), AK(9, 9), AN(9, 9), GAM(9)
1     CALL MATRED (AM, 9, 9)
      CALL MATRED (AK, 9, 9)
      CALL MATINV (AM, AN, 9)
      CALL MATMUL (AN, AK, H, 9, 9, 9, 9)
      CALL MATEIG (H, GAM, 9)
      DO 6 I = 1, 9
6     FREQ(I) = SQRTF (GAM(I))
      PRINT 7, (FREQ(I), I = 1, 9)
7     FORMAT (////19HNATURAL FREQUENCIES/8F9.4)
      GO TO 1
      END
```

7.22   Plotting the Output and Tabulated Data

Very often the printed results of a digital computer calculation for an engineering problem are plotted, so that the engineer can better visualize the trends of these results. The interpretation of the computed results is an important step in the solution of an engineering problem. This plotting of the computer results is often done manually, and if this plotting is performed by a non-professional subordinate who is unfamiliar with the problem, then the computer program should be designed to print the results in as clear a manner as possible in order to minimize the plotting errors. Many large computers have provisions for plotting data points on a cathode-ray tube, and a photograph of the cathode-ray tube furnishes the engineer with a graph of the computed results. As noted in Chap. 6, analog computers have the advantageous feature that the output data can be easily plotted and the inputs can be easily modified for the next case according to the results of the plotted output. Fortran IV has features that permit the efficient digital-computer programming of a Fortran subprogram that plots tabulated data. Some of these subprograms are available at computer installations for use in other programs. Some of these Fortran subprograms can plot several variables on one graph.

Let us now consider how tabulated data may be plotted using the Fortran II system for the IBM 1620 computer. We shall discuss two rather crude means for doing this, where we wish to plot the computed values of displacement x versus time t. The x-axis will be directed horizontally across the page, while the t-axis will be directed vertically and downward on the page. Thus, the user will have to turn the output sheet $90°$ to interpret this plot. We shall assume that the t_t-and x_t-values are stored in arrays T and X (i.e., the program sets $T(I) = t_t$ and $X(I) = x_t$) just before they are printed in the main program. If lack of adequate storage is a problem, we would not need the T array since the t_t-values can be recalculated in the plotting subprogram. This is because the interval for the t_t-values is constant and equals $(\Delta t)_{\text{print}}$. One of the problems involved in writing a plotting subprogram is the scaling of the plot. This is not a big problem for the vertical t-direction if one does not mind using a lot of output paper. Most printers can print up to six vertical characters per inch. Scaling is a problem in the horizontal y-direction, because we are limited to the width of the output paper and by the number of print bars. Most printers can print ten horizontal characters per inch. We shall simplify this scaling problem by normalizing the value of x to be between zero and unity. A plot of these normalized values will still allow us to see the trends of the output variation and to visually interpret the results. Thus, this plotting subprogram will first normalize the values of x. This can be done by using a DO loop to find the maximum and minimum values of x, which we shall call x_{max} and x_{min}, and then by using another DO loop to calculate the normalized values and store them in array R, by utilizing the following equation:

$$R(I) = \frac{X(I) - x_{\text{min}}}{x_{\text{max}} - x_{\text{min}}}$$

The last part of this subprogram must plot these normalized values. We shall also print the NP-values of T(I), X(I), and R(I) on the same horizontal line that we plot the T(I), R(I) data-point. This is done so that the user can determine the actual magnitudes of this plotted data. This part of the subprogram can be done by using one DO loop containing many IF and FORMAT statements to print an asterisk whose location denotes the size of quantity R(I) (and hence Y(I)) as follows:

DO 55 I = 1, NP
IF (R(I) − 0.95) 42, 42, 41

```
41      PRINT 111, T(I), X(I), R(I)
111     FORMAT (2F10.4, F10.6, 2X1H*)
        GO TO 55
42      IF (R(I) — 0.90) 44, 44, 43
43      PRINT 112, T(I), X(I), R(I)
112     FORMAT (2F10.4, F10.6, 4X1H*)
        GO TO 55
44      IF (R(I) — 0.85) 46, 46, 45
                    ⋮
55      CONTINUE
```

Note in the previous program portion how the data-point asterisk is printed in a horizontal location according to the size of quantity R(I). That is, each FORMAT statement has a different coefficient for symbol X. Two blanks are printed after the value of R (and before the * plot symbol) if $0.95 \leq R < 1.00$, four blanks are printed after the value of R if $0.90 \leq R < 0.95$, etc. We can also use slash marks in the FORMAT statements to obtain a wider vertical spacing if desired.

The following DO loop is a much shorter means for plotting the normalized values in array R. The plot that results will be of a cruder form since the plot symbol will be the digit 1 and since this plot symbol will be followed horizontally by zeroes all the way to the right side of the page. This plot symbol is plotted by printing the value of the calculated term P, whose size varies with the value of R(I). For example, if $R(I) = 0.930$, $V = 37.20$, $NV = 37$, and $P = 10^3$. Thus, statement 8 will print the value of P as 1000. If $R(I) = 0.882$, $V = 35.28$, $NV = 35$, and statement 8 will print the value of P as 100000. This example illustrates the usefulness of the truncation feature in fixed-point arithmetic.

```
        DO 8 I = 1, NP
        V = 40.0 * R(I)
        NV = V
        P = 10.0 ** (40 — NV)
8       PRINT 10, T(I), X(I), R(I), P
10      FORMAT (2F10.4, F9.6, F43.0)
```

The following SUBROUTINE subprogram uses a DO loop to compute x_{min} and x_{max} (represented by SMALX and BIGX) and another DO loop to both compute and plot R(I) by the method described in the preceding paragraph. NP is the number of tabulated data points (i.e., the number of elements in arrays X and T). This subprogram could be used to show the trends for the time-varying output problems in Secs. 7.14, 7.15, 7.24, and 7.26.

```
          SUBROUTINE PLOT (T, X, R, NP)
          DIMENSION T(30), X(30), R(30)
          BIGX = X(1)
          SMALX = X(1)
          DO 2 I = 2, NP
          IF (X(I) — BIGX) 3, 4, 4
    4     BIGX = X(I)
    3     IF (X(I) — SMALX) 5, 2, 2
    5     SMALX = X(I)
    2     CONTINUE
          DIFFX = BIGX — SMALX
          DO 8 I = 1, NP
          R(I) = (X(I) — SMALX)/DIFFX
          V = 40.0 * R(I)
          NV = V
          P = 10.0 ** (40 — NV)
    8     PRINT 10, T(I), X(I), R(I), P
   10     FORMAT (2F10.4, F9.6, F43.0)
          RETURN
          END
```

7.23   Interpolation of Tabulated Data

The data for a nonlinear damper could be furnished as a table or curve for damping coefficient c versus velocity $\dot{x}$, and the data for a nonlinear spring could be furnished as a table or curve for spring constant k versus displacement x. Similarly the applied force during a forced vibration can be more generally represented by a curve of force F_E versus time t. We would use such curves or tabulated data when the relationship is not easy (and perhaps impossible) to describe mathematically. Many problems in engineering practice require the use of tabulated or plotted data, and the tabulated data may have been obtained experimentally. Suppose that we wish to use such a curve as input for a Fortran program. Let us assume that this curve can be approximated by a series of straight lines as shown in Fig. 7–6. To approximate a curve by a series of straight lines, take a ruler (or some other straight-edge) and line this ruler along the curve. Mark the straight-line segments such that there is very little daylight between the curve and these segments. Thus, the points do not have to be equally spaced.

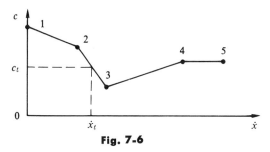

Fig. 7-6

In Ill. Exs. 7.12 and 7.13, we found that a curve can be represented in a Fortran program by two one-dimensional arrays. Let us represent a $c = f(\dot{x})$ curve by arrays C and DX, and let each array consist of NC points. That is, $c = f(\dot{x})$ is represented by $C = f(DX)$, or C(I) is the damping coefficient value at velocity DX(I). Thus, in Fig. 7-6, point 1 is represented by C(1) and DX(1), point 2 is represented by C(2) and DX(2), etc. (i.e., the array DX is arranged in ascending values of velocity). It is wise to tabulate the values of these NC points in a table. In Sec. 7.16, we found that this curve can be read into the computer by a statement such as

READ 11, NC, (DX(I), C(I), I = 1, NC)

Suppose that in a time-varying problem we have just calculated velocity $\dot{x}_t$ at time t_t and that we wish to compute the damping coefficient c_t, at time t_t, as a function of $\dot{x}_t$. That is, we want to compute, at each time interval, the value of c_t for a specific value of velocity $\dot{x}_t$. We can do it in the program by *linear interpolation*, because we have tabulated the points from the graph so that this is valid. Before we can interpolate, we must find the straight-line segment in which velocity $\dot{x}_t$ lies (e.g., look at the dotted lines in Fig. 7-6). The following Fortran instructions can be used to calculate c_t (which is represented by the symbol CT) at velocity $\dot{x}_t$ (which is represented by the symbol DXT).

```
8     DO 2 I = 1, NC
      IF (DXT − DX(I))3, 2, 2
2     CONTINUE
      I = NC
3     SLPE = (C(I) − C(I − 1))/(DX(I) − DX(I − 1))
      CT = C(I − 1) + SLPE * (DXT − DX(I − 1))
```

The first three statements of these Fortran instructions (i.e., the DO loop) scan the DX array to find the interval in which time DXT (i.e., $\dot{x}_t$) lies. This is explained in detail in Ill. Ex. 7.16 in Sec. 7.17. Thus, at statement 3 we know that velocity DXT lies between (or equals) DX(I − 1) and DX(I) in value. The I = NC statement takes care of the situation when DXT ≥ DX(NC) (i.e., the last point on the curve). Statement 3 calculates the slope m (represented by the symbol SLPE) of the straight-line segment using the following equation:

$$m = \frac{c_i - c_{i-1}}{\dot{x}_i - \dot{x}_{i-1}}$$

The last statement calculates the damping coefficient c_t using the equation

$$c_t = c_{i-1} + m(\dot{x}_t - \dot{x}_{i-1})$$

Note that the previous equations are valid if velocity DXT equals DX(I − 1) or DX(I). If velocity DXT exceeds the last velocity point DX(NC), then extrapolation (instead of interpolation) results. This extrapolation might be a valid approximation. *Extrapolation* is the evaluation of a function $f(x)$ for $x = x_k$ when the interpolated x_k-point lies outside the range of the x-data-points for $f(x)$. If velocity DXT is less in value than the first point DX(1), then an erroneous result will occur. The engineer should be notified if either condition occurs, and this is done in the following linear-interpolation subroutine which first prints a warning message along with the values of DXT, DX(1), and DX(NC). Statement 8 is the first statement of the linear interpolation procedure, and this procedure interpolates a Y = f(X) curve to approximate the value of Y (i.e., YA) at X = XA. That is, this SUBROUTINE subprogram calculates YA = f(XA) from the N points for a Y = f(X) curve. This interpolation procedure for the Y = f(X) curve is the same as that previously discussed for the C = f(DX) curve. The reason that we had programmed a linear interpolation procedure as a SUBROUTINE subprogram is because many vibration and other engineering problems require interpolations from several different curves for a specific problem. An example of such a problem is furnished in Sec. 7.24.

```
     SUBROUTINE LINTRP(X, Y, XA, N, YA)
     DIMENSION X(20), Y(20)
     IF (XA − X(1))3, 7, 6
6    IF (XA − X(N))7, 7, 3
3    PRINT 4, XA, X(1), X(N)
4    FORMAT (31HPOINT IS OUTSIDE RANGE OF
         CURVE/3F11.3)
```

```
7    DO 8 I = 1, N
     IF (XA − X(I))9, 8, 8
8    CONTINUE
     I = N
9    SLPE = (Y(I) − Y(I − 1))/(X(I) − X(I − 1))
     YA = Y(I − 1) + SLPE * (XA − X(I − 1))
     RETURN
     END
```

Frequently, the approximation of a curve by straight-line segments, as shown in Fig. 7-6, will not be a very accurate representation, especially in regions where the curvature is large. That is, it is not always easy to approximate all portions of a curve with straight-line segments. There are several interpolation methods available which represent $(n + 1)$ successive points on a curve by a polynomial of degree n and which interpolate along this polynomial curve. Thus, this means that the interpolated curve should be smooth (e.g., have no discontinuities) along these $(n + 1)$ points, and therefore caution must be exercised when a polynomial interpolation subroutine is utilized. Also note in Fig. 7-7 that the high-degree polynomial $P(x)$ fits (i.e., goes through) the tabulated points, but does not fit the function $f(x)$. Thus, a lower-degree polynomial (which has less maxima and minima) should be used. Since most curves that have regions of fairly large curvature can be adequately represented by successive quadratic polynomials, we shall be very interested in interpolation methods that represent three successive points on the curve by a quadratic polynomial. For this type of interpolation, the curve must be smooth along three successive points, and, thus, the choice of these three points can be important. When in professional practice, the reader will find many SUBROUTINE subprograms available that can perform polynomial interpolations of plotted curves and calculated or tabulated data (e.g., test results).

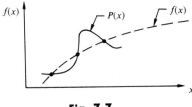

Fig. 7-7

There are several methods for the polynomial interpolation of data where these data points are equally spaced. Some of these equal-interval, polynomial-interpolation methods, which are often applied by using finite-difference tables, are the *Newton-Gregory*, *Newton-Gauss*, *Bessel*, *Stirling*, and *Everett* methods. Space does not allow us to discuss these methods, but the details of these methods can be found in numerical analysis textbooks. Besides, engineers will usually find much more value in interpolation methods where the tabulated points do not have to be equally spaced. For example, test data is often tabulated at unequal intervals. It should be mentioned that each of these polynomial interpolation formulas passes a polynomial of degree n through $(n + 1)$ equally spaced points. Thus, although the appearances of these formulas are quite different, application of any of these methods will give the same result *if* the same tabulated points are used for each interpolation. This is because each of these interpolation formulas would use the same polynomial approximation equation, but each is written in a different way to denote a difference sequence of finite-difference type of calculations. The first three terms of the forward Newton-Gauss formula can be expressed as

$$f(x_A) = f_i + r(f_{i+1} - f_i) + \left(\frac{r^2 - r}{2}\right)(f_{i+1} - 2f_i + f_{i-1})$$

which can be rewritten as the following three-point quadratic polynomial:

$$f(x_A) = f_i + \frac{r}{2}(f_{i+1} - f_{i-1}) + \frac{r^2}{2}(f_{i+1} - 2f_i + f_{i-1}) \qquad (7\text{-}2)$$

where $r = (x_A - x_i)/h$, h is the size of the equal interval (i.e., $h = x_{i+1} - x_i = x_i - x_{i-1}$), $f_{i-1} = f(x_{i-1})$, $f_i = f(x_i)$, and $f_{i+1} = f(x_{i+1})$.

For data whose points are unequally spaced, *Lagrange's interpolation formula* is very frequently used, and it is given by the following equation where the x-coordinates for the data are $x_0, x_1, x_2, \ldots,$ and x_n:

$$f(x_A) = \frac{(x_A - x_1)(x_A - x_2) \cdots (x_A - x_n)}{(x_0 - x_1)(x_0 - x_2) \cdots (x_0 - x_n)} f(x_0) \qquad (7\text{-}3)$$
$$+ \frac{(x_A - x_0)(x_A - x_2) \cdots (x_A - x_n)}{(x_1 - x_0)(x_1 - x_2) \cdots (x_1 - x_n)} f(x_1) + \cdots$$
$$+ \frac{(x_A - x_0)(x_A - x_1) \cdots (x_A - x_{n-1})}{(x_n - x_0)(x_n - x_1) \cdots (x_n - x_{n-1})} f(x_n)$$

This Lagrange formula passes a polynomial of degree n through $(n + 1)$ successive points. Point x_A can be located anywhere in the interval from x_0 to x_n. Equation (7–3) can be used for extrapolation if point x_A lies outside this range. When $n = 2$, Eq. (7–3) becomes

$$f(x) = \frac{(x - x_1)(x - x_2)}{(x_0 - x_1)(x_0 - x_2)}f(x_0) + \frac{(x - x_0)(x - x_2)}{(x_1 - x_0)(x_1 - x_2)}f(x_1) \quad \text{(7-4)}$$
$$+ \frac{(x - x_0)(x - x_1)}{(x_2 - x_0)(x_2 - x_1)}f(x_2)$$

where a quadratic polynomial is provided along the three successive points x_0, x_1, and x_2, and where we replaced $f(x_A)$ by $f(x)$ and x_A by x. We could derive Eq. (7-4), but instead let us just verify it. It can be seen that Eq. (7-4) is a polynomial of degree two in x since each of the three terms is a polynomial of degree two in x. That is,

$$f(x) = f(x_0)P_0(x) + f(x_1)P_1(x) + f(x_2)P_2(x)$$

where $P_0(x)$, $P_1(x)$, and $P_2(x)$ are all polynomials of degree two. For example, $P_0(x) = (x - x_1)(x - x_2)/(x_0 - x_1)(x_0 - x_2)$. Illustrative Example 7.25 verifies this in a more complete manner. Also when $x = x_0$, it can be seen that the first term in Eq. (7-4) equals $f(x_0)$ and the other two terms equal zero. Similarly, it can be shown that Eq. (7-4) gives $f(x) = f(x_1)$ when $x = x_1$ and $f(x) = f(x_2)$ when $x = x_2$. Equation (7-3) can be verified in the same manner that we verified Eq. (7-4).

Illustrative Example 7.25.   Write the three-point Lagrange interpolation formula in the quadratic form

$$f(x) = a_0 + a_1 x + a_2 x^2$$

Solution.   Rewrite Eq. (7-4) as follows:

$$f(x) = e(x - x_1)(x - x_2) + g(x - x_0)(x - x_2) + h(x - x_0)(x - x_1)$$

where constant $e = f(x_0)/(x_0 - x_1)(x_0 - x_2)$, constant $g = f(x_1)/(x_1 - x_0)(x_1 - x_2)$, and constant $h = f(x_2)/(x_2 - x_0)(x_2 - x_1)$. Multiplication of the factors in the terms of the $f(x)$-equation gives

$$f(x) = [ex^2 - e(x_1 + x_2)x + ex_1 x_2] + [gx^2 - g(x_0 + x_2)x + gx_0 x_2] + [hx^2 - h(x_0 + x_1)x + hx_0 x_1]$$

which can be rewritten in the following quadratic form:

$$f(x) = (e + g + h)x^2 + (-ex_1 - ex_2 - gx_0 - gx_2 - hx_0 - hx_1)x + (ex_1 x_2 + gx_0 x_2 + hx_0 x_1)$$

Illustrative Example 7.26.   Using an interval of $\pi/4$, approximate the value of $\cos(\pi/8)$ using the three-point, forward Newton-Gauss formula.

Solution.   We shall choose the three tabulated (i.e., known) points to be: $x_{i-1} = 0$, $x_i = \pi/4$, and $x_{i+1} = \pi/2$. Thus, $f_{i-1} \equiv \cos 0 = 1.000$, $f_i \equiv \cos(\pi/4) = 0.707$, and $f_{i+1} \equiv \cos(\pi/2) = 0$. The value of r is given by

$$r = \frac{x - x_i}{h} = \frac{\frac{\pi}{8} - \frac{\pi}{4}}{\frac{\pi}{4}} = -\frac{1}{2}$$

Application of Eq. (7-2) gives

$$f\left(\frac{\pi}{8}\right) \equiv \cos\left(\frac{\pi}{8}\right) = 0.707 + (-0.250)(0 - 1.000)$$
$$+ (0.125)(0 - 1.414 + 1.000) = 0.906$$

Since the correct value of $\cos(\pi/8)$ or $\cos 22.5°$ is 0.924, the error of this polynomial approximation is 0.018.

Illustrative Example 7.27. Solve Ill. Ex. 7.26 using the three-point Lagrange interpolation formula. Use the same three tabulated points.

Solution. For equal intervals, Eq. (7-4) (i.e., Lagrange's three-point formula) may be rewritten as follows:

$$f(x) = \frac{1}{2h^2}[(x - x_1)(x - x_2)f_0 - 2(x - x_0)(x - x_2)f_1 + (x - x_0)(x - x_1)f_2]$$

Since $x_0 = 0$, $x_1 = \pi/4$, and $x_2 = \pi/2$, $f_0 = 1.000$, $f_1 = 0.707$, and $f_2 = 0$. Substitution in the previous $f(x)$-equation gives

$$f\left(\frac{\pi}{8}\right) \equiv \cos\left(\frac{\pi}{8}\right) = \frac{1}{2\left(\frac{\pi}{4}\right)^2}\left[\left(-\frac{\pi}{8}\right)\left(-\frac{3\pi}{8}\right)(1.000) - 2\left(\frac{\pi}{8}\right)\left(-\frac{3\pi}{8}\right)(0.707)\right.$$
$$\left. + \left(\frac{\pi}{8}\right)\left(-\frac{\pi}{8}\right)(0)\right] = 0.906$$

which is the same result as that in Ill. Ex. 7.26. This is to be expected since the two quadratic-approximation formulas were applied over the same three tabulated points.

Illustrative Example 7.28. Write a SUBROUTINE subprogram that will perform a three-point Lagrange interpolation.

Solution. The subprogram that follows is exactly the same as the linear interpolation subprogram in this section, except for the first statement (which contains the title of the subprogram) and for some statements that follow statement 8. Thus, this section has already described the details of how the subprogram locates the interval (i.e., the three points x_{i-1}, x_i, and x_{i+1}) in which to perform the Lagrange interpolation. The four statements that start with statement 9 apply Eq. (7-4), where $X(I - 1) = x_0$, $X(I) = x_1$, $X(I + 1) = x_2$, and array Y represents the tabulated $f(x_i)$-values. The $I = N - 1$ statement insures that there is an x_{i+1}-point in array X if the DO loop goes through N cycles (i.e., it sets $X(I + 1) = X(N)$).

```
SUBROUTINE LAGTRP (X, Y, XA, N, YA)
DIMENSION X(20), Y(20)
```

```
       IF (XA − X(1)) 3, 7, 6
6      IF (XA − X(N)) 7, 7, 3
3      PRINT 4, XA, X(1), X(N)
4      FORMAT (31HPOINT IS OUTSIDE RANGE OF CURVE/
          3F11.3)
7      DO 8 I = 1, N
       IF (XA − X(I)) 9, 8, 8
8      CONTINUE
       I = N − 1
9      TERM1 = (XA − X(I)) * (XA − X(I + 1)) * Y(I − 1)/((X(I − 1)
          − X(I)) * (X(I − 1) − X(I + 1)))
       TERM2 = (XA − X(I − 1)) * (XA − X(I + 1)) * Y(I)/((X(I) − X
          (I − 1)) * (X(I) − X(I + 1)))
       TERM3 = (XA − X(I − 1)) * (XA − X(I)) * Y(I + 1)/((X(I + 1)
          − X(I − 1)) * (X(I + 1) − X(I)))
       YA = TERM1 + TERM2 + TERM3
       RETURN
       END
```

7.24   Nonlinear Vibration with Graphed Inputs

Let us now consider a more general form of the nonlinear vibration problem in Sec. 7.15. That is, we shall represent data for the nonlinear damper by a $c = f(\dot{x})$ curve and in the Fortran program by arrays C and DXC so that C = f(DXC). We shall generalize this problem further by use of a $k = f(x)$ curve (and Fortran arrays CON and XK) to represent a nonlinear or linear spring and by use of a $F_E = f(t)$ curve (and Fortran arrays F and TF) to represent an external force that has an arbitrary variation with time. Note that it takes only two (x, k)-points to represent a linear spring for which k is constant. Since this problem will require the reading in of the tabulated data for three curves and will also require the interpolation of these three curves during each time cycle, we shall use the curve-reading subprogram in Ill. Ex. 7.19 and the linear interpolation subprogram in Sec. 7.23. If the data points for these three curves are not linearly tabulated, then the Lagrange interpolation subprogram in Ill. Ex. 7.28 should be used. The following is a calculation sequence to solve this nonlinear vibration problem, whose equation of motion is $\ddot{x} + c\dot{x} + kx = F_E$ and which is solved numerically by the Euler and trapezoidal methods. Compare this calculation sequence with the similar one in Sec. 7.15. Note that the last 12 steps of both calculation sequences are esentially the same.

(1) Load and print the F_E-versus-t curve.
(2) Load and print the k-versus-x curve.
(3) Load and print the c-versus-$\dot{x}$ curve.
(4) Load and print the single-term inputs: KODE, W, t_0, x_0, $\dot{x}_0$, Δt, t_{max}, and $(\Delta t)_{print}$.
(5) Set the initial value of t_{print} equal to t_0.
(6) Calculate $F_{E_t} = f(t_t)$ by the linear interpolation subprogram.
(7) Calculate $k_t = f(x_t)$ by the linear interpolation subprogram.
(8) $F_{s_t} = k_t x_t$.
(9) Calculate $c_t = f(\dot{x}_t)$ by the linear interpolation subprogram.
(10) $F_{d_t} = c_t \dot{x}_t$.
(11) $\ddot{x}_t = (F_{E_t} - F_{d_t} - F_{s_t})(386/W)$.
(12) Test $(t_t - t_{print})$: If ≥ 0, go to (13).
$\qquad\qquad\qquad$ If < 0, go to (18).
(13) New value of print-time t_{print} = old value + $(\Delta t)_{print}$.
(14) Test (KODE): If ≥ 0, go to (17).
$\qquad\qquad\qquad$ If < 0, go to (15).
(15) Print t_t, x_t, F_{E_t}, F_{d_t}, F_{s_t}, $\dot{x}_t$, and $\ddot{x}_t$.
(16) Go to (18).
(17) Print t_t, x_t, and F_{E_t}.
(18) $t_{t+\Delta t} = t_t + \Delta t$.
(19) Save the value of $\dot{x}_t$.
(20) $\dot{x}_{t+\Delta t} = \dot{x}_t + (\Delta t)\ddot{x}_t$.
(21) $x_{t+\Delta t} = x_t + (\Delta t/2)(\dot{x}_t + \dot{x}_{t+\Delta t})$.
(22) Test $(t_{t+\Delta t} - t_{max})$: If > 0, go to (1) for next case.
$\qquad\qquad\qquad$ If ≤ 0, go to (6) for next Δt-time-cycle.

The following Fortran program was written using the previous calculation sequence, and it utilizes the curve-reading and linear-interpolation subprograms mentioned in this section. The $F_E = f(t)$ curve is represented by arrays F and TF (which have NF elements); the $k = f(x)$ curve is represented by arrays CON and XK (which have NK elements); and the $c = f(\dot{x})$ curve is represented by arrays C and DXC (which have NC elements). The symbol FE represents F_{E_t}, the symbol CONT represents k_t, and the symbol CT represents c_t. The physical terms that are represented by the other Fortran terms are given in Sec. 7.15.

```
       DIMENSION TF(20), F(20), XK(20), CON(20), DXC(20),
         C(20)
1      CALL CURVRD (TF, F, NF)
       CALL CURVRD (XK, CON, NK)
       CALL CURVRD (DXC, C, NC)
       READ 12, KODE, W, T, X, DX, DT, TMAX, DTPRT
```

```
12      FORMAT (I5/7F8.4)
        PRINT 13, KODE, W, T, X, DX, DT, TMAX, DTPRT
13      FORMAT (////6HINPUTS/I5/7F12.4///6X4HTIME,
           3X9HDISPLMENT, 5X5HFORCE)
        TPRT = T
 5      CALL LINTRP (TF, F, T, NF, FE)
        CALL LINTRP (XK, CON, X, NK, CONT)
        FS = CONT * X
        CALL LINTRP (DXC, C, DX, NC, CT)
        FD = CT * DX
        D2X = (FE − FD − FS) * 386.0/W
        IF (T − TPRT) 4, 6, 6
 6      TPRT = TPRT + DTPRT
        IF (KODE) 2, 3, 3
 2      PRINT 14, T, X, FE, FD, FS, DX, D2X
14      FORMAT (F10.5, F12.5, 5F10.4)
        GO TO 4
 3      PRINT 14, T, X, FE
 4      T = T + DT
        DXB = DX
        DX = DX + DT * D2X
        X = X + (DT/2.0) * (DX + DXB)
        IF (T − TMAX) 5, 5, 1
        END
```

7.25   Use of the Rayleigh Method for Beams of Varying Area or Material

Suppose that we wish to compute the lowest natural frequency of the beam system shown in Fig. 7–8 using the Rayleigh method. We can compute the bending moment distribution $M(x)$, from the P_1-P_2-P_3 loads, using techniques that we used in our strength of materials class. The $M(x)$-versus-x curve will have no discontinuities, but the curve of EI versus x will have two sudden discontinuities because of the two sudden changes in beam area and beam material. Thus, since $(d^2y/dx^2)_x = M_x/(EI)_x$, the d^2y/dx^2-versus-x curve will have two sudden discontinuities at the two points where the beam area suddenly changes (i.e., where the EI-versus-x curve is discontinuous).

This beam system is an example of a system whose whose deflection

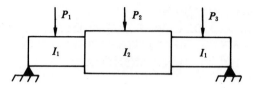

Fig. 7-8

curve $y(x)$ is difficult to estimate (i.e., it cannot be furnished easily as an equation), and we shall see that numerical or graphical integration methods must be employed in order to utilize the Rayleigh method. We could also use the graphical or numerical integration to determine the fundamental frequency of an airplane wing by treating it as a uniform-material cantilever beam whose EI-values continuously vary with x (since its cross-sectional area varies with x). Let the first part of a computer program to solve such a problem calculate d^2y/dx^2 versus x, at specific values of x, using the equation $d^2y/dx^2 = M/EI$. Once we have a tabulated set of d^2y/dx^2-versus-x values, the digital computer can use numerical integration techniques to compute the deflection distribution $y(x)$ from

$$\left(\frac{dy}{dx}\right)_x = \left(\frac{dy}{dx}\right)_{x=0} + \int_0^x \left(\frac{d^2y}{dx^2}\right) dx$$

$$y_x = y(0) + \int_0^x \left(\frac{dy}{dx}\right) dx$$

and the lowest natural frequency ω can be computed from the Rayleigh-method equation

$$\omega^2 = \frac{g \int_0^L (EI)_x \left(\frac{d^2y}{dx^2}\right)_x dx}{\int_0^L w_x y_x^2 dx}$$

where w is the weight per unit length of the beam. Note that four integrations (which can be done graphically or numerically) are needed to solve this problem.

As an alternate procedure, we can also iterate to find the deflection distribution $y(x)$ for a beam of variable EI, if it has no applied loads. That is, we can assume a tabulated beam deflection $y_1(x)$. Since the inertia loading per unit length is $\omega^2(w/g)y_1(x)$, we can use two numerical integrations to compute $V_x = \int_0^x w y_1 \, dx$ and $(d^2y/dx^2)_x = M_x/(EI)_x$ $= \int_0^x V_x \, dx/(EI)_x$, where the ω^2/g factor is omitted. We would use two more numerical integrations to compute $y'(x)$ and a new deflection distribution $y_2(x)$. The deflection distribution to use for the next iteration is the average value $[y_1(x) + y_2(x)]/2$. When $y_{i+1}(x) = y_i(x)$, we have converged to a solution for both $y''(x)$ and $y(x)$, which can be utilized in the Rayleigh-method equation to compute natural frequency ω, where two more numerical integrations are required.

Since we have shown their usefulness for applying the Rayleigh method to beams with variable EI-values, let us discuss some numerical integration techniques that might be applied. The problem here is to compute $\int f(x)\, dx$ when we have a tabulated set of $f(x)$-versus-x data. If we denote the ith tabulated point by (x_i, f_i), then the *trapezoidal method* can be expressed by the following formula, which assumes that the average value of $f(x)$ in the interval from x_i to x_{i+1} equals $(f_i + f_{i+1})/2$ and which represents the area under the $f(x)$-versus-x curve by a series of trapezoids as shown in Fig. 2–18 in Sec. 2. 14:

$$\int_{x_i}^{x_{i+1}} f(x)\, dx = \frac{1}{2}(f_i + f_{i+1})(x_{i+1} - x_i) \tag{7-5}$$

If should be noted that this equation does not require equally spaced points for its application. Also note from Fig. 2–18 that the trapezoidal method passes a first-degree polynomial (i.e., a straight line) through two successive points, and then finds the area under this straight-line segment.

The *Simpson one-third method* passes a quadratic (i.e., second-degree) polynomial through three equally spaced points, and then finds the area under this curve-segment. Denoting the interval size by h, the formula for this method is

$$\int_{x_i}^{x_{i+2}} f(x)\, dx = \frac{h}{3}(f_i + 4f_{i+1} + f_{i+2}) \tag{7-6}$$

If we pass a third-degree (i.e., cubic) polynomial through four equally spaced points and integrate the result, we will obtain the *Simpson three-eighths formlula*, which is

$$\int_{x_i}^{x_{i+3}} f(x)\, dx = \frac{3h}{8}(f_i + 3f_{i+1} + 3f_{i+2} + f_{i+3}) \tag{7-7}$$

Because of the sudden discontinuities that may occur in the integrand, we should not use either Simpson method to compute $(dy/dx)_x$, y_x, or the numerator of the Rayleigh-method formula by numerical integration. This is because the two Simpson methods assume a smooth polynomial curve through three and through four points. The Simpson methods can be used to to compute $\int_0^L w_x y_x^2\, dx$, the denominator of the

Rayleigh-method formula, because from physical considerations we know that the $y(x)$-versus-x deflection curve must be smooth (i.e., have no discontinuities). Since

$$\int_{x_1}^{x_N} f(x)\,dx = \int_{x_1}^{x_2} f(x)\,dx + \int_{x_2}^{x_3} f(x)\,dx + \cdots + \int_{x_{N-1}}^{x_N} f(x)\,dx$$

application of Eq. (7–5) gives

$$\int_a^b f(x)\,dx = \sum_{i=1}^{N-1} \frac{1}{2}\,(f_i + f_{i+1})(x_{i+1} - x_i)$$

for the trapezoidal method where $a = x_1$ and $b = x_N$. If N, the number of tabulated points, is an odd number, then

$$\int_{x_1}^{x_N} f(x)\,dx = \int_{x_1}^{x_3} f(x)\,dx + \int_{x_3}^{x_5} f(x)\,dx + \cdots + \int_{x_{N-2}}^{x_N} f(x)\,dx$$

and application of Eq. (7–6) gives

$$\int_a^b f(x)\,dx = \frac{h}{3}\,[(f_1 + 4f_2 + f_3) + (f_3 + 4f_4 + f_5) + \cdots$$
$$+ (f_{N-2} + 4f_{N-1} + f_N)]$$
$$= \frac{h}{3}\,(f_1 + 4f_2 + 2f_3 + 4f_4 + 2f_5 + \cdots$$
$$+ 2f_{N-2} + 4f_{N-1} + f_N)$$

Thus, Simpson's one-third method requires an even number of equal-size intervals or an odd number of equally spaced points. In the previous equation, note that the coefficients of f_1 and f_N equal unity and that the other coefficients alternate between 4 and 2. Illustrative Examples 7.29 and 7.30 furnish SUBROUTINE subprograms that perform a trapezoidal integration and a Simpson one-third method integration.

Illustrative Example 7.29. Suppose that we had N tabulated values of $f(x)$ versus x. Write a SUBROUTINE subprogram that utilizes the trapezoidal method to calculate the N integrals $\int_{x_1}^{x_i} f(x)\,dx$, where $i = 1$ to N.

Solution. In the following subprogram, arrays X and F contain the given x_i and $f(x_i)$ data, respectively, each of which contain NF points. Array SF contains the calculated values of $\int_{x_1}^{x_i} f(x)\,dx$. Thus, $X(4) = x_4$, $F(4) = f_4$, and $SF(4) = \int_{x_1}^{x_4} f(x)\,dx$. The third statement calculates $SF(1) \equiv \int_{x_1}^{x_1} f(x)\,dx = 0$, while statement 10 utilizes Eq. (7–5) to employ the trapezoidal method.

```
SUBROUTINE TRPINT (X, F, SF, NF)
DIMENSION X(50), F(50), SF(50)
```

```
      SF(1) = 0.0
      NFA = NF − 1
      DO 10 I = 1, NFA
10    SF(I + 1) = SF(I) + 0.50 * (F(I) + F(I + 1)) * (X(I + 1) − X(I))
      RETURN
      END
```

Illustrative Example 7.30.   Suppose in Ill. Ex. 7.29 that the tabulated values are equally spaced. Write a SUBROUTINE subprogram that utilizes Eq. (7-6) (i.e., the Simpson one-third method) to calculate

$$\int_{x_1}^{x_i} f(x)\,dx$$

Solution.   In the following Simpson's one-third formula subprogram, array FF contains the given $f(x_i)$-data, array SF contains the calculated values of $\int_{x_1}^{x_i} f(x)\,dx$, X1 is the value of x_1, DX is the value of the interval-size h, and NF is the number of tabulated $f(x_i)$-values. Arrays X and F are the values of x_i and $f(x_i)$ that correspond to array SF. This was done because the size of array SF will be about half the size of array FF. The third, fourth, and fifth statements calculate the first elements in arrays X, SF, and F, respectively. The DO loop calculates the rest of the values for these three arrays. The statement that calculates SF(J + 1) is based on the following equation:

$$\int_{x_1}^{x_{i+2}} f(x)\,dx = \int_{x_1}^{x_i} f(x)\,dx + \int_{x_i}^{x_{i+2}} f(x)\,dx$$

This statement uses Eq. (7-6), which is Simpson's one-third formula, to approximate the last integral in the previous equation. Statement 11 and those that follow are used to calculate the last integral by the Simpson three-eighths method if the number of points is even, as determined by the IF statement located just before statement 11.

```
      SUBROUTINE SIMINT (FF, SF, X, F, X1, DX, NF)
      DIMENSION FF(50), SF(30), X(30), F(30)
      X(1) = X1
      SF(1) = 0.0
      F(1) = FF(1)
      J = 1
      NFA = NF − 2
      DO 10 I = 1, NFA, 2
      X(J + 1) = X(J) + 2.0 * DX
      SF(J + 1) = SF(J) + (DX/3.0) * (FF(I) + 4.0 * FF(I + 1)
     + FF(I + 2))
      F(J + 1) = FF(I + 2)
10    J = J + 1
      IF (I − NFA) 11, 12, 11
```

11 X(J + 1) = X(J) + DX
  F(J + 1) = FF(NF)
  SF(J + 1) = SF(J − 1) + (0.375 ∗ DX) ∗ (FF(NF − 3)
   + 3.0 ∗ FF(NF − 2) + 3.0 ∗ FF(NF − 1) + FF(NF))
12 RETURN
  END

7.26 Vibration of One-Dimensional Distributed-Mass Systems

In Chap. 5, we found that the one-dimensional wave equation

$$\frac{\partial^2 \phi}{\partial x^2} = \frac{1}{a^2} \frac{\partial^2 \phi}{\partial t^2}$$

mathematically represents several engineering systems. It represents one-dimensional sound waves and the voltage and current flow in a lossless electric transmission line. It also represents the following four distributed-mass vibration systems:

(a) Vibration of an elastic string, where ϕ is the transverse displacement from the equilibrium position, a^2 equals T/m, T is string tension, and m is the mass per unit length.

(b) Torsional vibration of an elastic shaft, where ϕ is the angle of twist of the shaft as referenced to the equilibrium position, a^2 equals G/m, G is the shear modulus, and m is the mass per unit volume.

(c) Vibration of an elastic bar, where ϕ is the longitudinal displacement from the equilibrium position, a^2 equals E/m, E is the modulus of elasticity, and m is the mass per unit volume.

(d) Tidal waves in a long channel, where ϕ is the longitudinal displacement of the fluid, a equals $\sqrt{gh}$ and is the speed at which the tidal waves are propagated, g is the gravitational acceleration, and h is the depth of the fluid in the channel before it was disturbed.

We also saw that the one-dimensional wave equation can be numerically approximated by the following equation, where $\Delta t = \Delta x/a$:

$$\phi(x, t + \Delta t) = \phi(x + \Delta x, t) + \phi(x - \Delta x, t) - \phi(x, t - \Delta t)$$

To make this problem more general, we shall use four input curves. Note that we only need two curve points if a specific term is constant in value. We will have two curves for the initial values, $\phi(x, -\Delta t)$ versus x and $\phi(x, 0)$ versus x. We shall load $\phi(x, -\Delta t)$, $\phi(x, 0)$, and x

in arrays YB, Y, and X, respectively, in the Fortran program, where the x-values must start at $x = 0$ and must be spaced an interval Δx apart. We will also have two curves for the boundary values (i.e., the time-varying values at both ends of the system). We shall use the arrays Y0 and T0 to represent $y(0, t)$ versus time and the arrays YL and TL to represent $y(L, t)$ versus time, where L is the length of the one-dimensional system (e.g., string, rod, bar, etc.). The calculation sequence for the Fortran program in this section is as follows, where the curve-read and linear interpolation subroutines from Ill. Ex. 7.19 and Sec. 7.23 are utilized. If the data points for the curves are not linearly tabulated, we may wish to use the Lagrange interpolation subroutine in Ill. Ex. 7.28.

(1) Load and print the $\phi(0, t)$-versus-time curve.
(2) Load and print the $\phi(L, t)$-versus-time curve.
(3) Load and print the $\phi(x, -\Delta t)$-versus-x curve.
(4) Load and print the $\phi(x, 0)$-versus-x curve.
(5) Load and print the single-term inputs: Δx, a, $(\Delta t)_{print}$, and t_{max}.
(6) Print the x (i.e., horizontal-displacement) values.
(7) Set both t_t and t_{print} equal to zero.
(8) $\Delta t = \Delta x / a$.
(9) Test $(t_t - t_{print})$: If ≥ 0, go to (10).
        If < 0, go to (13).
(10) New value of t_{print} = old value + $(\Delta t)_{print}$.
(11) Print time t_t and the $\phi(x, t_t)$-values.
(12) Test $(t_t - t_{max})$: If ≥ 0, go to (1) for next case.
        If < 0, go to (13) for next time cycle.
(13) $t_{t+\Delta t} = t_t + \Delta t$.
(14) Calculate $\phi(0, t_{t+\Delta t})$ by the linear interpolation subprogram.
(15) Calculate $\phi(L, t_{t+\Delta t})$ by the linear interpolation subprogram.
(16) From $x = \Delta x$ to $x = (L - \Delta x)$, calculate $\phi(x, t + \Delta t) = \phi(x + \Delta x, t) + \phi(x - \Delta x, t) - \phi(x, t - \Delta t)$.
(17) Put the $\phi(x, t)$-values in the YB array and the $\phi(x, t + \Delta t)$-values in the Y array.
(18) Go to step (9).

In the following Fortran program, arrays YB, Y, and YA represent the $y(x, t - \Delta t)$-, $y(x, t)$-, and $y(x, t + \Delta t)$-values, respectively, and the symbols DX, A, DT, DTPRT, TMAX, T, TPRT, and N represent Δx, a, Δt, $(\Delta t)_{print}$, t_{max}, t_t, t_{print}, and the number of x-points, respectively.

```
        DIMENSION T0(20), Y0(20), TL(20), YL(20), X(20),
            YB(20), Y(20), YA(20)
```

```
1      CALL CURVRD (T0, Y0, N0)
       CALL CURVRD (TL, YL, NL)
       CALL CURVRD (X, YB, N)
       CALL CURVRD (X, Y, N)
       READ 12, DX, A, DTPRT, TMAX
12     FORMAT (4F8.4)
       PRINT 13, DX, A, DTPRT, TMAX
13     FORMAT (////6HINPUTS/4F12.4////)
       PRINT 14, (X(I), I = 1, N)
14     FORMAT (23HHORIZONTAL DISPLACEMENT/(F20.4,
          6F10.4))
       PRINT 15
15     FORMAT (//6X4HTIME, 8X11HDEFLECTIONS)
       T = 0.0
       TPRT = 0.0
       DT = DX/A
8      IF (T − TPRT) 3, 2, 2
2      TPRT = TPRT + DTPRT
       PRINT 16, T, (Y(I), I = 1, N)
16     FORMAT (7F10.4/(F20.4, 6F10.4))
       IF (T − TMAX) 3, 1, 1
3      T = T + DT
       CALL LINTRP (T0, Y0, T, N0, YA(1))
       CALL LINTRP (TL, YL, T, NL, YA(N))
       NA = N − 1
       DO 4 I = 2, NA
4      YA(I) = Y(I + 1) + Y(I − 1) − YB(I)
       DO 5 I = 1, N
       YB(I) = Y(I)
5      Y(I) = YA(I)
       GO TO 8
       END
```

PROBLEMS

7.1. Write each of the numbers -731, 231, 2.6×10^{-9}, and 7.8×10^{107} as a fixed-point constant and as a floating-point constant. State which are impossible to do and furnish a reason.

7.2. State why each of the following numbers are not proper floating-point constants: -718; $815{,}612.8$; $81E7$; $1.8E + 105$; and $7E - 8$.

7.3. State whether the following numbers are fixed-point constants or floating-point constants: $+85$; 0.0; $6{,}120$; -71; 81.5; $41E5$; 61507; $-5.8E4$; 0; $+4.8E - 8$; and $7{,}185.0$. If any of these numbers fall in neither category, state why.

7.4. State which of the following symbols can be used to represent a fixed-point variable and which can represent a floating-point variable: A4, IB, X$A, KN2, 3PA, N, D, K4A6, X $-$ 8, Y3.7, and MAXIMUM. If any of the preceding symbols cannot represent either variable, state why.

7.5. Write a Fortran expression to represent each of the following mathematical expressions:

a) $\dfrac{C}{D} - \dfrac{4A}{R}$

b) $7.8(X + Y^3)^{1.81}$

c) $\dfrac{XY - AB}{4.8E^4}$

d) $7X + \dfrac{13Y}{Z - 9.81}$

e) $\dfrac{4X + 9}{C + D^2}$

7.6. For each of the following Fortran expressions (if they are proper), write the corresponding mathematical expression. Also state the sequence of the calculations, and rewrite any of the Fortran expressions if any of them contain an unnecessary pair of parentheses. If any Fortran expression is improper, state why.

(a) $(X + Y)(A + 6.5)$

(b) $(C + 8.71) * (D/E)$

(c) $(X * Y + Z) ** (T + 9)$

(d) $X - 8.71 * R ** (I + 7)$

(e) $7.85 * X + P - K * T$

(f) $A - P * (R/(F/7.87)) * T ** (KP - 11)$

(g) $(K * IA) - (NPA ** C)$

(h) $S * (P - (T - R))$

7.7. Which of the following Fortran expressions represents the following mathematical expression? Write the mathematical expressions for the other incorrect Fortran expressions (if they are validly written). State the errors in the invalid Fortran expressions.

$$\frac{6X - 8}{Y + Z} - A^{I+5}$$

(a) $(6*X-8)/(Y+Z)-A**(I+5)$
(b) $(6.0*X-8.)/Y+Z-A**(I+5)$
(c) $(6.0*X-8.0)/(Y+Z)-A**(I+5)$
(d) $(6.0*X-8.0)/(Y+Z)-A**(I+5.0)$
(e) $6.0*(X-8.0)/(Y+Z)-A**(I+5)$
(f) $(6.0X-8.0)/(Y+Z)-A**(I+5)$
(g) $(6.0*X-8.0)/(Y+Z)-A**I+5$

7.8. Identify the error or errors in each of the following Fortran arithmetic statements.

(a) $Z=4*KR**B$
(b) $R+T=5*IN-KS**N$
(c) $7=A*B**6.8$
(d) $A=R*T**2-7*T**B$
(e) $-D=8.5*X+9.8*Y**2$
(f) $T=6.8*Y+(X-A)(Y-B)$
(g) $R=5.1*(Y-(A+B)$

7.9. State the value of Y or M stored for each of the following arithmetic statements if $J=3$, $K=8$, $L=5$, $B=2.0$, and $D=5.8$.

(a) $Y=K/J-L$
(b) $M=B*D**(L-J)$
(c) $Y=(K+L)/J$
(d) $M=K/J+L/J$
(e) $Y=D*B**(K/J)$
(f) $M=K+5*(L/J)$

7.10. Write the necessary Fortran statements to read quantities A, B, C, and D from an IBM card and to punch these same quantities on an IBM card for eventual printout. Read and punch these quantities in non-exponential form. For the input card, let each quantity occupy eight card columns with three digits after the decimal point. For the punched output card, let each quantity occupy twelve card columns with three digits after the decimal point. Let 12 be the number of the input FORMAT statement and 13 be the number of the output FORMAT statement.

7.11. Using only one READ statement, write the necessary Fortran statements to read the quantities A, B, and C from one IBM card and the quantities X, Y, and Z from another IBM card. Let each of these six quantities occupy seven card columns with two digits after the decimal point, and use the nonexponential form. Let 87 be the number of the FORMAT statement, and do not use the / symbol in this statement.

7.12. Write the necessary Fortran statements to print the quantities K, L, X, Y, and Z on one line. Let K and L occupy six print spaces,

X and Y occupy nine print spaces with two digits after the decimal point, and Z occupy fourteen spaces with three digits after the decimal point. Let Z be the only term printed in exponential form. Let 17 be the number of the FORMAT statement.

7.13. Let quantity D equal -814.7183. State how the output would appear if this quantity were printed by each of the following specifications: F9.4, F11.6, F8.3, F7.2, F8.4, F7.4, F5.3, and F5.0.

7.14. State whether the following forms can be properly used to represent the input quantity A, where $A = 348.5$, if this quantity is punched in exponential form on an IBM card as follows: 3.485E2, $34.85 + 1$, $348.5E + 00$, 3.485E02, or $3.485 + 01$.

7.15. Same as Prob. 7.12, except print quantities K and L on one line and X, Y, and Z on another line. Also write the FORMAT statement so that there will be two blank lines between these two printed lines.

7.16. Write the necessary Fortran statements to successively print the quantities MA, MB, IA, IB, A, B, C, D, E, F, G, and H on six lines. Print two quantities on each line, and let 31 be the number of the FORMAT statement. Use the specification 16 for each fixed-point quantity and the specification F11.3 for each floating-point quantity.

7.17. Write the necessary Fortran statements to print the following words, starting at the left-hand margin: QUANTITY A IS TOO LARGE. Let 32 be the number of the FORMAT statement.

7.18. Same as Prob. 7.12 except print the title OUTPUT DATA above the numerical values. Let this title start at the eleventh space from the left-hand margin.

7.19. Write the proper IF statement that will cause the program to go to statement 51 if $(A^2 - D) = R^{1.8}$, to statement 54 if $(A^2 - D) > R^{1.8}$, and to statement 47 if $(A^2 - D) < R^{1.8}$.

7.20. Write Fortran statements (including an IF statement) that will calculate the term R as follows:

$$R = 5.8X^{3.1} \qquad \text{if } X < (Y^2 - 7.2Z)$$
$$R = 3.7X^{1.2} \qquad \text{if } X = (Y^2 - 7.2Z)$$
$$R = -7.8X^{-0.7} \qquad \text{if } X > (Y^2 - 7.2Z)$$

7.21. Write a complete Fortran program to perform the following operations:

(a) Read quantities A and B using an F8.3 specification for each quantity.

(b) If $A < B$, print the words A IS LESS THAN B, and go to step (a) for the next case.

(c) If $A \geq B$, calculate quantity D, where $D = 4.8A^B$.

(d) Print quantities A, B, and D using an F12.3 specification for each quantity.

(e) Go to (a) for the next case.

7.22. Write enough Fortran instructions to perform the following calculations:

$$R = \left[\sum_{i=1}^{N} (A_i - B_i^2) \right]^{0.282}$$

7.23. Write a complete Fortran program to do the calculations of Prob. 7.22. Assume that each array consists of at most 70 elements.

7.24. Write enough Fortran instructions to perform the following calculations:

$$X_i = A_i B_i^2 \qquad \text{if } A_i < B_i$$
$$X_i = 0 \qquad \text{if } A_i = B_i$$
$$X_i = A_i^2 B_i \qquad \text{if } A_i > B_i$$
$$R = \sum_{i=1}^{N} X_i e^{B_i}$$

7.25. Write enough Fortran instructions to perform the following calculations:

$$\sum_{j=1}^{30} \sum_{i=1}^{40} (Y_{ij} \sin X_i + R_{ij})$$

7.26. Write a complete Fortran program to do the calculations of Prob. 7.25.

7.27. Write a complete Fortran program that will determine and print the maximum and minimum values of array A, which contains 200 elements.

7.28. Write a complete Fortran program to perform the following calculations using an arithmetic statement function:

$$X = A^2 - B^{0.86}/A$$
$$Y = C^2 - D^{0.86}/C$$
$$Z = E^2 - F^{0.86}/E$$

7.29. Write a complete Fortran program to perform the following matrix calculations and print the results. Make use of the subprograms in Sec. 7.20 All input matrices are square matrices of size N.

$$[T] = d[E] - [F] + [D][E]$$
$$[P] = [D][T][F] + [T][E]$$

7.30. Write and check out a Fortran program that computes the exact solution (i.e., the timewise displacement variation) for the damped spring-mass system in Fig. 1–19. Include the critically damped, overdamped, and underdamped cases.

7.31. Write and check out a Fortran program that computes the exact steady-state solution (i.e., timewise variation of the steady-state displacement) for the damped spring-mass system in Fig. 2–3 that that has an external force $F \sin \omega t$.

7.32. Modify the Fortran program in Sec. 7.15, that solves a nonlinear vibration problem, to use the midpoint-slope method to calculate $\dot{x}_{t+\Delta t}$ for $t \geq (t_0 + 2\Delta t)$. Use the Euler method to calculate the second starting value $\dot{x}(t_0 + \Delta t)$.

7.33. Same as Prob. 7.32, except use the Adams method instead of the midpoint-slope method.

7.34. Write a Fortran program to numerically solve the nonlinear vibration problem in Sec. 7.15 by using the second-order-accuracy Runge-Kutta method to compute both $\dot{x}_{t+\Delta t}$ and $x_{t+\Delta t}$.

7.35. Write a Fortran program to numerically solve the nonlinear, two-degree-of-freedom vibration problem in Sec. 3.11 using the Euler and trapezoidal methods. Use the calculation sequence given in that section.

7.36. Same as Prob. 7.35, except use the midpoint-slope method to compute velocities $\dot{x}_{1_{t+\Delta t}}$ and $\dot{x}_{2_{t+\Delta t}}$ for $t \geq (t_0 + 2\Delta t)$. Use the Euler method to compute these two velocities at time $(t_0 + \Delta t)$.

7.37. Same as Prob. 7.36, except use the Adams method instead of the midpoint-slope method.

7.38. Modify the Fortran program in Sec. 7.18 to compute $r^2(\text{M.F.})$ versus r and ζ.

7.39. Write and check out a Fortran program to compute transmissibility TR and the phase angle for vibration isolation versus r and ζ.

7.40. Modify the Holzer-method program in Sec. 7.19 so that it will also perform the calculations for a system that is rigidly supported at both ends. Use Ill. Exs. 4.4 to 4.6 as check cases for the modified program.

7.41. Expand the Holzer-method program in Sec. 7.19 so that it will iterate to find the zero roots of $y(\omega)$ by using the bisection method. Use Ill. Exs. 4.5 and 4.16 as check cases.

7.24. Expand the Holzer-method program in Sec. 7.19 so that it will iterate to find the zero roots of $y(\omega)$ by using the linear-slope or false-position method. Use Ill. Exs. 4.5 and 4.16 as check cases.

7.43. Write and check out a Fortran program to calculate the natural frequencies of a multiple-degree-of-freedom vibration system by using the Myklestad method.

7.44. Use the matrix-iteration method program in Sec. 7.21 to solve Ill.

Exs. 4.15 to 4.17 and Probs. 4.36, 4.37, 4.42, 4.43, and 4.44. Verify your results by using the Holzer-method program in Sec. 7.19.

7.45. Write and check out a Fortran SUBROUTINE subprogram that will invert a matrix of arbitrary size N by use of the Gauss-Jordan method. Use Eq. (4–30) to iterate your result.

7.46. Complete and check out the first plotting subprogram that is described in Sec. 7.22.

7.47. Check out the second plotting subprogram in Sec. 7.22 by using this subprogram in a Fortran program to plot $\sin x$, $\cos x$, and e^x from 0 to π for 51 equally spaced values of x.

7.48. Write and check out a four-point Lagrange interpolation subprogram. For these four points, let there be two tabulated points about the interpolated point if $X(2) < XA < X(N - 1)$.

7.49. Modify the Fortran program in Sec. 7.24 to use the three-point Lagrange interpolation subprogram in Ill. Ex. 7.28.

7.50. Modify the nonlinear vibration program in Sec. 7.24 to use the midpoint slope method to calculate $\dot{x}_{t+\Delta t}$ for $t \geq (t_0 + 2\Delta t)$. Use the Euler method to calculate the second starting value $\dot{x}(t_0 + \Delta t)$.

7.51. Same as Prob. 7.50, except use the Adams method instead of the midpoint-slope method.

7.52. Write a Fortran subprogram that calculates the roots of a polynominal using the Newton-Raphson method and the quadratic formula (for the last two roots). Write and incorporate a synthetic division subprogram to ease the Newton-Raphson calculations, as shown in Sec. 3.3.

7.53. Write a Fortran program to numerically calculate $\int \sin x \, dx$ from 0 to π for 11 and 21 equally spaced points (i.e., 10 and 20 intervals) using the trapezoidal- and Simpson-method subprograms in Ill. Exs. 7.29 and 7.30. Calculate the error and the percent error of all numerical calculations (i.e., for all upper limits of the running integral).

7.54. Write a Fortran program to compute the lowest natural frequencies of nonuniform beam systems by use of the Rayleigh method. Use the suggestions given in Sec. 7.25.

7.55. Solve Ill. Ex. 5.6 using the Fortran program in Sec. 7.26 and $\Delta x = 2.00$, 1.00, 0.500, 0.250, and 0.125 ft. Comment on the convergence of your results.

7.56. Modify the Fortran program in Sec. 7.26 so that it interpolates two curves to obtain the YB and Y initial value arrays. This interpolation is done with respect to the values in the X array, which is to be computed from inputs $XA \equiv X(1)$, Δx, and length L. This makes it easier to solve a problem for several values of Δx. Also modify this program so that it utilizes the three-point Lagrange interpolation subprogram in Ill. Ex. 7.28, rather than linear interpolation.

Appendix A-1

LAPLACE TRANSFORMS OF OPERATIONS

No.	$f(t)$ or $f(x, t)$	$L\{f(t)\} = F(s)$ or $L\{f(x,t)\} = F(x,s)$
1	$f(t)$	$\int_0^\infty e^{-st} f(t)\, dt = F(s)$
2	$\dfrac{df(t)}{dt}$	$sF(s) - f(0+)$
3	$\dfrac{d^n f}{dt^n}$	$s^n F(s) - [s^{n-1}f(0+) + s^{n-2}f'(0+) + \ldots + f^{(n-1)}(0+)]$
4	$\int_a^t f(t)\, dt$	$\dfrac{1}{s} F(s) + \dfrac{1}{s}\int_a^0 f(t)\, dt$
5	$\int_0^t f(t)\, dt$	$\dfrac{1}{s} F(s)$
6	$\int_0^t \int_0^T f(T)\, dT\, dt$	$\dfrac{1}{s^2} F(s)$
7	$f(t-a)u(t-a)$	$e^{-as} F(s)$
8	$tf(t)$	$-\dfrac{d}{ds} F(s)$
9	$\dfrac{1}{t} f(t)$	$\int_s^\infty F(s)\, ds$
10	$e^{at} f(t)$	$F(s-a)$
11	$a_1 f_1(t) + a_2 f_2(t)$	$a_1 F_1(s) + a_2 F_2(s)$
12	$\int_0^t f_1(t-T)f_2(T)\, dT$	$F_1(s) F_2(s)$

No.	$f(t)$ or $f(x, t)$	$L\{f(t)\} = F(s)$ or $L\{f(x, t)\} = F(x, s)$
13	$f(t) = f(t + T)$	$\left(\dfrac{1}{1 - e^{-Ts}}\right)\displaystyle\int_0^T e^{-st}f(t)\,dt$
14	$\lim\limits_{t \to 0} f(t)$	$\lim\limits_{s \to \infty} sF(s)$
15	$\lim\limits_{t \to \infty} f(t)$	$\lim\limits_{s \to 0} sF(s)$
16	$f(at)$	$\dfrac{1}{a}F\left(\dfrac{s}{a}\right)$
17	$f(x, t)$	$\displaystyle\int_0^\infty e^{-st}f(x, t)\,dt = F(x, s)$
18	$\dfrac{\partial f(x, t)}{\partial t}$	$sF(x, s) - f(x, 0+)$
19	$\dfrac{\partial^n f(x, t)}{\partial t^n}$	$s^n F(x, s) - \left[s^{n-1}f(x, 0+) + s^{n-2}\dfrac{\partial f(x, 0+)}{\partial t}\right.$ $\left. + \ldots + \dfrac{\partial^{n-1}f(x, 0+)}{\partial t^{n-1}}\right]$
20	$\dfrac{\partial f(x, t)}{\partial x}$	$\dfrac{\partial F(x, s)}{\partial x}$
21	$\dfrac{\partial^n f(x, t)}{\partial x^n}$	$\dfrac{\partial^n F(x, s)}{\partial x^n}$
22	$\dfrac{\partial^2 f(x, t)}{\partial x \partial t}$	$s\dfrac{\partial F(x, s)}{\partial x} - \dfrac{\partial f(x, 0+)}{\partial x}$

Appendix A-2

LAPLACE TRANSFORMS OF SPECIFIC FUNCTIONS

No.	$f(t)$	$L\{f(t)\} = F(s)$
1	1 or $u(t)$	$\frac{1}{s}$
2	$u(t-a)$	$\frac{1}{s} e^{-as}$
3	$u'(t-a)$	e^{-as}
4	$u''(t-a)$	se^{-as}
5	e^{at}	$\dfrac{1}{s-a}$
6	$\sin at$	$\dfrac{a}{s^2+a^2}$
7	$\cos at$	$\dfrac{s}{s^2+a^2}$
8	$\sinh at$	$\dfrac{a}{s^2-a^2}$
9	$\cosh at$	$\dfrac{s}{s^2-a^2}$
10	t	$\dfrac{1}{s^2}$
11	t^2	$\dfrac{2}{s^3}$
12	t^n	$\dfrac{n!}{s^{n+1}}$ where $n = 0, 1, 2, 3, \ldots$
13	$\dfrac{e^{-bt} - e^{-at}}{a-b}$	$\dfrac{1}{(s+a)(s+b)}$
14	$\dfrac{be^{-bt} - ae^{-at}}{b-a}$	$\dfrac{s}{(s+a)(s+b)}$
15	te^{at}	$\dfrac{1}{(s-a)^2}$
16	$t^n e^{at}$	$\dfrac{n!}{(s-a)^{n+1}}$ where $n = 1, 2, 3, \ldots$

No.	$f(t)$	$L\{f(t)\} = F(s)$
17	$e^{at} \sin bt$	$\dfrac{b}{(s-a)^2 + b^2}$
18	$e^{at} \cos bt$	$\dfrac{s-a}{(s-a)^2 + b^2}$
19	$t \sin at$	$\dfrac{2as}{(s^2 + a^2)^2}$
20	$t \cos at$	$\dfrac{s^2 - a^2}{(s^2 + a^2)^2}$
21	$t \sinh at$	$\dfrac{2as}{(s^2 - a^2)^2}$
22	$t \cosh at$	$\dfrac{s^2 + a^2}{(s^2 - a^2)^2}$
23	$J_0(t)$	$\dfrac{1}{\sqrt{s^2 + 1}}$
24	$J_1(t)$	$\dfrac{1}{(s + \sqrt{s^2 + 1})\sqrt{s^2 + 1}}$
25	$a^n J_n(at)$	$\dfrac{(\sqrt{s^2 + a^2} - s)^n}{\sqrt{s^2 + a^2}}$ where $n > -1$
26	$erf\left(\dfrac{t}{2a}\right)$	$\dfrac{1}{s} e^{a^2 s^2} [1 - erf\,(as)]$ where $a > 0$
27	$1 - erf\left(\dfrac{a}{2\sqrt{t}}\right)$	$\dfrac{1}{s} e^{-a\sqrt{s}}$ where $a \geq 0$
28	Rectangular pulse, magnitude M, duration a	$\dfrac{M}{s}(1 - e^{-as})$
29	Triangular pulse, magnitude M, duration $2a$	$\dfrac{M}{as^2}(1 - e^{-as})^2$
30	Sawtooth pulse, magnitude M, duration a	$\dfrac{M}{as^2}[1 - (as + 1)e^{-as}]$
31	Sinusoidal pulse, magnitude M, duration π/a	$\dfrac{Ma}{s^2 + a^2}(1 + e^{-\pi s/a})$

Appendix A-3

BIBLIOGRAPHY

Anderson, R. A. *Fundamentals of Vibrations*. New York: The Macmillan Company, 1967.

Bellman, R. *Perturbation Techniques in Mathematics, Physics, and Engineering*. New York: Holt, Rinehart, and Winston, Inc., 1964.

Bendat, J. S. *Principles and Applications of Random Noise Theory*. New York: John Wiley and Sons, Inc., 1958.

Biggs, J. M. *Introduction to Structural Dynamics*. New York: McGraw-Hill Book Co., Inc., 1965.

Chen, Y. *Vibrations: Theoretical Methods*. Reading, Massachusetts: Addison-Wesley Publishing Co., Inc., 1966.

Cheng, D. K. *Analysis of Linear Systems*. Reading, Massachusetts: Addison-Wesley Publishing Co., Inc., 1959.

Church, A. H. *Mechanical Vibrations*. New York: John Wiley and Sons, Inc., 1963.

Cramer, H. *The Elements of Probability Theory*. New York: John Wiley and Sons, Inc., 1955.

Crandall, S. H. *Random Vibration*. New York: John Wiley and Sons, Inc., 1959.

———— and W. D. Mark. *Random Vibration in Mechanical Systems*. New York: Academic Press, Inc., 1963.

Crede, C. E. *Vibration and Shock Isolation*. New York: John Wiley and Sons, Inc., 1951.

————. *Shock and Vibration Concepts in Engineering Design*. Englewood Cliffs, New Jersey: Prentice-Hall, Inc., 1965.

Den Hartog, J. P. *Mechanical Vibrations*. New York: McGraw-Hill Book Co., Inc., 1956.

Freberg, C. R. and E. N. Kemler. *Elements of Mechanical Vibration*. New York: John Wiley and Sons, Inc., 1949.

Gennaro, J. J. *Computer Methods in Solid Mechanics*. New York: The Macmillan Company, 1965.

Gere, J. M. and W. Weaver. *Matrix Algebra for Engineers*. Princeton, New Jersey: D. Van Nostrand Co., Inc., 1965.

370

Goldstein, S. *Classical Mechanics*. Reading, Massachusetts: Addison-Wesley Publishing Co., Inc., 1959.

Haberman, C. M. *Engineering Systems Analysis*. Columbus, Ohio: Charles E. Merrill Books, Inc., 1965.

————. *Use of Digital Computers for Engineering Applications*. Columbus, Ohio: Charles E. Merrill Books, Inc., 1966.

Hamming, R. W. *Numerical Methods for Scientists and Engineers*. New York: McGraw-Hill Book Co., Inc., 1962.

Hansen, H. M. and P. F. Chenea. *Mechanics of Vibration*. New York: John Wiley and Sons, Inc., 1956.

Harris, C. M. and C. E. Crede. *Shock and Vibration Handbook*. New York: McGraw-Hill Book Co., Inc., 1961.

Hayashi, C. *Nonlinear Oscillations in Physical Systems*. New York: McGraw-Hill Book Co., Inc., 1965.

Henrici, Peter. *Elements of Numerical Analysis*. New York: John Wiley and Sons, Inc., 1964.

Herriott, J. G. *Methods of Mathematical Analysis and Computation*. New York: John Wiley and Sons, Inc., 1963.

Hildebrand, F. B. *Introduction to Numerical Analysis*. New York: McGraw-Hill Book Co., Inc., 1956.

Holowenko, A. R. *Dynamics of Machinery*. New York: John Wiley and Sons, Inc., 1955.

Jacobsen, L. S. and R. S. Ayre. *Engineering Vibrations*. New York: McGraw-Hill Book Co., Inc., 1958.

James, M. L., G. M. Smith, and J. C. Wolford. *Analog and Digital Computer Methods in Engineering Analysis*. Scranton, Pennsylvania: International Textbook Co., 1964.

Jenness, R. R. *Analog Computation and Simulation: Laboratory Approach*. Boston, Massachusetts: Allyn and Bacon, Inc., 1965.

Johnson, C. L. *Analog Computer Techniques*. New York: McGraw-Hill Book Co., Inc., 1956.

Karman, T. von and M. A. Biot. *Mathematical Methods in Engineering*. New York: McGraw-Hill Book Co., Inc., 1940.

Karplus, W. J. and W. J. Soroka. *Analog Methods in Computation and Simulation*. New York: McGraw-Hill Book Co., Inc., 1959.

Korn, G. A. and T. M. Korn. *Electronic Analog Computers*. New York: McGraw-Hill Book Co., Inc., 1956.

Kuo, S. S. *Numerical Methods and Computers*. Reading, Massachusetts: Addison-Wesley Publishing Co., Inc., 1965.

Levine, Leon. *Methods for Solving Engineering Problems Using Analog Computers*. New York: McGraw-Hill Book Co., Inc., 1964.

MacDuff, J. N. and J. R. Currerri. *Vibration Control*. New York: McGraw-Hill Book Co., Inc., 1958.

McCormick, J. M. and M. G. Salvadori. *Numerical Methods in Fortran*. Englewood Cliffs, New Jersey: Prentice-Hall, Inc., 1964.

McCracken, D. D. and W. S. Dorn. *Numerical Methods and Fortran Programming*. New York: John Wiley and Sons, Inc., 1964.

Meirovitch, L. *Analytical Methods in Vibrations*. New York: The Macmillan Company, 1967.

Milne, W. E. *Numerical Solution of Differential Equations*. New York: John Wiley and Sons, Inc., 1953.

Minorsky, N. *Nonlinear Oscillations*. Princeton, New Jersey: D. Van Nostrand Co., Inc., 1962.

Morrow, C. T. *Shock and Vibration Engineering*, vol. 1. New York: John Wiley and Sons, Inc., 1963.

Morse, P. M. *Sound and Vibration*. New York: McGraw-Hill Book Co., Inc., 1948.

Myklestad, N. O. *Fundamentals of Vibration Analysis*. New York: McGraw-Hill Book Co., Inc., 1956.

Newmark, N. M. "A Method of Computation for Structural Dynamics," *Trans. ASCE*, vol. 127, 1962.

Pelton, W. J. *Computers and the Teaching of Engineering Mathematics*. Report Number SP-2052/000/01, System Development Corporation, Santa Monica, California, May, 1965.

Pennington, R. H. *Introductory Computer Methods and Numerical Analysis*. New York: The Macmillan Company, 1965.

Pipes, L. A. *Matrix Methods for Engineering*. Englewood Cliffs, New Jersey: Prentice-Hall, Inc., 1963.

Ralston, A. *A First Course in Numerical Analysis*. New York: McGraw-Hill Book Co., Inc., 1965.

Scanlan, R. H. and R. Rosenbaum. *Aircraft Vibration and Flutter*. New York: The Macmillan Company, 1951.

Seto, W. W. *Theory and Problems of Mechanical Vibrations*. New York: Schaum Publishing Co., 1964.

Smith, G. M. and G. L. Downey. *Advanced Dynamics for Engineers*. Scranton, Pennsylvania: International Textbook Co., 1960.

Soroka, W. W. *Analog Methods in Computation and Simulation*. New York: McGraw-Hill Book Co., Inc., 1954.

Stoker, J. J. *Nonlinear Vibrations in Mechanical and Electrical Systems*. New York: John Wiley and Sons, Inc., 1950.

Thomson, W. T. *Laplace Transformation*. Englewood Cliffs, New Jersey: Prentice-Hall, Inc., 1960.

———. *Mechanical Vibrations*. Englewood Cliffs, New Jersey: Prentice-Hall, Inc., 1953.

———. *Vibration Theory and Applications*. Englewood Cliffs, New Jersey: Prentice-Hall, Inc., 1965.

———. *Introduction to Space Dynamics*. New York: John Wiley and Sons, Inc., 1963.

Timoshenko, S. and D. H. Young. *Advanced Dynamics*. New York: McGraw-Hill Book Co., Inc., 1948.

———. *Vibration Problems in Engineering*. New York: D. Van Nostrand Co., Inc., 1956.

Tobias, S. A. *Machine-Tool Vibration*. New York: John Wiley and Sons, Inc., 1965.

Tong, K. N. *Theory of Mechanical Vibration*. New York: John Wiley and Sons, Inc., 1960.

Tse, F. S., I. E. Morse, and R. T. Kinkle. *Mechanical Vibrations*. Boston, Massachusetts: Allyn and Bacon, Inc. 1963.

Vernon, J. B. *Linear Vibration Theory: Generalized Properties and Numerical Methods*. New York: John Wiley and Sons, Inc., 1967.

———. *Linear Vibration and Control System Theory, Computer Applications*. New York: John Wiley and Sons, Inc., 1967.

Vierck, R. K. *Vibration Analysis*. Scranton, Pennsylvania: International Publishing Co., 1967.

Volterra, E. and E. C. Zachmanoglou. *Dynamics of Vibrations*. Columbus, Ohio: Charles E. Merrill Books, Inc., 1965.

Wilson, W. K. *Practical Solution of Torsional Vibration Problems*. New York: John Wiley and Sons, Inc., 1948.

Wylie, C. R. *Advanced Engineering Mathematics*. New York: McGraw-Hill Book Co., Inc., 1960.

Index

373